PAPERS IN GAME THEORY

JOHN C. HARSANYI
Professor of Business Administration, and of Economics,
University of California, Berkeley

PAPERS IN GAME THEORY

D. REIDEL PUBLISHING COMPANY
DORDRECHT : HOLLAND / BOSTON : U.S.A.
LONDON : ENGLAND

Library of Congress Cataloging in Publication Data

CIP

Harsanyi, John C.
Papers in game theory.

(Theory and decision library ; v. 28)
Includes index.
1. Game theory–Addresses, essays, lectures. I. Title.
II. Series.
QA269.H368 519.3 81-23413
ISBN 90-277-1361-8 AACR2

Published by D. Reidel Publishing Company,
P.O. Box 17, 3300 AA Dordrecht, Holland

Sold and distributed in the U.S.A. and Canada
by Kluwer Boston Inc., Lincoln Building,
190 Old Derby Street, Hingham, MA 02043, U.S.A.

In all other countries, sold and distributed
by Kluwer Academic Publishers Group,
P.O. Box 322, 3300 AH Dordrecht, Holland

D. Reidel Publishing Company is a member of the Kluwer Group

Printed in The Netherlands

To Anne

TABLE OF CONTENTS

SECTION D / NONCOOPERATIVE MODELS FOR COOPERATIVE GAMES

SECTION E / GAME-THEORETICAL ANALYSIS OF MORAL DECISIONS

ACKNOWLEDGEMENTS

The author, the editor, and the publisher are grateful to the following journals and institutions for permission to reprint the papers included in this volume:

"Approaches to the Bargaining Problem Before and After the Theory of Games: A Critical Discussion of Zeuthen's, Hicks', and Nash's Theories", *Econometrica* **24** (1956), 144–157.
Copyright 1956 The Econometric Society.

"On the Rationality Postulates Underlying the Theory of Cooperative Games", *Journal of Conflict Resolutions* **5** (1961), 179–196.
Copyright 1961 Sage Publication, Beverly Hills.

"A Simplified Bargaining Model for the *n*-Person Cooperative Game", *International Economic Review* **4** (1963), 194–220.
Copyright 1963 International Economic Review.

"Games with Randomly Disturbed Payoffs: A New Rationale for Mixed-Strategy Equilibrium Points", *International Journal of Game Theory* **2** (1973), 1–23.
Copyright 1973 Physica-Verlag.

"Oddness of the Number of Equilibrium Points: A New Proof", *International Journal of Game Theory* **2** (1973), 235–250.
Copyright 1973 Physica-Verlag.

"Games with Incomplete Information Played by 'Bayesian' Players. Part I: The Basic Model", *Management Science* **14** (November 1967), 159–182.
Copyright 1967 The Institute of Management Sciences.

"Games with Incomplete Information Played by 'Bayesian' Players. Part II: Bayesian Equilibrium Points", *Management Science* **14** (January 1968), 320–334.
Copyright 1968 The Institute of Management Sciences.

"Games with Incomplete Information Played by 'Bayesian' Players. Part III: The Basic Probability Distribution of the Game", *Management Science* **14** (March 1968), 486–502.
Copyright 1968 The Institute of Management Sciences.

"Uses of Bayesian Probability Models in Game Theory", in D. H. Mellor (ed.), *Science, Belief and Behaviour: Essays in Honour of R. B. Braithwaite,*

Cambridge University Press, Cambridge, England, 1980, pp. 189–201.
Copyright 1980 Cambridge University Press.

"An Equilibrium-Point Interpretation of Stable Sets and a Proposed Alternative Definition", *Management Science* **20** (1974), 1472–1495.
Copyright 1974 The Institute of Management Sciences.

"A New General Solution Concept for Both Cooperative and Noncooperative Games", in *Papers of the Rhineland-Westphalian Academy of Sciences* **287** (1979), Westdeutscher Verlag, Opladen, West Germany, 7–28.
Copyright 1979 Westdeutscher Verlag.

"Rule Utilitarianism, Rights, Obligations and the Theory of Rational Behavior", *Theory and Decision* **12** (1980), 115–133.
Copyright 1980 D. Reidel Publishing Co.

INTRODUCTION

This volume contains twelve of my game-theoretical papers, published in the period of 1956–80. It complements my *Essays on Ethics, Social Behavior, and Scientific Explanation*, Reidel, 1976, and my *Rational Behavior and Bargaining Equilibrium in Games and Social Situations*, Cambridge University Press, 1977.

These twelve papers deal with a wide range of game-theoretical problems. But there is a common intellectual thread going though all of them: they are all parts of an attempt to generalize and combine various game-theoretical solution concepts into a *unified solution theory* yielding one-point solutions for both cooperative and noncooperative games, and covering even such 'non-classical' games as games with incomplete information.

SECTION A

The first three papers deal with bargaining models. The first one discusses Nash's two-person bargaining solution and shows its equivalence with Zeuthen's bargaining theory. The second considers the rationality postulates underlying the Nash–Zeuthen theory and defends it against Schelling's objections. The third extends the Shapley value to games without transferable utility and proposes a solution concept that is at the same time a generalization of the Shapley value and of the Nash bargaining solution.

SECTION B

The next two papers deal with equilibrium points. The first of these shows that mixed -strategy equilibrium points can be interpreted as limits of pure-strategy equilibria in games with small random fluctuations in their payoffs, and argues that this model provides a new justification for use of mixed-strategy equilibrium points. The other paper presents a new proof for Wilson's theorem according to which in 'almost all' games the number of equilibrium points is finite and odd.

SECTION C

The four papers following these discuss some uses of probabilistic models in game theory. Three papers are parts of a longer article on games with incomplete information (which are games in which some or all players have less than full information about one another's payoff functions and/or strategy spaces). Classical game theory had no way of dealing with such games. These three papers argue that such games can be brought within the scope of game theory by introducing suitably chosen random variables (random moves) in our mathematical representation for these games. It is shown that the concept of equilibrium points can be naturally extended to these probabilistic game models for incomplete-information games but that cooperative solution concepts and, in particular, the Nash solution cannot be directly extended to them. Finally, the problem of mutual consistency between the probability distributions entertained by the different players is discussed.

The fourth paper in this section discusses two uses of subjective probabilities in game theory: One is the use of probabilistic models in analyzing games with incomplete information (see above). The other is the tracing procedure, which is a mathematical model for convergence of rational players' expectations to a specific equilibrium point in an n-person non-cooperative game.

SECTION D

The next two papers discuss noncooperative-game models for cooperative games. The first of these describes a noncooperative bargaining model for von Neumann-Morgenstern stable sets. The second discusses the problem of modelling cooperative games by means of noncooperative bargaining models in general terms. It also outlines the Harsanyi–Selten solution theory, which provides a one-point solution for every noncooperative game – and, of course, also for every cooperative game remodelled as a noncooperative game. (The new solution theory will be described in full detail in a forthcoming book by Reinhard Selten and the present writer.)

SECTION E

The last paper uses game-theoretical concepts to clarify the difference between act utilitarian and rule utilitarian moral theories. It is argued that act utilitarianism interprets moral behavior as a *noncooperative* game in *extensive* form whereas rule utilitarianism interprets it as a *cooperative* game in *normal* form, in which the concept of a rational commitment plays an essential role.

SECTION A

BARGAINING MODELS

CHAPTER 1

APPROACHES TO THE BARGAINING PROBLEM BEFORE AND AFTER THE THEORY OF GAMES: A CRITICAL DISCUSSION OF ZEUTHEN'S, HICKS', AND NASH'S THEORIES

It is proposed[1] to show that Professor Zeuthen's solution of the bargaining problem[2] (antedating the von Neumann-Morgenstern theory of games[3] by more than a decade) is mathematically equivalent to Mr. Nash's solution[4] (based on the theory of games), except that Zeuthen regards the two parties' mutual bargaining threats as given while Nash furnishes a solution also for the problem of selecting optimal threats. At the same time, it will be submitted that Zeuthen's approach in turn supplements Nash's more abstract treatment in an important way by supplying a plausible psychological model for the actual bargaining process. However, certain minor changes in the basic postulates of Zeuthen's theory and a simplication of Nash's method for finding the optimal threats will be suggested. Comparing Nash's theory with Professor Hicks' theory of collective bargaining[5] (published shortly after Professor Zeuthen's), it will be argued that the former rests on assumptions both more realistic and more consistent with the postulate of rational behaviour. Finally, the economic significance of the Zeuthen-Nash theory of bargaining will be discussed.

1. THE INDETERMINACY OF THE BARGAINING PROBLEM IN ORTHODOX ECONOMICS

As is well known, ordinary economic theory is unable to predict the terms on which agreements tend to be reached in cases of oligopoly, of bilateral monopoly, and in general in situations where settlements must be arrived at by means of explicit or implicit bargaining. Still less is current economic theory (even with the help of current political theory) able to predict on what terms agreements tend to be reached in situations where one or both of two bargaining parties use political pressure as one of the bargaining weapons.

All that orthodox economic theory can say in these cases it that any agreement tends to fall within a certain range, called by Professor Pigou the 'range of practicable bargains', defined by the following two properties: (1) It is on the (Edgeworth) contract curve, i.e., the locus of situations where neither party's position can be improved without at the same time worsening the position of the other party. (2) It is on this curve between two limits, which may be called the 'maximum-concession points' of each party, and which are determined by the fact that neither party would accept an agreement that put him in a worse position than the conflict situation (not reaching an agreement). In other words, each party's maximum-concession point is that point at which all the net gain resulting from the agreement and the co-operation agreed upon would go wholly to the other party. But current economic theory does not explain how the position of the actual agreement point is determined *within* this range of practicable bargains, that is, how the division of the net gain is determined between the two bargaining parties.

At the same time, the theory of games itself, as developed originally by von Neumann and Morgenstern, is no more able than is orthodox theory to determine the exact position of the agreement point within the range of practicable bargains. Only on the basis of *additional* assumptions[6] does the theory of games furnish a determinate solution, as we shall see.

The unpredictability of the outcome of bargaining is, however, by no means a law of nature, but is only a gap in current economic and political theories. In practical life experienced businessmen, politicians, diplomats, etc., seem to be often quite able to predict the terms of different agreements with reasonable accuracy when they have sufficient information on the bargaining situation. Of course, information on the two parties' general economic and political strength alone may not suffice: the outcome may depend significantly on such 'accidental' factors as the bargaining skill of the two parties' actual representatives.[7] But it is a perfectly legitimate question (and one to which we are entitled to an answer from economics and political science) to ask what sort of agreement tends to result from a given objective economic and political balance of power between two parties (as well as from given subjective attitudes on their part) if all disturbing forces are assumed away[8] (e.g., by postulating perfect rationality and perfect knowledge). In the absence of a theory yielding determinate predictions as to the outcome of bargaining, even the mere *description* of many social situations in a satisfactory way becomes impossible.[9] Thus a determinate solution for the bargaining problem – as attempted by the three theories to be discussed – would fill a significant gap in both economics and political science.

2. BARGAINING WITH, AND WITHOUT, A CHOICE AMONG ALTERNATIVE THREATS

An important advance over earlier authors has been the clear distinction in Nash's theory between bargaining situations that do and those that do not allow each party a choice among alternative threats by which to bring pressure to bear upon the other party.

Bargaining situations with one possible threat by each party arise either when the two parties can achieve a certain gain by co-operation but when each of them can threaten to withhold his co-operation unless a profit-sharing agreement satisfactory to him is reached, or when one or both parties are able to inflict one particular sort of positive damage on the other party and use this possibility as a threat. It is, however, often a matter of definition what we regard as mere non-co-operation and what we regard as positive damage, and so it is desirable to bring both cases under the same terminology by concentrating on the fact that in either case an agreement will mean at least a *relative* gain to both parties over the conflict situation resulting from the absence of agreement. In both cases agreement means saving the costs of a conflict and this saving can be divided between the two parties.

But in more general situations the bargaining parties will have a choice among several possible threats, each involving different degrees of non-co-operative or positively damaging behaviour. Of course, if there is no obligation to carry out threats in the absence of agreement, the choice among alternative threats may become trivial since each party may then use the most formidable threat against his opponent as a matter of mere bluffing. To exclude this possibility, we may assume that each party is forced to carry out this threat if agreement fails.[10] Or alternatively we may assume – in accordance with the postulate of perfect knowledge – that each party can distinguish mere bluffs from serious threats (say, on the basis of Nash's theory of optimal threats). We may even admit the possibility that explicit threats are not used at all during the negotiations, for it suffices if both parties can form a reasonably accurate *estimate* of what retaliation would result in the event of a conflict. In any case, if bluffing is made impossible or useless, the choice among alternative threats becomes a very real problem since, in general, the more effective a certain non-co-operative or positively damaging policy is against the opponent the more expensive it tends to be also to the party using it.

We begin, however, with the simpler problem of bargaining with *given* threats, assuming that each party's threat is uniquely given either because he has only one possible threat or because the problem of choosing among several possible threats has already been solved. Accordingly, both parties will be assumed to know what kind of conflict situation will result if no agreement is reached.

3. NASH'S UTILITY-PRODUCT MAXIMIZATION RULE

The bargaining problem has an obvious determinate solution in at least one special case: viz., in situations that are completely symmetric with respect to the two bargaining parties. In this case it is natural to assume that the two parties will tend to share the net gain equally since neither would be prepared to grant the other better terms than the latter would grant him. For instance, everybody will expect that two duopolists with the same cost functions, size, market conditions, capital resources, personalities, etc., will reach an agreement giving equal profits to each of them.

Nash's theory of bargaining (with given threats) is fundamentally a generalization of this principle[11] for the more general case of asymmetric situations. This generalization is achieved by postulating that the general solution should have desirable mathematical properties.[12] Nash's conclusion is that, if both parties' utilities are measured by von Neumann–Morgenstern cardinal-utility indices, they will reach an agreement at that point of the range of practicable bargains at which the *product* of the utilities each party ascribes to his net gain (over the conflict situation) is *maximized.*

It is worth noting that this solution does not presuppose the interpersonal comparability of utilities since the index for each party's utility can be chosen independently of the other's (because the solution is invariant with respect to changes in the zero point or the unit of either party's utility index; in effect this invariance property is one of the postulates actually used by Nash in deriving his solution).

Moreover, it is easy to see that in the symmetric case this solution gives equal shares to both parties,[13] so that it is a generalization of the solution suggested above for this special case.

4. THE MATHEMATICAL EQUIVALENCE OF ZEUTHEN'S AND NASH'S THEORIES

We now show that Zeuthen's solution is equivalent to Nash's.

Zeuthen's approach is based on a direct analysis of the process of collective bargaining on the labour market, but is has general validity for any sort of bargaining situation.

Suppose that Party 1 would like to achieve the terms A_1, but has been offered only the less favourable terms A_2 by Party 2. Will he accept A_2 or will he insist on obtaining A_1? This will obviously depend on Party 1's view of the probability that Party 2 would definitely reject the terms A_1, and that his own insistence on them would lead to a conflict.

Let $U_1(A_1)$ and $U_1(A_2)$ be the net utility gains over the conflict situation that Party 1 would derive from A_1 and A_2, respectively. By assumption, $U_1(A_1) > U_1(A_2)$. Moreover, let p_2 be the probability that Party 2 would finally reject the terms A_1. Then, if Party 1 accepts A_2 he will obtain $U_1(A_2)$ with certainty, while if he rejects A_2 and insists on the better terms A_1 he will have the probability $(1 - p_2)$ of obtaining the higher utility $U_1(A_1)$ and the probability p_2 of obtaining nothing. Therefore, on the assumption that Party 1 tries to maximize his expected utility,[14] he will accept the terms A_2 if $U_1(A_2) > (1 - p_2) \cdot U_1(A_1)$, that is, if $[U_1(A_1) - U_1(A_2)]/U_1(A_1) < p_2$, and will reject A_2 and insist on the better terms A_1 in the opposite case. Consequently, the utility quotient $[U_1(A_1) - U_1(A_2)]/U_1(A_1)$, which may be written briefly as $\Delta U_1/U_1$, expresses the maximum risk (maximum probability of conflict) that Party 1 is prepared to face in order to secure the better terms A_1 instead of the less favourable terms A_2.

On the other hand, denoting the net utility gains Party 2 would derive from the two situations by $U_2(A_1)$ and $U_2(A_2)$ (where $U_2(A_1) < U_2(A_2)$, it can similarly be shown that the maximum risk Party 2 would take in order to achieve the terms A_2 more favourable to him is equal to the utility quotient $\Delta U_2/U_2 = [U_2(A_2) - U_2(A_1)]/U_2(A_2)$. Thus the two utility quotients $\Delta U_1/U_1$ and $\Delta U_2/U_2$ – which in a sense measure the relative advantage of A_1 over A_2 to Party 1 and the relative advantage of A_2 over A_1 to Party 2, respectively – decide the strength of each party's 'determination' to insist on the alternative more favourable to himself.

At this point Professur Zeuthen introduces the further assumption[15] that each party will make a concession to his opponent once he finds that the latter's determination is firmer (i.e., the latter's readiness to risk a conflict is greater) than is his own. Thus Party 1 will always make a further concession if

$$(1) \qquad [U_1(A_1) - U_1(A_2)]/U_1(A_1) < [U_2(A_2) - U_2(A_1)]/U_2(A_2)$$

while Party 2 will always make a concession in the opposite case. A concession need not be the complete acceptance of the other party's last offer, but it must be large enough to reverse the inequality sign in (1) if the utilities corresponding to the new offer are substituted. (Zeuthen also assumes that if both parties' utility quotients happen to be exactly equal, *both* of them will make some concession in order to avoid a deadlock.) This process will continue until the two parties meet somewhere in the middle. It will end after a finite number of steps because the indivisibility of the smallest monetary unit (and other technical or psychological indivisibilities) sets a lower limit to the size of admissible concessions.

Zeuthen has also shown that this process will actually lead, in the essentially symmetric case assumed by him, to an equal division of the net gain. This is, of course, in agreement with Nash's theory. It is easy to see, however, that Zeuthen's model leads to the same solution as Nash's theory in the general case as well. In effect, inequality (1) can also be written in the form:

$$(2) \qquad U_1(A_1) \cdot U_2(A_1) < U_1(A_2) \cdot U_2(A_2)$$

so that we can restate Zeuthen's assumption by saying that the party whose last offer is associated with a lower value of the utility product, $U_1 \cdot U_2$, than is his opponent's last offer will always make a further concession and put forward a new offer associated with an even higher value of the product $U_1 \cdot U_2$ than was his opponent's last offer. (If both parties' last offers gave the same value for $U_1 \cdot U_2$, both will make concessions and put forward offers giving higher values of $U_1 \cdot U_2$.) Thus, each new offer made during the negotiations is bound to increase the value of the product $U_1 \cdot U_2$ and in the end the two parties will arrive[16] at terms that maximize[17] the value of $U_1 \cdot U_2$ – which is exactly Nash's solution.

5. DERIVATION OF ZEUTHEN'S BASIC ASSUMPTION FROM MORE GENERAL POSTULATES

We propose to derive now, from a few simpler and more general postulates, Zeuthen's assumption that each bargaining party will be prepared to make a concession to his opponent whenever he finds that the latter's readiness to risk a conflict is at least as great as his own.

For this purpose we need the following postulates:

(1) *Symmetry.* The bargaining parties follow identical (symmetric) rules of behaviour (whether because they follow the same principles of rational behaviour or because they are subject to the same psychological laws).[18]

(II) *Perfect knowledge.* Each party can estimate correctly the probability that the other party will definitely reject a certain offer.

(III) *Monotonity.* Suppose that Party 1 has received the offer A_2 from Party 2. Then, the probability p_1 of Party 1's rejecting A_2 and insisting on the more favourable terms A_1 is a monotone nondecreasing function[19] of the difference $U_1(A_1) - U_1(A_2)$ if all other independent variables are kept constant.

(IV) *Expected-utility maximization.* Unless both parties agree to make simultaneous concessions, each party will make a concession if and only if his making this concession will give him a prospect with higher expected utility than his refusing the concession would.

(IV′) *Efficiency.* However, the two parties will agree to make simultaneous concessions if this policy gives both of them higher expected utilities than they would obtain otherwise.

We shall first assume that Postulate (IV) applies without the qualifications expressed in the qualifying clause ("Unless ...) and in Postulate (IV′), and shall later consider what modifications are needed in our conclusions in view of these qualifications.

We have seen above (p. 7) that Party 1 will or will not insist on the more favourable terms A_1 according as the utility quotient $\Delta U_1/U_1$ is larger or smaller than p_2. To derive this result we used only the utility maximization postulate (corresponding to Postulate (IV). This result can also be expressed as follows:

(3) $$p_1 = 0 \text{ if } p_2 > \Delta U_1/U_1, \text{ and } \quad p_1 = 1 \text{ if } p_2 < \Delta U_1/U_1,$$

where p_1 denotes the probability that Party 1 will reject the terms A_2.

By the same token the analysis of Party 2's behaviour gives us:

(4) $$p_2 = 0 \text{ if } p_1 > \Delta U_2/U_2, \quad \text{and} \quad p_2 = 1 \text{ if } p_1 < \Delta U_2/U_2.$$

On the other hand, it is easy to see that (3) and (4) together require that one of the following three possibilities should be true:

(5)
(a) Either $p_1 = 0$ and $p_2 = 1$, or
(b) $p_1 = 1$ and $p_2 = 0$, or
(c) $p_1 = \Delta U_2/U_2$ and $p_2 = \Delta U_1/U_1$.

But we have still to find out when each of these three cases applies. Now, (3) makes p_1 a function of p_2 and $\Delta U_1/U_1$. We can therefore write:

(6) $$p_1 = F(p_2, \Delta U_1/U_1).$$

Similarly, on the basis of (4) as well as the Symmetry Postulate (1), we have:

(7) $$p_2 = F(p_1, \Delta U_2/U_2).$$

(6) and (7) are two simultaneous equations in the two unknowns, p_1 and p_2. Since they are symmetric with respect to these unknowns their solution

must be of the form:

(8) $p_1 = G(\Delta U_1/U_1, \Delta U_2/U_2)$ and $p_2 = G(\Delta U_2/U_2, \Delta U_1/U_1)$.

But the function G is uniquely determined by Postulate III and property (5), so that we have:

(9.1) $p_1 = 0$ and $p_2 = 1$, if $\Delta U_1/U_1 < \Delta U_2/U_2$,

(9.2) $p_1 = p_2 = \Delta U_1/U_1$, if $\Delta U_1/U_1 = \Delta U_2/U_2$,

(9.3) $p_1 = 1$ and $p_2 = 0$, if $\Delta U_1/U_1 > \Delta U_2/U_2$.

(9.2) is, however, inconsistent with Postulate (IV′) since it will result in wasteful conflicts with probability $(\Delta U_1/U_1)^2$, which can be avoided if the two parties agree to make concessions simultaneously, i.e., adopt the rule of behaviour:

(9.2′) $p_1 = p_2 = 0$ if $\Delta U_1/U_1 = \Delta U_2/U_2$.

Thus we obtain the result that:
Party 1 will make a concession if $\Delta U_1/U_1 < \Delta U_2/U_2$;
Party 2 will make a concession if $\Delta U_1/U_1 > \Delta U_2/U_2$; and
Both will make concessions if $\Delta U_1/U_1 = \Delta U_2/U_2$, as desired.

6. PSYCHOLOGICAL MODELS AND THE AXIOMATIC METHOD

The somewhat unexpected identity of the solutions yielded by Nash and by Zeuthen is a most encouraging sign because on the surface their approaches appear rather different. Many economists as well as other empirical social scientists will probably find Zeuthen's reasoning more convincing as it is based on a farily plausible psychological model of the bargaining process, and, at the same time, they may look somewhat askance at Nash's game-theoretical method, which relies on abstract mathematical postulates whose empirical relevance may be less obvious. But, in actual fact, the two approaches are complementary.

The abstract axiomatic method should primarily be regarded as a heuristic device. If it can be shown that there is, consistent with peoples' behaviour in a certain situation, one and only one hypothesis satisfying certain attractive general postulates (whether formal postulates based on considerations of mathematical simplicity, continuity, symmetry, determinateness, etc.,[20] or 'material' postulates suggested by observation or introspection), this is a strong argument[21] *for* this particular hypothesis – or at least *against* any alternative ones – even if no convincing psychological model is known which would explain the behaviour predicted by this hypothesis. Of course, if a satisfactory psychological model is also found, this will further increase our confidence in the hypothesis. But the converse is equally true. Our confidence in a given psychological model will certainly increase if the pattern of behaviour predicted by it is found to be one and the only one compatible with certain general postulates that are otherwise attractive to us.

Thus Zeuthen's and Nash's theories corroborate each other. At the same time, Zeuthen's approach supplements Nash's in at least two important respects. It furnishes a more detailed analysis of the actual bargaining process. Moreover, it explains how the result of the bargaining process depends on the two parties' cardinal utility functions in the von Neumann–Morgenstern sense, i.e., on their attitudes towards risk-taking – a dependence which in Nash's theory must be assumed (in order to obtain a determinate solution) without the possibility of being explained.

On the other hand, owing to its higher level of abstraction and its mathematical rigour, Nash's theory has the important advantage of complete generality, which assures us that no particular case falling within the scope of the theory will possibly be overlooked. While Zeuthen's argument in its original form is restricted to symmetric bargaining situations, Nash's theory automatically covers the asymmetric case also and even lends itself naturally to extension to the still more general case of bargaining with a choice among several alternative threats. Moreover, whereas Zeuthen's argument as it stands is limited to collective bargaining, Nash's theory at once covers all of bilateral monopoly, duopoly, and bargaining by means of political pressure, etc., without restriction.[22]

7. NASH'S THEORY OF OPTIMAL THREATS

Nash's theory of the selection of 'optimal' threats by each bargaining party naturally follows from his theory of bargaining with given threats.

He has shown that both parties always have optimal threats with maximin and minimax properties: i.e., if a given party uses an optimal threat, he will be assured of obtaining the best terms that can be exacted at all in the given situation from a rational opponent.[23] Nash also gives a geometric method by means of which both parties' optimal threats in this sense can be determined.[24] A simpler method can, however, be found on the basis of the following consideration.

Any threat aims at increasing the cost of a conflict to one's opponent without 'unduly' increasing the cost of a conflict to oneself. Nash's concept of optimal threats makes possible an exact definition of what 'unduly' means here. It can be shown that it is always profitable to increase the coast of a conflict to one's opponent if the cost to oneself either fails to increase at all or increases only in a smaller *proportion,*[25] and it is always profitable to allow the cost of a conflict to one's opponent even to decrease if the cost to oneself decreases in a still higher proportion. On this basis each party's optimal threat is easily found.

Not only will both parties follow the rule of behaviour expressed in this conclusion, but they will also expect each other to follow it. If either party attempted to put forward a threat that in the event of a conflict implied a policy disproportionately burdensome to himself in relation to its damaging effect on his opponent, the opponent would regard it as a mere bluff.

8. COMPARISON OF HICKS' THEORY OF BARGAINING WITH THE NASH–ZEUTHEN THEORY

It is instructive to compare Nash's above theory of optimal threats with Hicks' theory of collective bargaining on the labour market (1932).[26]

Hicks' theory can be summarized as follows. Each bargaining party will make a concession if, in his view, the strike resulting from his refusing this concession would cost him more than the concession would. (Of course, costs must be reckoned at their present values to make significant comparison possible.) The higher a given wage rate the more its acceptance will cost the employer and the longer the strike he will be prepared to endure rather than to accept this wage rate. Thus each wage rate will be associated in the employer's mind with a strike of a given length (with one which would be just as expensive as accepting this wage rate), higher wage rates being associated with longer strikes. On the other hand, the lower a given wage rate the more its acceptance will cost the trade union (in terms of disutility) and the longer the strike they will be prepared to undertake rather than to accept this wage rate. Thus each wage rate will be connected in the trade unionists' minds with a strike of a given length (one which would be just as bad as accepting the wage rate), but in this case lower wage rates will be connected with longer strikes. There will be a unique wage rate that both parties associate with a strike of the same length: this is the highest wage rate that the union can exact from the employer. The latter in effect will not accept a wage rate higher than this one since he knows that his refusal can at worst result only in a strike short enough to cost him less than accepting this wage rate would. On the other hand, for opposite reasons he will certainly be prepared to accept this particular wage rate or, of course, any lower one.

As is easy to see, the primary difference between Hicks' theory and Nash's is Hicks' assumption that either party will choose, or threaten to choose, a strike in preference to a concession only if this strike is expected to cost him less than the concession.[27] By contrast, in Nash's theory each party is ready to exert pressure on the other party by the threat of a strike whose actual occurrence, if agreement failed, would cost the threatening party *more* than a concession would, provided the threat is likely to exact better terms from the other party if agreement succeeds.[28]

I think it must be admitted that Nash's assumption is the more realistic one. Trade unions, employers, and bargaining parties in other fields, do use threats of retaliation that would, if carried out, entail losses for themselves no less than for their opponents – and it is rational for them to do so since they can thereby achieve better terms.[29]

9. ON THE ECONOMIC MEANING OF THE NASH–ZEUTHEN THEORY

What is the economic meaning of the Nash-Zeuthen theory of bargaining?

We have seen that this theory goes beyond orthodox economic theory in that it not only states that the agreement point will lie somewhere within the range of practicable bargains, but also predicts the actual position of this point within this range. Moreover, it predicts the two parties' behaviour even in the more general case where the limits of the range of practicable bargains themselves depend on the two parties' behaviour (viz., on the threats they use against each other).

According to this theory the most important factors determining the position of the agreement point within the range of practicable bargains are the two parties' attitudes towards risk-taking as expressed by their cardinal utility functions. A bargaining party will reach better terms the greater his risk preference, and the smaller the risk preference of his opponent. Thus orthodox economic theory is right in taking the view that, on the basis of the variables considered by the orthodox theory, the position of the agreement point is indeterminate within the range of practicable bargains: for it to become determinate it is necessary, though not yet sufficient, that the two parties' cardinal utility functions are also included among the variables of the theory.

The two parties' cardinal utility functions, however, are not the only variables influencing the position of the agreement point within the range of practicable bargains. This position will also depend on the conditions under which economic values (especially money) can be transferred from one party to the other: in particular, on whether, in the event of one party's making a concession to the other, the former's net loss will exceed, equal, or fall short of the latter's net gain. An obvious case where the net loss and the net gain will not be equal is a situation in which transfers are subject to transport costs. But the same will apply also if the two parties are subject to income taxation at different marginal rates. Again, in duopoly situations where direct side payments between the duopolists are not possible for one reason or another, but where concessions can take only the form of reducing output or allowing the opponent's output to rise, the loss in profits to one party will generally differ from the gain in profits accruing to the other party.[30] The conditions of transfer in this sense will obviously influence the readiness of both parties to make concessions and, therefore, the final outcome of the bargaining. We cannot here go into the details (which the reader can easily work out for himself), but it can be shown, for example, that each party's willingness to make concessions will be greater the more progressive the income tax to which he is subject[31] – a fact not unfamiliar to observers of wage determination by collective bargaining.

Another important corollary of the Zeuthen-Nash theory is the possibility of determining the exact economic value (nuisance value) of the ability to exert economic, political, or other pressure. Take the following simple example. Suppose that Party 1 tries to blackmail Party 2 by the threat of causing him a damage of $ 1000 unless Party 2 pays a certain amount. What is the amount they will tend to agree upon?

Some people would probably argue that this figure would be just a little below $ 1000 since Party 2 would be prepared to pay anything short of the full $ 1000 in order to avert Party 1's damaging action. But this would be no more valid than to argue that Party 1 would be ready to accept any amount, however small, since, if no agreement could be reached, he would receive nothing at all. In actual fact, these two arguments prove only that any figure between $ 0 and $ 1000 would be within the range of practicable bargains, but they tell us nothing about which figure will in fact be agreed upon. To determine this figure one must know the two parties' attitudes towards risk-taking as expressed by their cardinal utility functions.

The conclusion reached has general validity. The value of the ability to exert pressure on people by economic, political, or other threats is in general less than the value of the loss which the victim would suffer if these threats were carried out. The exact ratio between these two values, on the other hand, is primarily determined by the two parties' attitudes towards risk (and, of course, by the extent to which the threatened party can use effective counter-threats).

It may be noted that the bargaining theory considered here can be rendered more dynamic by allowing for the fact that, in deciding on their policies, the bargaining parties may also consider the effects of their present behaviour on their future bargaining positions. A tough policy followed in the present may increase the chance of a conflict in the present transaction, but may at the same time help to gain the reputation of toughness and thereby help to obtain better terms in other transactions. Taking account of this fact increases the relative desirability of obtaining better terms in the present deal and has the same effect on a bargainer's behaviour as an increase in his risk preference. The final terms reached will be most affected when this consideration has a much different importance for the two parties, e.g., if one of them is a large economic or political unit expecting to have to settle many similar transactions by bargaining in the future, while the other party is a small unit expecting to be involved in few or no similar transactions. In this case, the greater the difference between the sizes and transaction volumes of the two units, the tougher the policy that the larger unit will find it profitable to follow; and if this difference is considerable, the smaller unit may be forced back virtually to its maximum-concession point – which means, of course, the disappearance of actual bargaining in the usual sense. Markets with one large monopolist (or monopsonist) on one side and a great number of small atomistic competitors on the other and oligopolistic markets with one large price leader and a greater number of followers are both instances of such a situation.[32]

The main applications of the Zeuthen-Nash theory of bargaining are obviously in the fields of duopoly, of bilateral monopoly (including the problem of collective bargaining on the labour market), and of bargaining with the help of political pressure. (Another of its applications is in the analysis of political equilibrium between two political units, but this is beyond the scope of this paper.) But in no other field, perhaps, has its superiority over erarlier theories been more conspicuous than in the field of duopoly, which has so notoriously defied so many of the great names of economic science.[33]

On the other hand, the most important limitation[34] of the Zeuthen-Nash theory lies in the fact that it is restricted to two-party bargaining situations ("two-person games"), so that it is, for instance, inapplicable to oligopoly situations with more than two oligopolists. The removal of this restriction depends on finding a satisfactory determinate solution for the general n-person game,[35] and a solution to this problem will do much more than make a general theory of oligopoly possible; it will open the way for a general theory of the distribution of income and of power within society.

University of Queensland

[1] The author would like to thank Professor H. W. Arndt, of Canberra University College, and Mr. W. R. Lane, of the University of Queensland, for reading the manuscript of this paper and for making helpful comments.

[2] F. Zeuthen, *Problems of Monopoly and Economic Warfare,* London, 1930, Chap. IV.

[3] The first edition of the *Theory of Games and Economic Behavior* was published in 1944, the second in 1947.

[4] J. F. Nash, Jr., 'The Bargaining Problem', *Econometrica* **18** (1950), 115–162, and 'Two-Person Co-operative Games', *ibid.,* **21** (1953) 128–140.

[5] J. R. Hicks, *The Theory of Wages,* London, 1932, Chap. VII.

[6] Besides Nash's approach to be discussed, an alternative method for finding determinate solutions has been proposed by Mr. Raiffa, but his solutions involve arbitration, not bargaining. See H. Raiffa, 'Arbitration Schemes for Generalized Two-Person Games', University of Michigan, Report No. M720-1, R 30, June 1951; reprinted (in a shortened form) in *Contributions to the Theory of Games,* ed. by H. W. Kuhn and A. W. Tucker, Vol. II, Princeton, 1953, pp. 361–387.

[7] Often the bargaining skill of these representatives is not really an accidental factor because the ability to hire more skilled negotiators itself depends on greater economic or political strength.

[8] Abstraction from disturbing forces in constructing theoretical models is no more objectionable in the theory of bargaining than it is in other parts of social science, and only requires the same sorts of precautions in applying conclusions to the real world.

[9] For example, such descriptive terms as the 'balance of power' between two parties (or even the term economic or political 'power' itself), the degree of a bargaining party's 'determination' or 'toughness', the strength of the economic or political 'pressure' one party can exert against the other, etc., can be given a definite meaning only on the basis of such a theory.

[10] In an abstract model we may imagine, as Nash does, that the carrying-out of threats is enforced by some kind of special authority. In the real world, of course, the enforcing agency might simply be the fear of discrediting oneself and spoiling the efficacy of one's threats in the future.

[11] The present writer feels that this aspect of Nash's approach deserved greater emphasis than it received in Nash's original exposition.

[12] Nash's second paper also gives an alternative derivation. See his 'Two-Person Cooperative Games', *loc, cit.,* esp. pp. 131–134.

[13] This applies, of course, only if both parties' cardinal utility functions (i.e., their attitudes towards risk-taking) are also similar (symmetric).

[14] That is, the actuarial value of his (cardinal) utility. Actually, Zeuthen measures the values of A_1 and A_2 to each party, not in terms of utility, but in terms of money, and assumes that each party tries to maximize the actuarial value of his net money gain. This assumption, however, is valid only in the special case when the marginal utility of money

is constant to both parties. In the more general case we must measure all values in terms of von Neumann-Morgenstern cardinal utility.

[15] The intuitive plausibility of this assumption has been questioned (see J. Pen, 'A General Theory of Bargaining', *American Economic Review* **XLII**, No. 1 (March, 1952), esp. 33–34). We shall, however, see that this asseumption can be dispensed with as an independent postulate (see Section 5 below).

[16] Actually, they arrive at slightly different terms if at the end of the negotiations one party 'overshoots the mark' by an over-generous concession. Strictly speaking, we have to assume that each party always makes the smallest concession with suffices to reverse the inequality (2). This assumption is permissible since the sizes of concessions are assumed to have a lower bound.

[17] If the upper-right boundary of the set of the pairs of real numbers (U_1, U_2) (which represent the possible values of the two parties' net utility gains) is a smooth curve without an angular point, the same argument can also be put in the following way. At the equilibrium point the inequality (2) must be replaced by an equality for small displacements and so we must have $d(U_1 \cdot U_2)/dU_1 = d(U_1 \cdot U_2)/dU_2 = 0$. But this means that the product $U_1 \cdot U_2$ must have a maximum at this point since the set of the pairs (U_1, U_2) is a convex set.

[18] This postulate is here equivalent to the assumption which we would express on the common-sense level by saying that each party will make a concession at a given stage of the negotiations if and only if he thinks he has at least as much 'reason' as his opponent has to yield ground at that point.

[19] We cannot make p_1 a monotone strictly increasing function of this difference since p_1 cannot increase further once it has reached unity, however much this difference further increases.

[20] The history of the natural sciences – especially in the last few decades – has forcibly attested to the heuristic value of the axiomatic method and, in particular, of formal mathematical simplicity, etc., requirements in the selection of scientific hypotheses.

[21] But, obviously enough, it is never a conclusive argument. However plausible a certain postulate seems to be at first sight, there is always the possibility that it will have to be dropped if further empirical research finds it inconsistent with the actual facts – or even if further theoretical analysis finds it inconsistent with some other logical or mathematical postulates of a more fundamental nature.

[22] No economic theory relying on the concepts of orthodox economics alone could possibly cover such prima-facie different phenomena as duopoly, bilateral monopoly, etc., by the same argument; this is possible only on the higher level of abstraction adopted by the theory of games.

[23] It is, of course, assumed throughout that, after the choice of threats by each party, the outcome of bargaining is determined according to Nash's utility-product maximization rule.

[24] See his 'Two-Person Co-operative Games', *loc. cit.*, esp. pp. 134–136.

[25] This can be shown as follows:

Suppose that in the original position each party's net gain connected with the equilibrium agreement point (as above the conflict situation) is $\bar{U}_1$ and $\bar{U}_2$ respectively. Obviously $\bar{U}_1$ and $\bar{U}_2$ will also express the cost of a conflict to each party. Moreover, we know that

(a) $\bar{U}_1 \cdot \bar{U}_2 > U_1 \cdot U_2$,

where U_1 and U_2 are utility gains connected with any alternative terms. Now let the cost of a conflict to Party 1 and Party 2 increase by $a \cdot \bar{U}_1$ and $b \cdot \bar{U}_2$ respectively, where

(b) $a < b$.

Then the net utility gains connected with any terms will also increase by $a \cdot \bar{U}_1$ and $b \cdot \bar{U}_2$ for each party. Suppose the net utility gains connected with the new equilibrium agreement point (as above the new conflict situation) are $(\bar{U}'_1 + a \cdot \bar{U}_1)$ and $(\bar{U}'_2 + b \cdot \bar{U}_2)$, while the net utility gains connected with the same point in the original position (as above the original conflict situation) were $\bar{U}'_1$ and $\bar{U}'_2$. According to the utility maximization rule, in the new position the utility product *now* connected with the *new* equilibrium point must be larger than the utility product now connected with the *old* equilibrium point, so that

(c) $(\bar{U}'_1 + a \cdot \bar{U}_1) \cdot (\bar{U}'_2 + b \cdot \bar{U}_2) > (\bar{U}_1 + a \cdot \bar{U}_1) \cdot (\bar{U}_2 + b \cdot \bar{U}_2)$.

On the other hand, (a) implies that

(d) $\bar{U}'_1 \cdot \bar{U}'_2 < \bar{U}_1 \cdot \bar{U}_2$.

(c), (d) and (b) taken in conjunction give us:

(e) $\bar{U}_1 \cdot (\bar{U}'_2 - \bar{U}_2) > \bar{U}_2 \cdot (\bar{U}_1 - \bar{U}'_1)$.

It will be shown that (e) entails that

(f) $\bar{U}'_1 \geqslant \bar{U}_1$ and $\bar{U}'_2 \leqslant \bar{U}_2$.

For, suppose that $\bar{U}'_1 < \bar{U}_1$, so that $\bar{U}'_2 > \bar{U}_2$, then (e) could be written as

(g) $(\bar{U}'_2 - \bar{U}_2)/(\bar{U}_1 - \bar{U}'_1) > (\bar{U}_2/\bar{U}_1)$.

Let us denote the upper-right boundary curve of the set of all possible pay-off points (U_1, U_2) by C. (The curve C represents, of course, the contract curve between the two parties, and both the point (U_1, U_2) and the point (U'_1, U'_2) will lie on C.) Then two cases may be distinguished.

(1) If the curve C has no angular point at $(\bar{U}_1, \bar{U}_2)$ then the tangent drawn to C at $(\bar{U}_1, \bar{U}_2)$ will have the backward slope $\bar{U}_2/\bar{U}_1$ because $(\bar{U}_1, \bar{U}_2)$ is the point at which the product $U_1 \cdot U_2$ takes its maximum value. Therefore, in view of (g), the point $(\bar{U}'_1, \bar{U}'_2)$ would lie above and to the right of this tangent, which is impossible since the set of the points (U_1, U_2) is a convex set.

(2) If the curve C does have an angular point at $(\bar{U}_1, \bar{U}_2)$, two tangents can be drawn to it at this point. The tangent, however, that is drawn from the side nearer to the point $(\bar{U}'_1, \bar{U}'_2)$ will, in general, have a slope even smaller than $\bar{U}_2/\bar{U}_1$ so that the point $(\bar{U}'_1, \bar{U}'_2)$ will all the more lie above and to the right of this tangent, which is once more impossible because of the convexity of the set of points (U_1, U_2).

Consequently, (f) must be true, and the new equilibrium agreement point $(\bar{U}'_1, \bar{U}'_2)$ must be more favourable than (or at worst as favourable as) the old equilibrium agreement point $(\bar{U}_1, \bar{U}_2)$ was to Party 1 – which is what we have wanted to show.

26 *Op. cit.* (See note 5 above).

[27] Professor Hicks admits that a union may occasionally embark upon strikes not profitable in terms of immediate results and costs in order to keep the employer aware of the union's power (*op. cit.*, p. 146). But he does not discuss the use of threats of such 'unprofitable' strikes as bargaining weapons.

[28] Thus in Nash's theory it is not required that this strike should itself be a profitable venture, but only that its cost to this party should not be out of proportion to the losses this strike would cause to the other party.

[29] Professor Hicks' theory yields a determinate solution only because this special assumption just discussed reduces the range of practicable bargains to one unique point. If this assumption is dropped, indeterminacy returns and recourse must be made to some other theory of bargaining to obtain a determinate solution.

[30] For an analysis of a duopoly situation on the basis of Nash's theory and a comparison of the results with other theories of duopoly, see J. P. Mayberry, J. F. Nash, and M. Shubik, 'A Comparison of Treatments of a Duopoly Situation', *Econometrica* **21** (1953), 141–154.

[31] The reason is that, other things being equal, a progressive income tax tends to decrease the value of the utility quotient $\Delta U/U$ because it raises the marginal rate of taxation relative to its average rate, and this is the case all the more, the more progressive this income tax is.

[32] The last part of our argument, of course, has carried us over from the field of two-person games to the field of n-person games (with a large value for n), a field we are not qualified to enter in strict logic (but see note 35).

[33] See note 30.

[34] Another important limitation is the restrictive assumption of perfect knowledge, in particular the assumption that each bargaining party perfectly well knows the other party's subjective preferences (including his attitude towards risk-taking) as well as his objective damaging power. This assumption excludes the use of secrecy, deception and bargaining skill, so important in the real world. But it can be shown that this restrictive assumption might be dropped without essentially affecting the validity of the theory. Only, the position of the agreement point must now be made dependent not only on the two parties' *actual* cardinal utility functions but also on each party's *estimate* of the other party's utility function, and even on each party's estimate of what the other party's estimate is of his own (the first party's) utility function, etc. (which is the traditional problem of 'conjectural variation'). But the discussion of this problem must be left for another occasion.

[35] For a solution of this problem in relation to a certain important but restricted class of n-person games, see L. S. Shapley, 'A Value for n-Person Games', in *Contributions to the Theory of Games*, ed. by H. W. Kuhn and A. W. Tucker, Vol. II, Princeton, 1953, pp. 307–317. (It may be mentioned that the same value for this class of n-person games was found independently by the present writer before Mr. Shapley's paper became available in Australia.) A more general solution, extending the Zeuthen-Nash concept of bargaining equilibrium to the n-person co-operative game, will be described in a forthcoming paper of the present writer. (See Chapter 3 of this volume.)

CHAPTER 2

ON THE RATIONALITY POSTULATES UNDERLYING THE THEORY OF COOPERATIVE GAMES

The purpose of this paper is to re-state and re-examine the rationality postulates underlying the theory of cooperative games. It is also proposed to discuss T.C. Schelling's recent criticism of some game-theoretical postulates, in particular the symmetry postulate [15, 16].

I

The theory of games aims at developing criteria for rational behavior in situations that involve a total or partial conflict of interests[1] between two or more individuals. Its primary concern is with the case where a rational individual is faced with an opponent who himself also behaves rationally (or with several such opponents), although its interest also extends to the case where a rational individual has to deal with an opponent (or opponents) liable to more or less irrational behavior.

If a given player could regard his opponents' strategies as *given*, the problem of rational behavior for him would reduce to the selection of a strategy that maximizes his expected utility, i.e., the expected value (or actuarial value) of his utility function. But the point is precisely that in general he cannot regard his opponents' strategies as given independently of his own. For, if his opponents behave rationally, then their strategies will depend on the strategy they expect *him* to follow, in the same way as his own strategy depends on the strategies he expects *them* to follow. Hence the fundamental problem of the theory of games is to establish criteria for defining what sorts of mutual expectations can be *rationally* entertained by the players of a game concerning one another's strategies.

II

In some special cases one can derive satisfactory criteria for rational behavior (rational decision rules) without making stronger assumptions than that: (1) each player tries to maximize his own expected utility (postulate of *individual utility maximization* or of *individual rationality*); and that (2) each player expects (and acts on the expectation) that the other player(s) will also try to maximize his (their) own expected utility (postulate of *mutually expected rationality*).

These two postulates we shall call the *weak* postulates of *rationality.* They involve only assuming that all players act rationally and also expect each other to act rationally, but do not involve any stronger assumptions about what sort of strategies the players expect each other to follow.

In particular, von Neumann and Morgenstern have shown [11, ch. 3] that in the case of the two-person zero-sum (or constant-sum) game, if mixed strategies are admitted, these two assumptions make it a rational decision rule for both players to use maximin strategies. This is so because neither player can obtain more than his maximin payoff if the other player plays rationally – as he is expected to under the expected-rationality postulate. On the other hand, by using a maximin strategy, each player can make sure to obtain at least that much; while if he uses some other strategy he may run the risk of obtaining even less than that.

In some other games, *dominance*[2] considerations make it possible to derive well-defined rational decision rules without making use of stronger assumptions than the postulates of utility maximization and of expected rationality.

By the utility-maximization postulate, a rational player will always give preference to a dominant strategy over a strategy dominated by it, and so will never use a dominated strategy at all (assuming that he has some undominated strategies). On the other hand, by the expected-rationality postulate, the other players will be aware of this and will choose their own strategies on this expectation.

Hence, to derive a rational decision rule for the players, we can replace the original game by a smaller one, by means of eliminating for each player all strategies dominated by some other strategy. Then, we eliminate all dominated strategies from this new smaller game, etc. In games where this procedure leaves us finally with one unique undominated strategy for each player, the rational decision rule for each player will be obviously to use this one undominated strategy.[3]

An important application of this principle is to *ultimatum games,* i.e., to bargaining games where one of the players can firmly commit himself in advance under a heavy penalty that he will insist under all conditions upon a certain specified demand (which is called his *ultimatum*). Let us call a penalty *adequate* if it is large enough to make it unprofitable for the first player to break his commitment, whatever strategy his opponent may use. (Sometimes the loss of face the player would suffer if he broke his commitment may be an adequate penalty). Moreover, let us call a demand by the first player *admissible* if its acceptance would still leave the second player better off than he would be in the conflict situation that would arise if no agreement could be reached. It can be shown by dominance considerations that it will be rational for the second player to accept any admissible demand to which the first player has committed himself under an adequate penalty. Consequently, it will be rational for the first player to commit himself to his *maximum demand*,[4] i.e., to the most extreme admissible demand he can make.[5, 6]

III

However, apart from such special cases, the weak rationality postulates of individual utility maximization and of expected rationality are in themselves quite insufficient for establishing well-defined decision rules for the players, and additional assumptions have to be made. In what follows we shall restrict ourselves to *cooperative games*,[7] where the players are allowed free pre-play communication and, what is more important, where *agreements* between the players are fully *enforced.* More particularly, we shall concentrate on two-person *simple bargaining games* because once we have a satisfactory theory for this important class of cooperative games, our results can be easily extended to the general two-person cooperative game and even to the n-person cooperative game.[8]

By a simple bargaining game we mean a situation where two individuals can obtain certain prizes if they can agree on how these should be divided between them: while if they cannot agree, then a conflict situation will arise with specified payoffs to both parties (e.g., in the absence of agreement both parties may simply retain their existing positions; or may have to pay certain specified fines; or may receive certain consolation prizes). Communication between the two players is completely free and the communication facilities are not biased in favor of either party. The players are also assumed to know each other's utility payoff functions.[9]

A simple bargaining game differs from a general cooperative game in that in the former the players' conflict payoffs are completely fixed by the rules of the game, while in the latter, these conflict payoffs are in general dependent on the retaliatory strategies that the players would actually use against each other in the conflict situation. A simple bargaining game thus corresponds to the special case where each player has only one possible retaliatory strategy (e.g., plain non-cooperation) but has no choice among several alternative retaliatory strategies.

A simple bargaining game differs from an ultimatum game (see Section II) in providing equal communication facilities for both players, so that neither player can unilaterally decide the outcome of the game by putting forward an ultimatum before the other player could take any effective counteraction. Given equal communication facilities, if ultimatums are permitted, both parties, if they are rational, will put forward an ultimatum simultaneously at the very moment the game starts – or alternatively, they may rule out ultimatums altogether by common consent (or even by simple refusal to consider any ultimatum the other party may try to make).

Accordingly, the conclusions we have reached for ultimatum games have no application to simple bargaining games. In ultimatum games it was a rational strategy for one of the players to put forward an ultimatum containing the most extreme admissible demand he could make. But in simple bargaining games, if ultimatums are allowed at all, both parties will make an ultimatum; but then both parties will have an incentive to moderation because, if the two players' ultimatums contain incompatible demands, a conflict will result, unprofitable to both parties. Thus, in a simple bargaining game it is not a rational strategy against a rational opponent to insist on an extreme demand. The rational strategy is rather to be ready for some sort of *compromise*; and the main task of game theory in the case of simple bargaining games is to decide *what kind* of compromise will represent a rational strategy for both players.[10]

At the same time, the rational compromise that game theory tries to define is a compromise of a rather special sort. It is a compromise that two rational bargainers will tend to reach if their behavior is determined solely by *self-interest* (or, more generally, by those personal preferences and value judgments that enter into each player's utility function) as well as by consideration of what concessions they are likely to obtain from the other party. That is, altruism, fairness, and other moral considerations can come in only so far as they are included in each party's (or both parties') utility function(s). But the emergence of a compromise in the model is not to be made dependent on the two parties' sharing a common set of moral principles (or on their willingness to accept the moral principles[11] of a third party, say, an arbitrator). One of the purposes of our analysis is to show – contrary to statements sometimes made in the literature – that it is *possible* to define a unique rational compromise purely in terms of the two parties' self-interest (as expressed by their utility functions) and in terms of certain rationality criteria, without making use of any moral criteria at all.

That is, for the sake of conceptual clarity, among the various theories dealing with bargaining situations, we must clearly distinguish between *bargaining theories* in a narrower sense, and *arbitration theories.* The former are parts of game theory proper and try to define solutions based on the rational pursuit of self-interest (or, more generally, on the rational pursuit of individual utility), while the latter are parts of ethics or welfare economics and try to define solutions satisfying certain moral criteria. In this sense Braithwaite's and Raiffa's theories [1; 12, pp. 361-87] are arbitration theories while the Zeuthen-Nash theory (which we shall discuss) is a bargaining theory proper.[12]

IV

As the postulates of individual utility maximization and of expected rationality do not define a unique determinate solution for the simple bargaining game, Nash has suggested certain further postulates which achieve this purpose [9]. As I have shown elsewhere [3], the Nash solution can also be obtained from weaker and, I believe, more fundamental postulates if one makes use of the Zeuthen model of the bargaining process. I now propose to present a revised form of my original argument, with a view to more clearly bringing out the logical function of each postulate. (I have also tried to make other improvements over my earlier exposition.)

The purpose of the Zeuthen model is to find a rational decision rule which tells each bargainer at any given stage of the bargaining process whether he should himself make the next concession or should wait for his opponent to make the next concession.[13]

We shall use the following six postulates:

1. Individual Utility Maximization (as in Section II)

Definitions. Let R_1 and R_1' be alternative decision rules for player 1 while R_2 and R_2' be rules for player 2. We shall call the pair (R_1, R_2) more efficient than the pair (R_1', R_2') if the use of R_1 and R_2 would yield higher payoffs to both players than would the use of R_1' and R_2'. We shall call the pair (R_1, R_2) *most efficient* within a given set S of decision rules if there is no pair more efficient than (R_1, R_2) within the set S.

2. Efficiency

(A) Out of the set of all decision rules consistent with postulates 1, 3, 4, 5, and 6, the players will select a pair (R_1, R_2), of most efficient decision rules. (B) However, if there is an alternative pair (R_1', R_2') of decision rules even more efficient than (R_1, R_2), then the players will adopt (R_1', R_2') instead of (R_1, R_2) so long as R_1' and R_2' are consistent with postulates 3, 4, 5, and 6, even if they are *not* consistent with postulate 1.

3. Acceptance of Higher Payoffs

Whenever a given player makes a certain offer to the other player, this means that he will accept the terms specified in this offer and, *a fortiori,* any terms even more favorable to him.

4. Symmetry

The decision rules of the two players have the same mathematical form and differ only in having the variables associated with the two players interchanged.

5. Restriction of Variables

The decision rules of the players involve only such variables as have to be included by virtue of our present postulates.

6. Mutually Expected Rationality

The players expect each other to follow decision rules consistent with the present postulates.

These six postulates we shall call the *strong* postulates of *rationality.*

Among the new postulates, the efficiency postulate is a natural extension – and also a natural qualification – of the utility maximization postulate. Rational pursuit of self-interest by the two players not only requires that they should individually maximize their utilities and should choose efficient individual strategies. It also requires that they should choose individual strategies which, taken in conjunction, represent an efficient *joint* strategy. Indeed, self-interest demands even that the players should suspend independently maximizing their utilities if both can benefit by switching to a more efficient joint strategy not based on individual utility maximization. (We shall use this efficiency postulate only for excluding individual utility-maximizing strategies which would lead to unnecessary conflicts between the two players.)

The postulate on the acceptance of higher payoffs can also be regarded as a *rationality postulate* concerning the rational pursuit of self-interest by the players.[14] Or, alternatively, it may be taken as an *institutional assumption* about the nature of the bargaining process envisaged by our model. That is, we may assume that the rules of our bargaining game require that at any given stage of the bargaining process each player must state his *minimum* demand, which then automatically commits him to accept, not only the terms actually specified in this demand, but also any alternative terms even more favorable to him.

While postulates 1, 2, and possibly 3 have the function of stating what the rational pursuit of self-interest by the players demands, postulates 4, 5, and 6 have the function of stating, in terms more specific than the expected-rationality postulates of Section II, what expectations rational players can consistently entertain about each other's behavior.

Postulate 6, like the expected-rationality postulate of Section II, is a direct expression of the principle that the two players expect each other to act rationally. But now this means that they must expect each other to act in conformity not only with the utility-maximization postulate but with our other, newly adopted, postulates as well.

Postulate 4 is also based on the principle that the two players expect each other to act rationally. Its purpose is to exclude bargaining strategies that a given player could adopt only on the expectation that the other player would follow a different, and less rational, bargaining strategy. In particular, it wants to rule out that a player should act on the expectation that a rational opponent would make a concession in a situation in which he himself, following his own criteria of rational behavior, would refuse to make a concession.

Finally, postulate 5 serves to exclude, from the players' decision rules, irrelevant variables extraneous to our rationality postulates. If we dropped postulate 5, our remaining five postulates would not rule out some quite arbitrary, or even silly, decision rules. For instance, it would be quite compatible with our other five postulates that the players should divide any joint profit in proportion to their telephone numbers, or in proportion to the logarithms of their waist measurements, etc.

The only reason why a rational player could possibly accept such an arbitrary decision rule would be the expectation that his opponent would insist on an agreement along these lines. Likewise the only reason why a rational opponent could possibly accept such a decision rule would be the expectation that the first player would insist on it. But, assuming that a fully rational decision rule exists, neither player can have any reason to expect that the other player will adopt such an arbitrary and, on the face of it, quite irrational decision rule – at least if he thinks that the other player is also a rational individual.

The question is, of course, whether in actual fact there is a sufficiently definite, fully rational decision rule for the players to follow. This question we shall be able to answer in the affirmative only when we have shown (as we propose to do in Section V) that our rationality postulates do furnish a well-defined decision rule for the players, which provides a unique solution for the game.

That is, the final justification for excluding extraneous variables from the players' decision rules lies in the fact that it is possible to define a satisfactory rational decision rule not involving any extraneous variables. This means that, in strict logic, our acceptance of postulate 5 at this stage must remain tentative. We can fully accept it as a basic assumption of our theory only when we have satisfied ourselves in Section V that postulate 5, together with our five other postulates, does yield a uniquely defined decision rule for the players.

We shall come back again to postulates 4 and 5 in Sections VI and VII when comparing our own theory with Schelling's theory of bargaining.

To sum up what has been said about the last three postulates, postulate 6 directly states that the two players expect each other to act rationally; postulate 4 excludes bargaining strategies which depend on the expectation that the opponent will use a less rational bargaining strategy than the player concerned himself does; and postulate 5 excludes bargaining strategies based on the expectation that the opponent's strategy will depend on irrelevant variables extraneous to the criteria of rational behavior.

V

At a given stage of the bargaining process, let A_1 be player 1's last offer and A_2 be player 2's last offer. Suppose that the terms of A_1 would yield the utility $U_1(A_1)$ to player 1 and the utility $U_2(A_1)$ to player 2, whereas the terms of A_2 would yield the utilities $U_1(A_2)$ and $U_2(A_2)$ to the two players. Let $U_1(C)$ and $U_2(C)$ be the utilities that the two players would assign to the conflict situation which would arise if no agreement could be reached. Finally, let p_{12} be the subjective probability that player 1 attaches to the possibility that player 2 will firmly stick to his last offer A_2 and will make no further concession; and let p_{21} be the subjective probability that player 2 attaches to the possibility that player 1 will firmly stick to his last offer A_1 and will make no further concession.

What decision rules will the two players follow as to making further concessions?

First let us make the provisional simplifying assumption that both players restrict their choices to the extreme strategies of *either* fully accepting their opponent's last offer *or* fully keeping to their own last offer.

Under this assumption, by postulate 1, player 1 will stick to his own last offer A_1 if

$$p_{12} \quad U_1(C) + (1 - p_{12}) \cdot U_1(A_1) > U_1(A_2)$$

that is, if

$$\text{(a)} \qquad p_{12} < q_1 = \frac{U_1(A_1) - U_1(A_2)}{U_1(A_1) - U_1(C)}$$

and will accept his opponent's last offer A_2 if the inequalities are reversed.

By similar reasoning, player 2 will stick to his last offer A_2 if

$$\text{(b)} \qquad p_{21} < q_2 = \frac{U_2(A_2) - U_2(A_1)}{U_2(A_2) - U_2(C)}$$

and will accept his opponent's last offer A_1 if the inequality is reversed.

Comparison of the two sides of inequality (a) could serve as a decision rule for player 1 if he could attach a numerical value to p_{12}.

By postulate 6, he will be aware of the fact that player 2's behavior is governed by inequality (b). Consequently, in view of the definition of p_{12}, he will know that his assessment of p_{12} will be realistic only if it satisfies the condition

$$\text{(c)} \qquad \begin{array}{l} p_{12} = 0 \text{ if } p_{21} > q_2 \text{ and} \\ p_{12} = 1 \text{ if } p_{21} < q_2 \end{array}$$

Hence he will look for a decision rule consistent with condition (c).

On the other hand, in view of postulate 4, by symmetry, if a decision rule satisfies condition (c) it must also satisfy

$$\text{(d)} \qquad \begin{array}{l} p_{21} = 0 \text{ if } p_{12} > q_1 \text{ and} \\ p_{21} = 0 \text{ if } p_{12} < q_1 \end{array}$$

Conditions (c) and (d) taken together allow only three possible cases:

(e) *Either*

(i) $p_{12} = 0$ and $p_{21} = 1$,

or (ii) $p_{12} = 1$ and $p_{21} = 0$,

or (iii) $p_{12} = q_1$ and $p_{21} = q_2$

But it still has to be decided when each of these three cases applies.

Now, (c) makes p_{12} a step function of p_{21} and q_2, so that one can write

$$\text{(f)} \qquad p_{12} = F(p_{21}, q_2)$$

In view of postulate 4, by symmetry

$$\text{(g)} \qquad p_{21} = F(p_{12}, q_1)$$

(f) and (g) are two simultaneous equations in the two unknowns, p_{12} and p_{21}. By symmetry their solution must be of the form

(h) $$p_{12} = G(q_1, q_2) \text{ and}$$
$$p_{21} = G(q_2, q_1)$$

Thus our postulates make p_{12} and p_{21} functions of the variables q_1 and q_2. But they do not make them functions of any other variables. Hence, by postulate 5, they are in fact not functions of any other variables.

By postulates 3 and 6, p_{12} must be a monotone nondecreasing function of $U_1(A_1)$ if all other utilities are held constant while p_{21} must be a monotone nondecreasing function of $U_2(A_2)$. Therefore G must be a monotone nondecreasing function of its first argument.

All functions G satisfying this monotonity requirement and at the same time also satisfying postulate 4 as well as condition (e) must be of the following form. They must be defined over the square $0 \leqslant q_1 \leqslant 1; 0 \leqslant q_2 \leqslant 1$, dividing this square into three regions corresponding to cases (i), (ii), and (iii) of condition (e). (See diagram.) Region (i) must include the q_1 axis while region (ii) must include the q_2 axis. Region (iii) must include the line $q_1 = q_2$ and must be symmetric with respect to this line. Any straight line running parallel to the q_2 axis in the positive direction must traverse regions (i), (iii), and (ii) in this order, whereas any straight line running parallel to the q_1 axis in the positive direction must traverse these regions in opposite order.

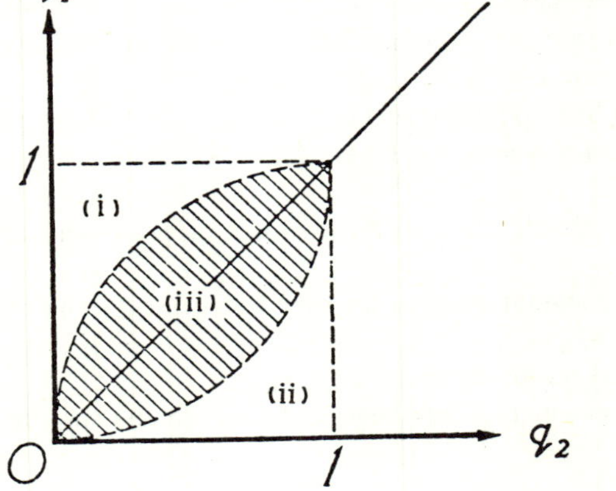

Any such function G corresponds to a possible decision rule for the players to the effect that:

in region (i) player 1 should stick to his last offer A_1, and player 2 should yield;

in region (ii) player 2 should stick to his last offer A_2, and player 1 should yield;

whereas:

in region (iii) the players should use mixed strategies, so that player 1 is sticking to his last offer A_1 with probability q_1 and is yielding with probability $(1 - q_1)$ while player 2 is sticking to his last offer A_2 with probability q_2 and is yielding with probability $(1 - q_2)$.

Different functions G, of course, will define different regions (i), (ii), and (iii), and so will furnish different decision rules. Let S be the set of all decision rules corresponding to any admissible function G. All decision rules in this set will be consistent with postulates 1, 3, 4, 5, and 6.

However, adoption by the players of any decision rule in set S will result in conflicts between them in region (iii), with probability $\bar{p} = q_1 q_2$. Therefore the most efficient decision rule in set S is the one which, of all members of S, reduces region (iii) to the smallest possible size, i.e., which reduces region (iii) to the line $q_1 = q_2$ itself. This means that all points satisfying the inequality $q_1 > q_2$ will be assigned to region (i) while all points satisfying $q_1 < q_2$ will be assigned to region (ii). Let us call this decision rule R^*. By part A of postulate 2, the players cannot adopt any other decision rule in set S than (possibly) R^*.

At the same time, R^* is still not a perfectly efficient decision rule because its adoption would still result in conflicts between the players in the case $q_1 = q_2$, with probability $\bar{p} = q_1 q_2 = q_1{}^2$. These conflicts could be avoided if in region (iii), i.e., on the line $q_1 = q_2$, both players would *always* yield instead of adopting a mixed strategy. Let this modified form of R^* be called R^{**}. By part B of postulate 2, R^{**} supersedes R^*.[15]

So we obtain decision rule R^{**} as the only decision rule consistent with all our rationality postulates. It says that:

player I should stick to his own last offer A_1 and player 2 should yield if $q_1 > q_2$;

player 2 should stick to his own last offer A_2 and player 1 should yield if $q_1 < q_2$; and

both players should yield if $q_1 = q_2$.

This decision rule has been obtained on the simplifying assumption that the two players restrict their choices to the two extreme strategies of *either* fully accepting their opponent's last offer *or* fully sticking to their own last offer.

Now let us drop this simplifying assumption. Suppose that the two players' last offers A_1 and A_2 are such that $q_1 > q_2$. It will still be true that, so long as neither player makes a new offer, our rationality postulates will require player 2 to accept player 1's last offer A_1 in full. But now player 2 can avoid this extreme move, by making a new offer A_2' intermediate between his own last offer A_2 and his opponent's last offer A_1. As making this new offer will be preferable for him to full acceptance of A_1, it will be the rational course for him to take. (On the other hand, player 1 will have no incentive to make a new offer and, even less, to accept player 2's last offer A_2).

Hence, if $q_1 > q_2$, it will be player 2 who has to make a new concession (although he need not go as far as fully accepting his opponent's last offer).

By similar reasoning, if $q_1 < q_2$, it will be player 1 who has to make the next concession (but again he need not go as far as fully accepting his opponent's last offer).

Finally, if $q_1 = q_2$ then both players will have to make some concessions.

This is the final form of the decision rule we obtain.

It can be shown [3] that if the two players follow this decision rule they will tend to reach an agreement at the point A where the product

$$P = [U_1(A) - U_1(C)] \cdot [U_2(A) - U_2(C)]$$

takes its maximum value. This point, however, is unique and is the Nash solution of the bargaining game.

VI

In my earlier paper I did not introduce the restriction-of-variables postulate as a separate assumption distinct from the symmetry postulate. But now I think that the logical structure of our model comes out more clearly if these two assumptions are made separate postulates.

Out of these two postulates, in my view, the restriction-of-variables postulate is the really crucial assumption. This is because our model differs from alternative theories of bargaining, such as Raiffa's, Braithwaite's, and Schelling's, mainly in making the two parties' bargaining strategies dependent only on a restricted set of variables, viz., on those variables which are directly relevant for the rational pursuit of individual utility by the two parties – whereas Raiffa and Braithwaite make the two parties' strategies dependent on certain ethical criteria, and Schelling makes them dependent on certain psychological considerations. In our model these ethical and psychological variables are excluded by the restriction-of-variables postulate.

Unlike the restriction-of-variables postulate, the symmetry postulate is by no means peculiar to our model. On the contrary, the symmetry postulate has to be satisfied, as a matter of sheer logical necessity, by *any* theory whatever that assigns a unique outcome to the bargaining process. This is true irrespective of whether the theory in question means to make a realistic prediction, or means to set up a normative standard – and, in the latter case, whether this normative standard is based on individual rationality or is based on moral criteria. In any case such a theory must make the outcome (the dependent variables) a function of some particular independent variables, such as the variables describing the players themselves, their physical environment, their rights and duties within the game, etc. As any theory must apply to *both* players, if the two players happen to be equal with respect to all relevant *independent* variables they must be assigned full equality also with respect to the *dependent* variables, i.e., with respect to the outcome. But this is precisely what the symmetry postulate says. Different theories of bargaining may differ in what variables they regard as the *relevant* independent variables – but, if the two players are equal on *all* variables regarded by the theory as relevant, the theory must allot both players the same payoffs.[16]

Although Schelling objects to the symmetry postulate, his own theory, also, cannot avoid conforming to this postulate if it proposes to make any definite predictions, even in a mere probabilistic sense, about the outcome of the bargaining process [16]. For instance, he argues that, even in an otherwise symmetric bargaining game, rational players may come to divide their joint profit in proportion 80:20 (instead of 50:50 as predicted by the Zeuthen-Nash theory in such cases) because some object or event in their present environment or in their past history may catch their fancy and may suggest a division of 80:20 [16, p. 220]. (For example, they may see a sign on a blackboard suggesting this particular proportion, or they may hear of a precedent where two other players adopted this proportion, etc.) This account shows that for Schelling's theory the set of potentially relevant independent variables is much larger than for our own theory. He would include certain environmental variables (e.g., blackboard signs, precedents, etc.) which we would regard as irrelevant for decision-making by rational players. However, given this larger set of relevant variables under Schelling's theory, the general psychological laws determining the outcome must be the same for the two players, and therefore the outcome (or the probability distribution of possible outcomes) must be a symmetric function of the independent variables if *all* relevant independent variables are taken into account. If the blackboard sign suggested that player B should get 80% and player A 20%, instead of suggesting the opposite (and if all other relevant psychological and non-psychological variables were also interchanged between A and B) then the outcome predicted by the theory would also have to be exactly the opposite, with B now getting the 80% and A now getting the 20%. (In a stochastic model the relevant probabilities, rather than the outcome itself, would be reversed, with the proportion 20:80 now being assigned the same probability as was assigned to the proportion 80:20 in the original case, etc.) Schelling's theory could avoid this conclusion only if it were to assume that human behavior in bargaining situations were not governed by any kind of general laws at all – not even by probabilistic laws.

Thus, Scheling cannot really avoid the symmetry postulate if he is to propose any definite theory of bargaining at all; and the real difference between his theory and our own lies not in our symmetry postulate, but rather in our restriction-of-variables postulate, which excludes many of Schelling's variables as being irrelevant for decision-making by rational players.

VII

The essential effect of our restriction-of-variables postulate is to exclude, from the players' decision rules, all variables which are not directly involved in the two players' choice of utility-maximizing bargaining strategies. According to inequalities (a) and (b) of Section V, the only variables relevant to the choice of utility-maximizing strategies are q_1, q_2, p_{12}, and p_{21}. Our subsequent argument in the same section then shows that, out of these four quantities, the variables p_{12} and p_{21} are dependent variables determined by q_1 and q_2. This leaves only q_1 and q_2 as the independent variables which must govern the two players' behavior at each particular stage of the game. Only when this conclusion had been established could the symmetry and the efficiency postulates be used to find the specific mathematical form that the two players' decision rules must take.

The rationale behind the restriction-of-variables postulate is the fact that *either* player's bargaining strategy would depend on additional variables (e.g., on those occurring in Schelling's theory) only if he had some reasons to believe that the *other* player's bargaining strategy would depend on these variables. For instance, by inequality (a), player 1's strategy would depend on the additional variable X (say, on the presence of a certain sign on the blackboard) only if he had reasons to think that p_{12} (which represents player 2's expected behavior) should depend on X. But, as both players are assumed to be rational individuals, in actual fact neither of them will have any reason to think that the other player will be influenced by X, or indeed by any other variable intrinsically unrelated to utility maximization.

The only reason why rational players could expect each other to be influenced by such extraneous variables would be the impossibility of defining a unique rational solution in terms of the directly relevant independent variables q_1 and q_2 alone. If no such solution could be found this would mean that no rational criteria existed to choose among alternative possible agreements. As the two players would still have a clear interest in reaching *some* agreement, i.e., in making *some* mutually consistent choices, it would be no doubt rational for them to bring in additional variables if this were thought to increase the chances of reaching an agreement. If any way of dividing one hundred dollars between the two players were equally rational (or equally arbitrary) then there could not be any objection on rationality grounds to the suggestion of dividing this amount in the proportion suggested by an accidental blackboard sign, or suggested by an arbitrary precedent, or, for that matter, even in the proportion of their telephone numbers. But the main result of our argument has been that it is *possible* to define a unique rational solution in terms of the directly relevant independent variables alone. Consequently, rational players can have no reasonable ground for making their bargaining strategies dependent on other variables intrinsically irrelevant for utility maximization.

VIII

This last conclusion can be further corroborated by comparing bargaining games with *coordination games.* Schelling has developed an ingenious theory of tacit coordination games and bases this theory of bargaining games on an assumed analogy with such tacit coordination games [14]. But I shall try to show that this analogy breaks down – precisely because tacit coordination games by definition always lack a unique rational solution, whereas bargaining games with free communication always possess a rational solution unique in all strategically relevant respects.[17]

By a pure coordination game we mean a game in which the two (or more) players have exactly identical interests. Both win certain fixed payoffs if they manage to choose coordinated strategies, and both lose if they fail to do so. For example, let us denote player 1's strategies by α_1, α_2 ..., and player 2's strateiges by β_1, β_2 ... Let α_1 and β_1, or α_2 and β_2, etc., denote coordinated strategies. Then both players will win if they use strategies with the same subscript, and both will lose if they use strategies with different subscripts. But so long as they use strategies with identical subscripts it makes no difference whether they use the pair (α_1, β_1) or the pair (α_1, β_2), etc. Consequently, the players have no rational criterion to choose between such alternative winning strategy pairs.

Nevertheless, if they are free to communicate, coordination of their strategies poses no real problem. Either player may suggest one particular winning strategy pair, and the other player will agree at once as he has no reason to prefer some alternative pair of strategies. But if the two players cannot communicate, i.e., in a *tacit* coordination game, the solution of the coordination problem may require real ingenuity, as each player must try to guess what strategy the other player will use. As there is no intrinsically rational criterion to choose among alternative potential winning strategies, the players may try to fall back on any arbitrary selection rule which picks out a unique strategy pair so long as they have any reason to expect that the other player may hit upon the same arbitrary selection rule.

For example, two individuals may have to meet in a large city without having agreed in advance on any particular meeting place. Both will 'win' if they go to the same place, and both will 'lose' if they go to different places. But so long as both choose the same place, any possible meeting place will do equally well.

Now suppose that both individuals have read a novel (and each knows) that the other has read this novel) in which two people had met in the same city at the Town Hall. By itself this fact would not be a good reason for them to meet at the same place. But, as they have no *more rational* criterion to choose a meeting place, it may be a good strategy for them to try the Town Hall, as each individual may reasonably hope that the novel they have both read will suggest this meeting place to the other individual as well.

But if one particular pair of strategies were clearly preferable to all others on some clear-cut rational ground, (e.g., if some possible meeting place were clearly much more convenient to both parties than any other) then no additional extraneous selection rule would be called for. Of course, by definition, in a coordination game this never happens: any particular potential winning strategy is always just as rational as any other. In contrast, in a bargaining game there is a unique rational selection rule, which yields the Zeuthen-Nash solution: therefore there is no real analogy between the two cases.

IX

This point can be brought out even more clearly if we consider *tacit bargaining games* which happen to involve *both* a bargaining and a coordination problem.

In such games, in order to avoid a conflict, the players have to agree not only on how to distribute the payoffs between them but also on some other, more or less independent, aspects of their joint strategies, and have to reach these agreements without any explicit communication. For instance, suppose there are two armies in a certain area. Both would like to control as much territory as possible, but both would like to avoid a battle, and to reach a tacit agreement on some sort of demarcation line between them.[18] This situation involves a (tacit) bargaining problem as far as both parties would like to increase their own territories at each other's expense, though without actual fight. But it also involves a coordination problem because, even if the two parties did reach an agreement about the proportion of the territory to be held by each side, there would still be infintely many ways of dividing up the total territory in this given proportion, and so agreement would have to be reached also in detail on who should get which particular areas. In general, a tacit bargaining game always involves a bargaining problem as far as the two parties have opposite interests as to the distribution of the payoffs; and it *may* also involve a coordination problem if there are several alternative strategies to achieve the same given payoff distributions, so that the choice among these alternative strategies does not involve a conflict of interest between the players.[19]

Now suppose that the two armies are at the beginning separated by a great river which flows across the whole area and divides it in two parts. Also suppose that the river is the only conspicuous landmark in the whole region. Then, as far as the problem of *tacit coordination* is concerned, it will be a natural idea to accept the river as the desired demarcation line because this may be the only solution that can be achieved without detailed explicit negotiations. So far we are in full agreement with Schelling's theory of the coordination problem. But, how readily this solution will find actual acceptance will also depend on the *bargaining* aspect of the game, i.e., on the relative bargaining positions (power positions) of the two parties – an aspect of the game completely neglected by Schelling's theory. Even if the river should represent the only possible demarcation line so long as the game remains a strictly *tacit* bargaining game, it will be the relative bargaining positions of the two parties, together with the costs of communication between them, which will decide how long the bargaining in fact remains restricted to the tacit level. If one of the two parties comes to feel that the area on its own side of the river is much too small in relation to its military strength as compared with the other party's, it will surely try to shift the demarcation line beyond the river, either by explicit negotiations, or by using sign language, or simply by putting pressure on the other party by means of small-scale military attacks, etc. – unless the costs and risks associated with any of these methods of communication happen to outweigh the value of the disputed territory. In any case, no satisfactory analysis of the situation is possible without recognizing *both* the bargaining considerations (balance of power considerations) *and* the coordination considerations involved, which may very well pull in opposite directions. A theory relying on coordination considerations alone, like Schelling's, cannot do justice to the influence that the two parties' bargaining positions, i.e., the balance of power between them, have on the outcome.

X

Our theory of bargaining is based on certain rationality postulates and describes actual bargaining behavior only to the extent to which this behavior is in fact rational (or to the extent to which possible deviations from perfect rationality have no significant effect on the outcome). For instance, we do not claim the people's actual bargaining behavior is never influenced by irrelevant considerations. We have only argued against Schelling that, in bargaining games with free communication (and also in tacit bargaining games that do not involve a coordination problem), if a player is influenced by irrelevant considerations this represents *irrational* behavior. That is, our contention is only that it is preferable – both on grounds of conceptual clarity and on grounds of analytical usefulness – to define rational behavior in such a way that it excludes behavior dependent on irrelevant considerations.

Models of rational bargaining behavior are meant to be used for three main analytical purposes:

(1) for the *prediction* or *explanation* of the outcome of actual bargaining behavior;

(2) for *conceptual clarification* of what sort of behavior really represents rational behavior in bargaining situations; and

(3) for defining a theoretical *bargaining equilibrium point* for any given bargaining situation, i.e., for determining that particular payoff distribution which would represent the stable equilibrium position between perfectly rational bargainers, and which therefore can be regarded as the payoff distribution corresponding to the real *balance of power* between the two parties.

Experience has shown in economics, political science, organization theory, and other fields that models based on rationality postulates can often be used with reasonable success to predict or explain actual social behavior. In the case of bargaining, of course, it is obvious enough that the outcome is often affected by irrational behavior on the part of either party (or both parties).

Indeed, what we call bargaining skill largely consists in exploiting the opponent's proclivities to such irrational behavior. But it seems reasonable to assume that, in many cases at least, economic, political, diplomatic, or other, bargaining tends to result in agreements that fairly closely correspond to the real balance of power between the two parties, with irrational influences being no doubt present but having only marginal significance. Similarly, the distribution of economic and other values in society between rival social classes or other social groups, even when it is decided without any explicit bargaining tends to correspond to the relative bargaining positions (power positions) of the parties concerned, and can be regarded as the outcome of a (tacit) bargaining game. Naturally, it is the task of detailed empirical research to determine in what particular fields, and under what particular conditions, our models based on rationality postulates give reasonably realistic predictions of the actual outcomes.

But even in cases where bargaining models based on rationality postulates have little direct predictive value they still have important analytical functions. Even when people's actual bargaining behavior is not particularly rational it is still an interesting theoretical problem what *would be* rational behavior in that particular situation. Both recent controversies on the theory of bargaining and the earlier history of bargaining problems (e.g., the earlier history of the duopoly problem[20]) clearly show that it is by no means a trivial task to develop a consistent idea of what rational behavior really involves in bargaining situations. How far does it involve insistence on extreme demands and how far does it involve readiness for a suitable compromise? What factors determine the actual terms of a rational compromise? How is a compromise based on a rational pursuit of self-interest related to a compromise based on certain ethical considerations? In the case of bargaining among several parties, to what extent does rational behavior involve the forming of two or more rival coalitions? Etc.

Herbert A. Simon has made out a strong case for the view that realistic models of social behavior must be based not on the assumption of 'unlimited' rationality, but rather on the assumption of some sort of 'limited rationality'.[21] But the setting-up of a fruitful model of *limited* rationality seems to presuppose a clear notion of what the assumption of *full* rationality would imply: We can best decide how to restrict our rationality postulates when we clearly know what predictions our unrestricted rationality postulates would entail and how these predictions deviate from the empirical facts.

Finally, irrespective of what the actual outcome of bargaining is in a given particular case, the bargaining models based on rationality postulates –

such as Nash's theory of the general two-person cooperative game [9, 10] and the present writer's theory of the n-person cooperative game [3, 5] – define a unique bargaining equilibrium point which represents the outcome theoretically corresponding to the real balance of power among the bargaining parties in that particular situation. The concept of bargaining equilibrium point thus gives an exact meaning to the balance of power notion, often used by social scientists for analytical purposes, but lacking a clear meaning in the absence of a unique definition of what outcome would theoretically correspond to the relative power position among the different parties. The concept of the bargaining equilibrium point also allows us to judge the influence of the two parties' relative bargaining skills on the actual outcome: the better the terms a given party has achieved, *as compared with the theoretical bargaining equilibrium point* between the two parties, the more his bargaining skill must have exceeded that of his opponent(s).

However, in order that a model of rational bargaining behavior can perform these analytical functions of prediction or explanation, of conceptual clarification, and of definition of a unique bargaining equilibrium point, it must be based on a very sharp and unambiguous concept of rational behavior. That is why we oppose Schelling's suggestion of using a broader and less sharply defined concept of rational behavior, based on weaker assumptions than our symmetry and our restriction-of-variables postulates. Even if intrinsically irrelevant variables do play some role in actual bargaining behavior (although it seems to me that Schelling greatly overstates their actual importance) this does not mean that such bargaining behavior represents a rational strategy against a rational opponent – at least in cases where rational bargaining strategies not involving such irrelevant variables are available.[22] A concept of rational behavior broad enough to include bargaining strategies based on irrelevant variables would be, in my view, too vague to be of much use for analytical purposes.

XI

To conclude, we have argued that the basic analytical problem of game theory is to determine what sorts of *expectations* rational players can consistently entertain about one another's behavior. In the case of the two-person zero-sum game and in some other cases no stronger assumption has to be made than that the players mutually expect each other to act rationally by maximizing their expected utilities. In other cases, however, stronger assumptions are required, specifying in greater detail what it means to expect one's opponent to act rationally.

In the case of simple bargaining games (and more generally in the case of cooperative games), under our own version of the Zeuthen-Nash theory of bargaining, these stronger assumptions include a symmetry postulate and a restriction-of-variables postulate. We have contended that the latter, rather than the former, is the really crucial assumption which distinguishes our own theory from alternative theories of bargaining, in particular from Schelling's. While Schelling claims that even rational players may be influenced by various factors intrinsically unrelated to the balance of power between the two parties, we have argued that being influenced by such irrelevant factors represents irratinal behavior – unless the concept of rational behavior is broadened to such an extent as to become useless as an analytical tool.

On the other hand, our own theory differs from Raiffa's and Braithwaite's theories of bargaining in making no use of ethical postulates and in assuming that the players are concerned only with the rational pursuit of their own individual utility. (Ethical considerations are assumed to affect the players' behavior only insofar as they are incorporated in each individual player's utility function.)

Finally, we have briefly discussed three main uses, in theoretical analysis, of bargaining models based on rationality postulates: their use for prediction or explanation of the outcome of actual bargaining behavior; for conceptual clarification of what strategies rational behavior involves in bargaining situations; and for definition of a bargaining equilibrium point which represents the outcome theoretically corresponding to the real balance of power between or among the bargaining parties. We have argued that our bargaining models can serve these analytical purposes only if they are based on a sharp and unambiguous concept of rational behavior as assumed by the Zeuthen-Nash theory of bargaining.

NOTES

[1] As a limiting case one may also include the study of situations that involve a full identity of interests between two or more individuals, in the sense of Marschak's *theory of teams* [7,8] or Schelling's *theory of pure coordination games* [14].

[2] We say that a given strategy of a certain player dominates another of his strategies if it yields a *higher* payoff than this other strategy would against at least *one* possible combination of strategies by the other players; provided that against any *other* combination of strategies by the other players it yields a payoff *no lower* than this other strategy.

[3] This method of arriving at rational decision rules corresponds to what Luce and Raiffa call a *solution in the complete weak sense for the game* [6, pp. 108-9]. They also show that this procedure works even if in the end several undominated strategies are left over for some players, provided that these strategies satisfy certain equivalence and interchangeability requirements.

4 For the sake of simplicity we are assuming that this maximum exists, e.g., owing to the indivisibility of the smallest monetary unit. If this maximum does not exist because the set of all possible demands is not a closed set, then the first player will have no optimal strategy in the strict sense at all, but will have only ϵ-optimal strategies. That is, the set of his possible payoffs will have an upper bound that he can never reach but can approximate as closely as desired. But from a practial point of view ϵ-optimal strategies are just as good as true optimal strategies.

5 On the application of the dominance method to this case, see Schelling [13; 15, pp. 240-4].

6 In the two-person zero-sum game, also, the solution can be said to depend on *dominance* considerations in a generalized sense.

7 However, I shall show in a later publication that these same principles, with minor modifications, can also be used to define decision rules for non-cooperative games.

8 On the two-person simple bargaining game, as well as on the two-person general cooperative game, see Nash [9, 10] and Harsanyi [3]. On the *n*-person cooperative game, see Harsanyi [5, pp. 325-55].

9 This would mean, for instance, in the case of a duopoly situation, that the two duopolists would know each other's cost functions and their joint demand function, as well as each other's attitudes towards risk-taking (as expressed by their von Neumann-Morgenstern cardinal utility functions for money). In many situations these assumptions will represent a reasonable approximation to the facts. But our model can be easily extended also to the more general case where the players do not know each other's utility payoff functions, if we assume that each player acts on the basis of his subjective *estimate* of his opponent's utility function. See Harsanyi [4].

10 I cannot agree with Schelling's contention that simple bargaining games, in which the symmetrical communication possibilities available to the players rule out the effective use of ultimatums, have less theoretical interest than ultimatum games have [16, p. 128]. I would rather argue for the very opposite of this view. Admittedly, ultimatum games may involve some interesting types of tactical moves (e.g., tampering with communication facilities) which are absent from simple bargaining games. But, at the level of basic theoretical analysis, simple bargaining games pose some rather deep theoretical problems of rational behavior (precisely because they have no determinate solution so long as we assume only utility maximization and expected rationality) – whereas ultimatum games have a rather trivial solution (based on dominance considerations), as we have seen. Moreover, the outcome of an ultimatum game may greatly depend on rather accidental features of the communication system between the players, while the outcome of a simple bargaining game always depends on the fundamental balance of power between the two parties (i.e., on each party's ability to inflict damage and to confer benefits on the other party). Hence, to determine the theoretical *bargaining equilibrium point* corresponding to the fundamental balance of power between the parties, we have to use a simple bargaining game model (see Section X below).

11 Of course, if both parties do happen to ascribe high utility to the observance of the same moral principles, it is reasonable to require that our model should furnish a solution which satisfies these principles – this follows from the efficiency postulate.

12 It is no accident that both Raiffa (in one version of his theory) and Braithwaite use interpersonal utility comparisons while Nash's theory explicitly excludes such comparisons. As I have argued elsewhere [2], no satisfactory theory of ethics or welfare

economics is possible without interpersonal comparisons of utility. On the other hand, such comparisons have no place in models based solely on the rational pursuit of individual utility.

13 Zeuthen's model envisages a simple bargaining game where no ultimatums are used [18, ch. 4]. In contrast, in Nash's model both players put forward ultimatums simultaneously. However, as I have shown [3], the two models are mathematically equivalent, and the Nash model can be regarded as a 'normalized' form of the Zeuthen model, itself considered as an 'extensive' game.

14 Admittedly, in real life it may occur that a given bargainer would be willing to accept a moderate concession by his opponent, but refuses to accept an unexpectedly large concession. This can happen because the first bargainer can take such an over-large concession for a 'sign of weakness' by his opponent. That is, it may induce him to revise his assumptions about the form of the opponent's utility payoff function, and/or about the latter's proficiency (rationality) as a bargainer, and so make him expect, and therefore insist upon, even larger further concessions. But this case cannot arise in our model, where the players are assumed to know each other's utility payoff functions, and to be firmly convinced of each other's rationality; under such conditions an offer of better terms can only increase, and can never decrease, one's willingness to accept.

15 It will be noted that in our argument the only *essential* use of the efficiency postulate occurs in the previous paragraph, where part A of the postulate is used to select the decision rule R^* out of the set S. In the present paragraph part B of the postulate is used to replace R^* by R^{**} – which is done only in order to deal with the very special case of $q_1 = q_2$, a case that can occur only with zero probability (i.e., corresponds to a set with measure zero).

16 The symmetry postulate must be satisfied not only by deterministic theories but also by probabilistic theories of bargaining. Of course, if the outcome is made a random variable, then the probability distribution of possible outcomes, rather than the actual outcome itself, will become a symmetric function, with respect to the two players, of the relevant independent variables.

17 For bargaining games the Zeuthen-Nash theory always determines a unique solution in the sense of defining unique equilibrium payoffs to the two players. On the other hand, there may be several alternative coordinated joint strategies by which the players can achieve these equilibrium payoffs. However, so long as there is free communication, the coordination of their strategies poses no problem to the players.

18 The example has been taken from Schelling [14, p. 24, example 6].

19 This means that the Zeuthen-Nash solution gives no criterion to choose among these alternative strategies.

20 Modern critics of the classical theories of duopoly agree that almost all of these theories at one stage or another attribute some highly irrational forms of behavior to the two duopolists.

21 For a recent summary of Simon's views, see [17].

22 A bargaining strategy involving intrinsically irrelevant variables may be, however, a perfectly rational strategy against an irrational opponent whom one expects to be himself influenced by these irrelevant variables.

REFERENCES

[1] Braithwaite, Richard B., *Theory of Games as a Tool for the Moral Philosopher.* Cambridge, England: Cambridge University Press, 1955.

[2] Harsanyi, John C., 'Cardinal Welfare, Individualistic Ethics, and Interpersonal Comparisons of Utility', *Journal of Political Economy,* **63** (1955), 309-21.

[3] Harsanyi, John, C., 'Approaches to the Bargaining Problem Before and After the Theory of Games: A Critical Discussion of Zeuthen's, Hicks', and Nash's Theories', *Econometrica* **24** (1956), 144-57.

[4] Harsanyi, John, C., 'Bargaining in Ignorance of the Opponent's Utility Function', *Cowles Foundation Discussion Paper,* No. 46, 1957.

[5] Harsanyi, John, C., 'Bargaining Model for the Cooperative n-Person Game', in A. W. Tucker and R. D. Luce (eds.), *Contributions to the Theory of Games,* Vol. 4, Princeton, N.J.: Princeton University Press, 1959.

[6] Luce, R. Duncan, and Raiffa, Howard, *Games and Decisions.* New York: John Wiley & Sons, 1957.

[7] Marschak, Jacob, 'Elements for a Theory of Teams', *Management Science* **1** (1955), 127–37. (Reprinted as *Cowles Foundation Paper, No. 94.)*

[8] Marschak, Jacob, 'Towards an Economic Theory of Organization and Information', in R. M. Thrall, C. H. Coombs, and L. R. Davis (eds.), *Decision Processes.* New York: John Wiley & Sons, 1954, pp. 187–220. (Reprinted as *Cowles Foundation Paper,* No. 95.)

[9] Nash, John, F., 'The Bargaining Problem', *Econometrica* **18** (1950), 155–62.

[10] Nash, John, F., 'Two-Person Cooperative Games', *ibid.,* **21** (1953), 128–40.

[11] von Neumann, John, and Morgenstern, Oskar, *Theory of Games and Economic Behavior.* Princeton, N.J.: Princeton University Press, 1953.

[12] Raiffa, Howard, 'Arbitration Schemes for Generalized Two-Person Games', in H. W. Kuhn and A. W. Tucker (eds.), *Contributions to the Theory of Games,* Vol. 2. Princeton, N.J.: Princeton University Press, 1953.

[13] Schelling, Thomas, C., 'Essay on Bargaining', *American Economic Review,* **46** (1956), 281–306.

[14] Schelling, Thomas, C., 'Bargaining, Communication, and Limited War', *The Journal of Conflict Resolution* **1** (1957), 19–36.

[15] Schelling, Thomas, C., 'The Strategy of Conflict: Prospectus for the Reorientation of Game Theory', *ibid.* **2** (1958), 203–64.

[16] Schelling, Thomas, C., 'For the Abandonment of Symmetry in Game Theory', *Review of Economics and Statistics* **41** (1959), 213–24.

[17] Simon, Herbert A., 'Theories of Decision-Making in Economics', *American Economic Review* **49** (1959), 253–83.

[18] Zeuthen, Frederik, *Problems of Monopoly and Economic Warfare.* London: George Routledge & Sons, 1930.

CHAPTER 3

A SIMPLIFIED BARGAINING MODEL FOR THE n-PERSON COOPERATIVE GAME*[1]

1. INTRODUCTION

THE BARGAINING MODEL I propose to discuss in this paper[2] is a simplified and, as I hope, improved version of my bargaining model published in the Princeton *Contributions* (see [3]). Like my earlier model, this one is also a generalization of the Nash solution for the two-person cooperative game [6, 7] to the n-person cooperative game; and it is at the same time also a generalization of (a somewhat modified form of) the Shapley value [8] from the transferable-utility[3] case to the general case.

Indeed my new model is even more closely related both to the Nash solution and to the Shapley value, and preserves many of their desirable mathematical properties that my old model did not preserve. As a result, the payoffs it yields are in much better agreement with our intuitive notions about the various players' power positions and threat possibilities in each specific game.[4] In particular, to any given game Γ (which of course in general will not have transferable utility) my new model assigns as payoffs the modified Shapley values[5] of that game $\bar{\Gamma}$ which, among all games with transferable utility, most closely approximates in an appropriate sense the original game Γ.

Conceptually the main difference between my present model and my earlier one is this. Since formally both models are generalizations of Nash's theory of two-person cooperative games, both models have to

* Manuscript received November 28, 1960, revised July 31, 1962.

[1] An earlier version of this paper was written while the author was at the Australian National University, Canberra, Australia. It was read at the Stanford meeting of the Econometric Society in August, 1960. The author is indebted to Professor Leonid Hurwicz, of the University of Minnesota, and to Professor Gerard Debreu, now of the University of California at Berkeley, for helpful comments.

[2] I have been informed by Dr. E. L. Kaplan, of the University of California Radiation Laboratory, Livermore, that he had independently developed a very similar model for payoff determination in the n-person cooperative game.

[3] Transferable utility means that the players are free to make side payments to each other, so that the sum of the two players' utility payoffs will remain unaffected, the payee's net utility gain being always exactly equal to the payer's net utility loss. If side payments are made in money, transferable utility may be the result of all players' having linear utility functions for money.

[4] The new model also overcomes the objections raised by Isbell [4] against the old model.

[5] For definition of the modified Shapley value, see Section 3 below. Cf. also [3, (353–54)].

reduce any n-person cooperative game ultimately to some combination of two-person subgames. But in my earlier model this reduction was accomplished in two steps. First any n-person cooperative game Γ was reduced to a set of several-person bargaining subgames Γ^S played by the s members of each possible subset (syndicate or coalition) S of players. Then each such s-person subgame Γ^S was itself reduced to a set of two-person bargaining subgames Γ^S_{ij} between each possible pair i, j among the s members of any given syndicate S.

In contrast, my new model avoids the unnecessary intermediate step of several-person bargaining subgames Γ^S, and reduces any n-person cooperative game Γ directly to a set of two-person bargaining subgames Γ_{ij} between each of the $n(n-1)/2$ possible pairs i, j among the n players. This approach results in a closer and more direct conceptual and mathematical relationship between my present model and the Nash solution for two-person bargaining games. My new model also has the advantage of giving rise to a system of simultaneous equations much simpler and easier to handle than were the equilibrium conditions of my model.

2. THE NASH SOLUTION IN THE CASE OF FIXED DISAGREEMENT PAYOFFS

Before describing my new model in greater detail, I should like to discuss certain properties of the Nash solution and of the Shapley value, which we shall need later on.

Let P be the set of all possible payoff pairs $x = (x_1, x_2)$ that the two players of a given two-person cooperative game Γ can achieve by any possible joint strategy (including jointly randomized strategies) available to them. P is called the *prospect space* of the game and is assumed to be a compact and convex set of full dimensionality (i.e. of 2 dimensions). Let H be the *upper-right boundary* of P, defined as the set of points in P undominated (even in a weak sense) by any point of P. The equation of H will be written as

$$H(x_1, x_2) = 0 .$$

As on the upper-right boundary H the variables x_1 and x_2 are mutually strictly decreasing functions of each other, we have[6]

$$\text{(2.1)} \qquad \left[\frac{dx_i}{dx_j}\right]_{H=0} \leqq 0 , \qquad i, j = 1, 2 , \quad i \neq j ,$$

[6] It will be throughout assumed, wherever convenient, that the function H and its n-dimensional analogue possess all first derivatives. This will result in no loss of generality as our results can be easily extended to any game in which H has corner points, by means of making Γ the limit of a sequence of certain games Γ_k in which H possesses first derivatives everywhere.

where the equality sign can hold at most at the end points of the curve H.

Let H_1 and H_2 be the first derivatives of the function H with respect to x_1 and x_2. By (2.1) H_1 and H_2 will have the same sign. Without loss of generality we can assume that

$$H_1, H_2 \geqq 0 . \tag{2.2}$$

The equality sign again can apply at most at the end points of H.

In view of the convexity of the prospect space P we also have

$$\left[\frac{d^2x_i}{dx_j^2}\right]_{H=0} \leqq 0 , \qquad i, j = 1, 2 , \quad i \neq j . \tag{2.3}$$

Let us first consider the special case of a game where the two players obtain certain fixed disagreement payoffs t_1 and t_2 if they cannot agree on the distribution of final payoffs between them. Then, according to Nash's theory [6, (159)] the two players will agree on the final payoffs $x_1 = u_1$, $x_2 = u_2$ such that

$$(u_1 - t_1)\cdot(u_2 - t_2) = \max_{\substack{(x_1, x_2) \in P \\ x_i \geqq t_i, i=1,2 \\ t_1, t_2 = \text{const.}}} (x_1 - t_1)\cdot(x_2 - t_2) . \tag{2.4}$$

By using Lagrange's method of multipliers one sees immediately that (2.4) is equivalent to the requirements:

$$H(u_1, u_2) = 0 , \tag{2.5}$$

$$a_1\cdot(u_1 - t_1) = a_2\cdot(u_2 - t_2) , \tag{2.6}$$

where

$$a_i = H_i(u_1, u_2) , \qquad i = 1, 2 . \tag{2.7}$$

The second-order requirements are taken care of by the convexity of the prospect space P.

Equations (2.5), (2.6) and (2.7) show that the Nash solution has the following important property:

Invariance with respect to commensurate increments or decrements in the two players' disagreement payoffs t_1 *and* t_2. Suppose that in a given game Γ both players' disagreement payoffs are increased (decreased) to $t_1' = t_1 + \Delta t_1$ and $t_2' = t_2 + \Delta t_2$, with

$$a_1\cdot \Delta t_1 = a_2\cdot \Delta t_2 ; \tag{2.8}$$

whereas the upper boundary H of the prospect space is left unchanged. Then the resulting new game will have the same solution $u = (u_1, u_2)$ as did the original game Γ.

This property can be used in the analysis of bargaining games embedded in larger bargaining games as follows.

The solution of a bargaining game imbedded in a larger bargaining game: Suppose a given two-person bargaining game Γ^* is embedded as a subgame in a larger bargaining game Γ between the same two players. More specifically, let P^* and P be the prospect spaces of Γ^* and of Γ respectively, with $P^* \subset P$. Let H^* and H be the upper-right boundaries of P^* and of P.

Suppose that the main game Γ is a composite bargaining game played in two stages. In stage 1 the players have to play subgame Γ^*. That is, they have to agree on some payoff vector $u^* = (u_1^*, u_2^*) \in P^*$. If they cannot agree then they receive the prearranged disagreement payoff vector $t = (t_1, t_2) \in P^*$.

In stage 2 the players have to agree on a final payoff vector $u = (u_1, u_2) \in P$. If they cannot reach agreement at this stage, then they receive, as their disagreement payoff vector for stage 2, the payoff vector u^* already agreed upon in stage 1.

Though game Γ is played in two stages, it is strategically equivalent to, and has the same normal form as, a one-stage bargaining game with the same prospect space P and the same disagreement payoff vector t. Therefore its Nash solution u will be defined by equations (2.5), (2.6) and (2.7). Consequently the two players, if they follow the rationality postulates of Nash's theory, must play the two-stage game Γ in such a way as to arrive at a final payoff vector u defined by these equations. But this implies that, during stage 1 of the game (corresponding to subgame Γ^*), the players must choose a payoff vector u^* such that by using this u^* as disagreement payoff vector during stage 2, they will arrive at the same final payoff vector u as they would arrive on the basis of the original disagreement payoff vector t. But by equation (2.8) this means that u^* must satisfy

(2.9) $$a_1 \cdot (u_1^* - t_1) = a_2 \cdot (u_2^* - t_2) ,$$

where

(2.10) $$a_i = H_i(u_1, u_2) , \qquad i = 1, 2 .$$

On the other hand, efficiency requires that u^* should satisfy the condition

(2.11) $$H^*(u_1^*, u_2^*) = 0 .$$

Intuitively, this argument amounts to saying that a rational player i in stage 1 of the game will not agree to a payoff vector u^* which would worsen his bargaining position with respect to his final payoff u_i. Therefore, if both players act rationally, then in stage 1 they will adopt a payoff vector u^* which will not alter their relative bargaining power with respect to the final outcome of the game.

Thus if a given bargaining game Γ^* is embedded as a subgame in a larger bargaining game Γ, then the solution of Γ^* will be in general different from what it would be if Γ^* were played as an independent bargaining game. In the former case the solution u^* of Γ^* will be defined by equations (2.9), (2.10), and (2.11). In the latter case, in view of equations (2.5), (2.6), and (2.7), the solution u^* would be defined by the equations

(2.12) $$a_1^* \cdot (u_1^* - t_1) = a_2^* \cdot (u_2^* - t_2)$$

(2.13) $$a_i^* = H_i^*(u_1^*, u_2^*) \qquad i = 1, 2$$

together with equation (2.11), which equally applies in both cases.

Extension of this analysis to composite bargaining games Γ played in several stages (say, m stages) is straightforward. Let $P_1 \subset P_2 \subset \cdots \subset P_m = P$, where P_j is the prospect space for stage j, $P_m = P$ being the prospect space for the whole game as well as for stage m. In each stage j the players agree on a payoff vector $u^j \in P_j$. In stage 1 the disagreement payoff vector is t. In each later stage j the disagreement payoff vector is the vector $u^{(j-1)}$ agreed upon in the previous stage. Then for every stage j

(2.14) $$a_1 \cdot (u_1^j - t_1) = a_2 \cdot (u_2^j - t_2) .$$

Hence

(2.15) $$a_1 \cdot \Delta u_1^j = a_2 \cdot \Delta u_2^j ,$$

where

(2.16) $$\Delta u_i^j = u_i^j - u_i^{(j-1)} , \qquad i = 1, 2 .$$

3. THE NASH SOLUTION IN THE CASE OF VARIABLE DISAGREEMENT PAYOFFS

In the general case the disagreement payoffs t_1 and t_2 of a given two-person cooperative game Γ will not be constants directly determined by the rules of the game. Rather, they will be *variables* depending on the actual retaliatory strategies ϑ_1 and ϑ_2 the two players would use against each other in case they could not agree on the final payoffs. Hence

(3.1) $$t_i = X_i(\vartheta_1, \vartheta_2) , \qquad i = 1, 2 ,$$

where X_i is the payoff function of player i. We shall also write

$$t = (t_1, t_2) = X(\vartheta_1, \vartheta_2) .$$

Under Nash's model the two players announce their retaliatory strategies ϑ_1 and ϑ_2 in advance at the beginning of the game and therefore these are called the *threat strategies* of the two players [7, (130–36)].

Against any given $\vartheta_2 = \vartheta_2^0$, player 1's *optimal threat strategy* will be that particular $\vartheta_1 = \vartheta_1^0$ which maximizes his final payoff u_1, subject to the constraints (2.5), (2.6) and (2.7). Similarly, against any given $\vartheta_1 = \vartheta_1^0$, player 2's optimal threat strategy will be that particular $\vartheta_2 = \vartheta_2^0$ which maximizes his own final payoff u_2, subject to the same constraints.

Nash [7, (135–36)] has shown that in any two-person cooperative game Γ there exists a pair of threat strategies ϑ_1^0 and ϑ_2^0 *mutually optimal* against each other. Though there may be several such pairs of mutually optimal threat strategies, all will lead to the same final payoffs u_1 and u_2.

We shall now prove

LEMMA 1. *Let*

$$t = (t_1, t_2) = X(\vartheta_1^0, \vartheta_2^0) \tag{3.2}$$

and let $u = (u_1, u_2)$ *be the Nash solution corresponding to this disagreement payoff vector* t, *so that*

$$H(u_1, u_2) = 0 \,, \tag{3.3}$$

$$a_1 \cdot (u_1 - t_1) = a_2 \cdot (u_2 - t_2) \,, \tag{3.4}$$

where

$$a_i = H_i(u_1, u_2) \,, \qquad i = 1, 2 \,. \tag{3.5}$$

Then, in order that threat strategy ϑ_i^0 *of player* i *be optimal against threat strategy* ϑ_j^0 *of player* j ($i, j = 1, 2$ *and* $i \neq j$) *it is both necessary and sufficient that*

$$a_i t_i - a_j t_j = \max_{\substack{\vartheta_i \in \theta_i \\ a_i, a_j = \text{const.}}} [a_i X_i(\vartheta_i, \vartheta_j^0) - a_j X_j(\vartheta_i, \vartheta_j^0)] \,, \tag{3.6}$$

where θ_i *is the set of all strategies available to player* i *in the game.*

PROOF. Take any point $u^* = (u_i^*, u_j^*)$ on the upper-right boundary H of the prospect space. Consider the set $T(u^*)$ of all possible disagreement payoff vectors $t^* = (t_i^*, t_j^*)$ such that would make this point u^* the Nash solution of the game. By (2.6) the set $T(u^*)$ is given by the intersection of the prospect space P and the straight line $L(u^*)$ whose points t^* satisfy the linear equation

$$H_i(u^*) \cdot t_i^* - H_j(u^*) \cdot t_j^* = H_i(u^*) \cdot u_i^* - H_j(u^*) \cdot u_j^* = \text{const.} \tag{3.7}$$

Now, any point t^* of the prospect space will belong to one and only one such set $T(u^*)$. Hence, the family of all sets $T(u^*)$, $u^* \in H$ represents a partitioning of the prospect space P into these sets $T(u^*)$

as equivalence classes. Clearly the mapping T of all points u^* of H into the family of all sets $T(u^*)$ is continuous, and so is the inverse mapping T^{-1}.

Comparing two possible disagreement payoff vectors $t = (t_1, t_2)$ and $t^* = (t_1^*, t_2^*)$, t^* is called *more favorable* than t for player i if $t \in T(u)$, $t^* \in T(u^*)$, and $u_i^* > u_i$.

Let $T^+(u)$ be the set of all points t^* in the prospect space P such that

$$H_i(u) \cdot t_i^* - H_j(u) \cdot t_j^* > H_i(u) \cdot u_i - H_j(u) \cdot u_j \,, \tag{3.8}$$

or equivalently such that

$$a_i t_i^* - a_j t_j^* > a_i u_i - a_j u_j = a_i t_i - a_j t_j \,, \tag{3.9}$$

where the equivalence of (3.8) and (3.9) follows from (3.5) and (3.4).

Let $u^* = (u_i^*, u_j^*)$ be any point on the upper-right boundary H such that

$$u_i^* > u_i \,. \tag{3.10}$$

Since $a_i \geqq 0$ we can write

$$a_i u_i^* \geqq a_i u_i \,. \tag{3.11}$$

By the convexity of the prospect space

$$a_i u_i + a_j u_j \geqq a_i u_i^* + a_j u_j^* \,. \tag{3.12}$$

By (3.11) and (3.12)

$$a_i u_i^* - a_j u_j^* \geqq a_i u_i - a_j u_j \,. \tag{3.13}$$

Hence the point u^* must lie either in the set $T(u)$ or in the set $T^+(u)$. But it cannot lie in $T(u)$ because then, in view of (3.7) and (2.1), we would have $u^* = u$, which would be inconsistent with (3.10). Hence

$$u^* \in T^+(u) \,. \tag{3.14}$$

Now, for player i the point $t^* = u^*$ as disagreement payoff vector would be certainly more favorable than any point t in $T(u)$ because clearly $u^* \in T(u^*)$ and by (3.10) $u_i^* > u_i$. But, in view of the continuity of the mapping T^{-1}, by the Bolzano-Weierstrass theorem if for player i any point u^* of $T^+(u)$ is more favorable than are the points in $T(u)$, then *all* points t^* of $T^+(u)$ must be more favorable than the points in $T(u)$.

Hence if player i had any threat strategy ϑ_i^* yielding a disagreement point $t^* = X(\vartheta_i^*, \vartheta_j^0)$ belonging to $T^+(u)$, i.e. satisfying the inequality $a_i t_i^* - a_j t_j^* > a_i t_i - a_j t_j$, then ϑ_i^0 could not be optimal against ϑ_j^0. Thus ϑ_i^0 can be optimal against ϑ_j^0 only if it maximizes the expression on the right-hand side of (3.6). This proves the lemma.

COROLLARY. *In order that* ϑ_i^0 *and* ϑ_j^0 *be mutually optimal against each other it is both necessary and sufficient that the point t should satisfy equations* (3.2) *to* (3.6) *as well as the additional equation*

$$(3.15)\qquad a_i t_i - a_j t_j = \min_{\substack{\vartheta_j \in \theta_j \\ a_i, a_j = \text{const.}}} [a_i X_i(\vartheta_i^0, \vartheta_j) - a_j X_j(\vartheta_i^0, \vartheta_j)] ,$$

where θ_j *is the set of all strategies available to player j.*

The corollary follows from the fact that minimizing the expression $(a_i t_i - a_j t_j)$ is equivalent to maximizing the expression $(a_j t_j - a_i t_i)$. Since Nash has shown that a pair of mutually optimal threat strategies ϑ_i^0 and ϑ_j^0 always exists, there is always a pair ϑ_i^0 and ϑ_j^0 satisfying equations (3.2) to (3.6) as well as (3.15). But if such a pair exists then it must have maximin and minimax properties, and therefore the quantity $\Delta_i = a_i t_i - a_j t_j$ will correspond to the maximin (as well as the minimax) value of the expression in square brackets on the right-hand side of (3.6) and (3.15).

LEMMA 2. *Assuming that conditions* (3.2) *to* (3.5) *are fulfilled, if strategies* ϑ_i^0 *and* ϑ_j^0 *will satisfy equations* (3.6) *and* (3.15), *then*

$$(3.16)\qquad a_i t_i - a_j t_j = \max_{\vartheta_i \in \theta_i} \min_{\substack{\vartheta_j \in \theta_j \\ a_i, a_j = \text{const.}}} [a_i X_i(\vartheta_i, \vartheta_j) - a_j X_j(\vartheta_i, \vartheta_j)] .$$

This statement remains true if we reverse the order of the max and min operators (i.e. if we use the maximin operator instead of the minimax operator).

PROOF. The lemma follows from the fact that equations (3.6) and (3.15) entail equation (3.16). Let

$$(3.17)\qquad F(\vartheta_i, \vartheta_j) = a_i X_i(\vartheta_i, \vartheta_j) - a_j X_j(\vartheta_i, \vartheta_j) .$$

Then for any $\vartheta_i \in \theta_i$

$$(3.18)\qquad \begin{aligned} \min_{\vartheta_j} F(\vartheta_i, \vartheta_j) \leqq F(\vartheta_i, \vartheta_j^0) &\leqq \max_{\vartheta_i} F(\vartheta_i, \vartheta_j^0) \\ &= F(\vartheta_i^0, \vartheta_j^0) = \min_{\vartheta_j} F(\vartheta_i^0, \vartheta_j) , \end{aligned}$$

where the last two equalities follow from (3.6) and (3.15). Consequently

$$(3.19)\qquad F(\vartheta_i^0, \vartheta_j^0) = \max_{\vartheta_i} \min_{\vartheta_j} F(\vartheta_i, \vartheta_j) ,$$

as desired. We can prove by a similar argument that

$$(3.20)\qquad F(\vartheta_i^0, \vartheta_j^0) = \min_{\vartheta_j} \max_{\vartheta_i} F(\vartheta_i, \vartheta_j) .$$

LEMMA 3. *If the variables* u_i, a_i, *and* t_i $(i = 1, 2)$ *satisfy equations* (3.3), (3.4), (3.5), *and* (3.16), *then the vector* $u = (u_1, u_2)$ *is the Nash solution of the game.*

Proof. Equation (3.4) can be written in the form

$$\Delta_i = a_i t_i - a_j t_j = a_i u_i - a_j u_j \ . \tag{3.21}$$

On the other hand, equations (3.3) and (3.5) do not contain the variables t_1 and t_2. Consequently the choice of t_1 and t_2 is important only as far as it determines the value of Δ_1 (or of $\Delta_2 = -\Delta_1$).

But, by Lemma 2, if equation (3.16) is satisfied, then the value of $\Delta_i (i = 1, 2)$ will be the same as the Δ_i value satisfying equations (3.6) and (3.15). Therefore, by Lemma 1, if equations (3.16) as well as (3.3), (3.4) and (3.5) are fulfilled, we shall obtain the same vector u as that corresponding to a pair of mutually *optimal* threat strategies ϑ_1^0 and ϑ_2^0, but this vector u is the Nash solution, as desired.

Lemmas 1, 2 and 3 establish the following theorem.

Theorem 1. *In the general case with variable disagreement payoffs t_1 and t_2, the Nash solution can be defined by the following conditions, which are both necessary and sufficient*:

(A) $$H(u_1, u_2) = 0 \ ;$$

(B) $$a_i = H_i(u_1, u_2) \ , \qquad i = 1, 2 \ ;$$

(C) $$a_1 \cdot (u_1 - t_1) = a_2 \cdot (u_2 - t_2) \ ;$$

(D) $$t_i = X_i(\vartheta_1^0, \vartheta_2^0) \ , \qquad i = 1, 2 \ ;$$

(E) $$\begin{aligned} a_1 t_1 - a_2 t_2 &= a_1 X_1(\vartheta_1^0, \vartheta_2^0) - a_2 X_2(\vartheta_1^0, \vartheta_2^0) \\ &= \max_{\substack{\vartheta_1 \in \theta_1 \\ a_1, a_2 = \text{const.}}} \min_{\vartheta_2 \in \theta_2} [a_1 X_1(\vartheta_1, \vartheta_2) - a_2 X_2(\vartheta_1, \vartheta_2)] \ , \end{aligned}$$

or, equivalently,

$$a_1 t_1 - a_2 t_2 = \min_{\substack{\vartheta_2 \in \theta_2 \\ a_1, a_2 = \text{const.}}} \max_{\vartheta_1 \in \theta_1} [a_1 X_1(\vartheta_1, \vartheta_2) - a_2 X_2(\vartheta_1, \vartheta_2)] \ .$$

Conditions (D) and (E) can be omitted in the special case where t_1 and t_2 are constants directly determined by the rules of the game.

In order to obtain conditions that do not assume the existence of the derivatives H_1 and H_2 at the solution point $u = (u_1, u_2)$, we can replace conditions (A) and (B) by the following conditions:

(AA) $$a_i \geqq 0 \ , \qquad i = 1, 2 \ ,$$

(BB) $$a_1 u_1 + a_2 u_2 = \max_{\substack{(x_1, x_2) \in P \\ a_1, a_2 = \text{const.}}} (a_1 x_1 + a_2 x_2) \ ,$$

while conditions (C), (D) and (E) can be retained unchanged. In cases where H_1 and H_2 exist, these new conditions are equivalent to conditions (A)–(E). In cases where these derivatives do not exist, we can justify the new conditions by the method outlined in footnote 6 above.

4. THE MODIFIED SHAPLEY VALUE

Let Γ be an n-person cooperative game with transferable utility. Then the Shapley value of game Γ to player i can be defined as

$$u_i = \sum_{\substack{S \subseteq N \\ S \ni i}} \frac{(s-1)!\,(n-s)!}{n!} [v(S) - v(\bar{S})] , \tag{4.1}$$

where N stands for the syndicate (or coalition) of all players, n for the number of players in N, s for the number of players in each syndicate $S \subseteq N$, and $\bar{S} = N - S$ for the syndicate complementary to S. This definition is clearly equivalent to the one given by Shapley [8, (311, equation (11))].

Under Shapley's original definition the symbol $v(S)$ denotes the von Neumann-Morgenstern characteristic function of the game, corresponding to the *maximin* value of the joint payoff $U^S = \sum u_i^S$ $(i \in S)$ of each syndicate S.

However, as I have argued elsewhere [3, (353–54)], if the players' final payoffs are in fact determined according to equation (4.1), and if the players are free to make binding threats to one another, then one must redefine the symbol $v(S)$ in equation (4.1) as follows.

By equation (4.1), the final payoff u_i of each member i of syndicate S is maximized if the difference $D^S = U^S - U^{\bar{S}}$ is maximized. On the other hand, the final payoff u_j of any member j of the complementary syndicate $\bar{S}$ is maximized if $-D^S = U^{\bar{S}} - U^S$ is maximized, that is, if $D^S = U^S - U^{\bar{S}}$ itself is minimized. Hence the choice of threat strategies (conflict strategies) ϑ^S and $\vartheta^{\bar{S}}$ by the two complementary syndicates S and $\bar{S}$ can be regarded as a two-person zero-sum subgame, in which one side tries to maximize the same quantity D^S that the other side tries to minimize. This subgame will have a unique solution, corresponding to the maximin (as well as the minimax) value of this quantity D^S.

Therefore in equation (4.1) the symbol $v(S)$ should be re-interpreted as denoting, not the von Neumann-Morgenstern characteristic function of the game, but rather a modified characteristic function. It should be made to correspond, not to the maximin value of the joint payoff U^S itself, but rather to the maximin value of the difference $D^S = U^S - U^{\bar{S}}$. This gives the definition

$$\begin{aligned} v(S) &= U^S(\vartheta_0^S, \vartheta_0^{\bar{S}}) , \\ v(\bar{S}) &= U^{\bar{S}}(\vartheta_0^S, \vartheta_0^{\bar{S}}) , \end{aligned} \tag{4.2}$$

with

$$D^S = U^S(\vartheta_0^S, \vartheta_0^{\bar{S}}) - U^{\bar{S}}(\vartheta_0^S, \vartheta_0^{\bar{S}})$$
$$= \max_{\vartheta^S \in \theta^S} \min_{\vartheta^{\bar{S}} \in \theta^{\bar{S}}} [U^S(\vartheta^S, \vartheta^{\bar{S}}) - U^{\bar{S}}(\vartheta^S, \vartheta^{\bar{S}})]$$
$$= \min_{\vartheta^{\bar{S}} \in \theta^{\bar{S}}} \max_{\vartheta^S \in \theta^S} [U^S(\vartheta^S, \vartheta^{\bar{S}}) - U^{\bar{S}}(\vartheta^S, \vartheta^{\bar{S}})] ,$$

where θ^S and $\theta^{\bar{S}}$ are the sets of all joint strategies available to S and to $\bar{S}$.

If in (4.1) we define the symbol $v(S)$ in accordance with equation (4.2) then the resulting quantity u_i is called the *modified* Shapley value of the game to player i.

This modification also overcomes Luce and Raiffa's objection [5, (137–38)] to the original Shapley value, based on the latter's insufficient sensitivity to the players' threat possibilities.

Our own solution for the general n-person cooperative game will be a generalization of this modified Shapley value.

5. ADMISSIBLE STRATEGIES AND THE PROSPECT SPACE

We shall now resume discussion of our new bargaining model for the general n-person cooperative game.

We shall make the usual assumption that in a cooperative game Γ the n players are free to use any *jointly randomized* mixed strategy ϑ^N. The set of all joint strategies ϑ^N available to the n players (i.e. to the all-player syndicate N) will be denoted by θ^N. In the case of a conflict between two complementary syndicates S and $\bar{S}$, it will be assumed that each side is free to use a jointly randomized mixed strategy ϑ^S or $\vartheta^{\bar{S}}$ respectively, where now joint randomization is restricted to the members of the relevant syndicate S or $\bar{S}$. The set of all joint strategies ϑ^S available to syndicate S will be denoted by θ^S.

Let P be the set of all payoff vectors $x = (x_1, \cdots, x_n)$ that the n players can achieve by any joint strategy $\vartheta^N \in \theta^N$. P will again be called the *prospect space* of the game. P will be obviously a convex set. We shall also assume that it is a compact set of full dimensionality (i.e. of n dimensions).

Let $\tilde{P}$ be the set of all points $\tilde{x}$ in the utility space U^n of the n players, such that $\tilde{x}$ either is itself a point of P or is at least weakly dominated by some point $x \in P$. Clearly, $\tilde{P} \supset P$. $\tilde{P}$ will be called the *extended prospect space*. We shall assume that any player i is free to decrease voluntarily his own payoff x_i by any amount. This essentially amounts to making $\tilde{P}$ the effective prospect space of the game.[7]

[7] The motivation for this assumption will be explained later (Section 9). Of course, when our model is applied to the special case of games with transferable utility (where

This assumption also enlarges the strategy space θ^S available to each syndicate $S \subseteq N$ to an *extended strategy space* $\tilde{\theta}^S \supset \theta^S$.

The set H of all points x in the original prospect space P such that are not dominated, even weakly, by any other point $x^* \in P$ is called the *upper boundary* of the prospect space. On the other hand, the set $\tilde{H}$ of all points $\tilde{x} \in \tilde{P}$ such that are not strongly dominated by some other point $\tilde{x}^* \in \tilde{P}$ is called the *extended upper boundary*.

Obviously, $\tilde{H} \supset H$. Both H and $\tilde{H}$ are hypersurfaces of $(n-1)$ dimensions, convex to above.

The equation of $\tilde{H}$ (and of course of H) will be written as

$$H(x_1, \cdots, x_n) = 0 . \tag{5.1}$$

H_i will again denote the first partial derivative of the function H with respect to x_i.[8]

Since at any given point these partial derivatives must all have the same sign, we can assume without loss of generality that

$$H_i \geqq 0 , \qquad i = 1, \cdots, n . \tag{5.2}$$

At any point $x \in \tilde{H}$ at least one H_i must be nonzero. Indeed, at any point of H itself, except for the points on the boundary of H, all H_i must be nonzero.

6. DEFINITIONS AND ASSUMPTIONS

The purpose of my present model, like that of my original one [3], is to define a set of payoffs that represent the equilibrium outcome of bargaining among rational players. We shall assume that the final payoffs of the game are determined by a whole network of various agreements among the players; and we shall try to define the equilibrium conditions of mutual consistency and interdependence that these various agreements have to satisfy. That is, we shall assume that each such agreement will represent a bargaining equilibrium situation between the participants if all other agreements between the players are regarded as given.

As my earlier model, the present one is also based on the fundamental assumption that every possible subset (syndicate or coalition) S of the players will agree to cooperate in protecting their common interests

---

it always yields as payoffs the modified Shapley values of the game) the compactness assumption must be further relaxed. For in that case the prospect space will contain, as subsets, unbounded hyperplanes of the form $\sum x_i = \text{const.}$ But in the general case it is convenient to assume the compactness of at least the set P itself, which generates the noncompact set $\tilde{P}$.

[8] See Footnote 6 above.

against the rest of the players. More particularly, it will be assumed that the members of each syndicate S, including both the all-player syndicate N and each of the sectional syndicates $S \subset N$, publicly guarantee certain payments, called *dividends*, to one another and agree to cooperate in securing these dividends to all members of the syndicate. These dividends will be assumed to be additional to the dividends that each member of the syndicate may receive from other syndicates. A syndicate may also distribute negative dividends, i.e. may reduce the payoffs of its members by certain amounts. Any decision by a syndicate is supposed to require the consent of all its members. The dividend of player i from syndicate S will be denoted by w_i^S.

The final payoff of a given player i from the whole game will be denoted by u_i. It will simply represent the sum of the dividends to i from all syndicates of which he is a member. That is

$$u_i = \sum_{\substack{S \ni i \\ S \subseteq N}} w_i^S \,, \qquad \text{for all } i \in N \,. \tag{6.1}$$

We shall require that the final payoff vector $u = (u_1, \cdots, u_n)$ of the game should be efficient (Pareto-optimal), so that

$$u = (u_1, \cdots, u_n) \in H \,. \tag{6.2}$$

It will also be assumed that each sectional syndicate S backs the dividends it guarantees to its members, by announcing a threat strategy ϑ^S, i.e. a joint strategy that the members of S would use against the other players if they could not agree with them on the distribution of final payoffs. The threat strategies ϑ^S and $\vartheta^{\bar{S}}$ announced by a given syndicate S and by the complementary syndicate $\bar{S}$ fully determine the conflict payoffs u_i^S and $u_j^{\bar{S}}$ that each member i of S and each member j of $\bar{S}$ (i.e. all players) would receive in case of a conflict between S and $\bar{S}$. Thus we can write

$$u_i^S = X_i(\vartheta^S, \vartheta^{\bar{S}}) \qquad \text{for each } S \subset N \text{ and } i \in S \,, \tag{6.3}$$

where X_i denotes the payoff function of the game.

We want to exclude dividend-guaranteeing agreements which are in themselves unrealistic or are inconsistent with the other commitments of the participants. Therefore we shall require that the dividends guaranteed to any player i by a given syndicate S and by all its subsets taken together should not exceed the conflict payoff u_i^S, which is the largest payoff that the cooperative effort of the members of S could secure for him in the event of a conflict between S and $\bar{S}$. On the other hand, we also want to exclude dividend-guaranteeing agreements which are inefficient (not Pareto-optimal), and therefore also require that the sum of these dividends should not fall short of this conflict payoff, either. Therefore we postulate

(6.4) $$u_i^S = \sum_{\substack{R \ni i \\ R \subseteq S}} w_i^R \qquad \text{for each } S \subset N \text{ and } i \in S\,.$$

For the sake of notational symmetry, we also define the symbol u_i^N as

(6.5) $$u_i^N = \sum_{\substack{R \ni i \\ R \subseteq N}} w_i^R = u_i \qquad \text{for each } i \in N\,,$$

where the last equality holds by (6.1).

7. THE DIVIDEND-PROPORTIONALITY RULE

We shall first discuss the case where in a given game Γ all dividends w_i^S from the various syndicates S are nonnegative.

Consider the bargaining subgame Γ_{ij} in which players i and j have to agree on the distribution of dividends between them for all syndicates S of which both of them are members. That is, they have to choose dividends w_i^S and w_j^S for all $S \ni i, j$ and $\subseteq N$. We shall assume that for the purposes of subgame Γ_{ij} all other dividends within the game are regarded as given. Thus we regard as given all dividends w_i^S of player i from syndicates S of which player j is not a member; all dividends w_j^S of player j from syndicates S of which player i is not a member; and all dividends w_k^S of all other players $k \neq i, j$ from any syndicate S whatever. (In view of (6.1) this implies that the final payoffs u_k of all other players k are also regarded as given.)

With respect to subgame Γ_{ij}, the disagreement payoffs t_i and t_j of players i and j are obviously

(7.1) $$t_i = \sum_{\substack{S \subseteq N \\ S \not\ni j}} w_i^S\,, \qquad t_j = \sum_{\substack{S \subseteq N \\ S \not\ni i}} w_j^S\,.$$

In view of equation (6.1), the payoffs of the two players i and j from subgame Γ_{ij} are simply their final payoffs u_i and u_j from the main game Γ as a whole. Applying conditions (A), (B) and (C) of the Nash solution (see Section 3 above) to subgame Γ_{ij}, the payoffs u_i and u_j must satisfy:

(7.2) $$H(u_1, \cdots, u_n) = 0\,,$$

(7.3) $$a_i \cdot (u_i - t_i) = a_j \cdot (u_j - t_j)\,,$$

(7.4) $$a_m = H_m(u_1, \cdots, u_n)\,, \qquad m = i, j\,,$$

where only u_i and u_j are to be regarded as variables since all the other u_k's are considered as given.

(7.3), in view of (7.1), can also be written as

$$a_i \cdot \sum_{\substack{S\subseteq N\\ S\ni i,j}} w_i^S = a_j \cdot \sum_{\substack{S\subseteq N\\ S\ni i,j}} w_j^S . \tag{7.5}$$

But we are assuming that in subgame Γ_{ij} players i and j have to agree not only on the sums $\sum w_i^S$ and $\sum w_j^S$ for all syndicates S of which both are members, but also separately on each pair of dividends w_i^S and w_j^S for each particular syndicate $S \ni i, j$. Agreement on each such pair w_i^S and w_j^S can be regarded as a smaller bargaining subgame Γ_{ij}^S embedded in the larger subgame Γ_{ij}, which itself can be considered as a composite bargaining game. Hence, by (2.16) in each subgame Γ_{ij}^S separately the two players must agree on dividends w_i^S and w_j^S such that

$$a_i w_i^S = a_j w_j^S \tag{7.6}$$

where a_i and a_j are defined by (7.4). This requirement applies to every syndicate S of which both i and j are members. Thus if $R \ni i, j$ then also

$$a_i w_i^R = a_j w_j^R ,$$

where a_i and a_j are the same quantities as in (7.6). Consequently, for any two syndicates $S \subseteq N$ and $R \subseteq N$, if both players i and j are members of both S and R, we must have

$$\frac{w_i^S}{w_j^S} = \frac{w_i^R}{w_j^R} = \frac{a_j}{a_i} = \frac{H_j(u_1, \cdots, u_n)}{H_i(u_1, \cdots, u_n)} . \tag{7.7}$$

In view of this fact we shall call equation (7.6) the *dividend-proportionality rule.*

The n proportionality factors $a_1 = H_1(u_1, \cdots, u_n), \cdots, a_n = H_n(u_1, \cdots, u_n)$ determining the distribution of dividends within each syndicate S we shall call the *weights* of the game.

We shall now consider the case where some of the dividends w_i^S in the game are negative. We define the quantities

$$t_i^N = \sum_{\substack{S\ni i\\ S\subseteq N}} w_i^S , \tag{7.8}$$

$$t_i^S = \sum_{\substack{R\ni i\\ R\subset S}} w_i^R . \tag{7.9}$$

By (6.4) and (6.5)

$$t_i^N = u_i^N - w_i^N = u_i - w_i^N , \tag{7.10}$$

$$t_i^S = u_i^S - w_i^S . \tag{7.11}$$

It may happen that the dividends w_i^S that the various players i

receive from the sectional syndicates S are so large that the vector

$$t^N = (t_1^N, \cdots, t_n^N) \notin P . \tag{7.12}$$

In such cases requirement (6.2) can be satisfied only if at least some of the dividends $w_1^N, \cdots, w_n^N$ are negative.

Likewise, it may happen that the dividends w_i^R of player i from the various subsets R of S are so large that

$$t_i^S > u_i^S = X_i(\vartheta^S, \vartheta^{\bar{S}}) . \tag{7.13}$$

In such cases requirement (6.4) can be satisfied only if player i's dividend w_i^S from syndicate S itself is negative.

We shall assume that such negative dividends w_i^N or w_i^S must also satisfy the dividend-proportionality rule (7.6). (As the a_i's are all nonnegative, this implies that, if one dividend w_i^S, $S \subseteq N$, is negative from some given syndicate S, all other dividends w_j^S from this syndicate S must also be negative or zero.) We propose to justify this extension of requirement (7.6) as follows.

We have seen that the Nash solution remains unchanged if the two players' disagreement payoffs t_1 and t_2 are increased or decreased by commensurate increments or decrements Δt_1 and Δt_2 such that

$$a_1 \cdot \Delta t_1 = a_2 \cdot \Delta t_2 \tag{7.14}$$

[equation (2.8)]. We have inferred from this fact that at each stage in several-stage composite bargaining games the players will agree on commensurate payoff increments Δu_1 and Δu_2 such that

$$a_1 \cdot \Delta u_1 = a_2 \cdot \Delta u_2 \tag{7.15}$$

[equation (2.15)], so that their relative bargaining positions with respect to their final payoffs will remain unchanged.

We now postulate a similar principle in cases where the rules of the game require the two players (or at least one of them) to accept decreases in their payoffs, but where it is left to these players themselves how they distribute these decreases between them. We shall assume that in such cases the players will agree on commensurate payoff decrements satisfying (7.15)—which will mean, intuitively speaking, payoff decrements such that leave their relative bargaining positions unchanged. Applying this principle to the case where, for a given syndicate $S \subseteq N$, at least one of the two dividends w_i^S or w_j^S must be *negative*, we obtain the requirement

$$a_i w_i^S = a_j w_j^S , \tag{7.16}$$

extending the dividend-proportionality rule (7.6) also to negative dividends.

8. THE GENERALIZED SHAPLEY-VALUE EXPRESSIONS

As can be verified by substitution, the inverse relationship to (6.4) and to (6.5) is

$$w_i^S = \sum_{\substack{R \ni i \\ R \subseteq S}} (-1)^{s-r} u_i^R \,, \tag{8.1}$$

where s and r are the number of players in syndicate S and in syndicate R respectively.

We introduce the new quantities

$$Y^S = \sum_{i \in S} a_i w_i^S \,, \tag{8.2}$$

$$Z^S = \sum_{i \in S} a_i u_i^S \,. \tag{8.3}$$

Then by (8.1), (8.2) and (8.3)

$$Y^S = \sum_{R \subseteq S} (-1)^{s-r} Z^R \,. \tag{8.4}$$

By (7.6), (8.2) and (8.4)

$$a_i w_i^S = \frac{1}{s}\, Y^S = \sum_{R \subseteq S} \frac{1}{s} (-1)^{s-r} Z^R \,. \tag{8.5}$$

By (8.5), and in view of (6.4) and (6.5),

$$a_i u_i^S = \sum_{\substack{R \ni i \\ R \subseteq S}} \sum_{Q \subseteq R} \frac{1}{r} (-1)^{r-q} Z^Q \,, \tag{8.6}$$

where q is the number of players in syndicate Q.

The result of this double summation is

$$a_i u_i^S = \sum_{\substack{R \ni i \\ R \subseteq S}} \frac{(r-1)!\,(s-r)!}{s!} \cdot (Z^R - Z^{\bar{R}}) \,, \tag{8.7}$$

where $\bar{R} = N - R$ is the syndicate complementary to R, and where we define

$$Z^{\bar{N}} = 0 \,. \tag{8.8}$$

If we now take the special case $S = N$, and set $u_i^N = u_i$; $a_1 = \cdots = a_n = 1$; and $Z^R = v(R)$; $Z^{\bar{R}} = v(\bar{R})$; then this equation becomes identical to equation (4.1) defining the Shapley value. We shall call the right-hand side of (8.7) a *generalized Shapley-value expression.*

9. THE OPTIMAL THREAT STRATEGIES

Equations (7.2), (6.1), (6.5), (7.6), and (7.4) give us the following requirements concerning the variables u_i^S and w_i^S ($S \subseteq N$ and $i \in S$), as well as $a_i (i \in N)$:

$$(9.1) \qquad H(u_1^N, \cdots, u_n^N) = 0 ;$$

$$(9.2) \qquad u_i^S = \sum_{\substack{R \ni i \\ R \subseteq S}} w_i^R , \qquad i \in S,\ S \subseteq N ;$$

$$(9.3) \qquad a_i w_i^S = a_j w_j^S , \qquad i, j \in S,\ S \subseteq N ;$$

$$(9.4) \qquad a_i = H_i(u_1^N, \cdots, u_n^N) , \qquad i \in N .$$

We have all together $n \cdot 2^{n-1}$ variables of form u_i^S; the same number of variables of form w_i^S; and n variables of form a_i. On the other hand, we have 1 equation of form (9.1); $n \cdot 2^{n-1}$ equations of form (9.2); $(s-1)$ independent equations of form (9.3) for each syndicate $S \subseteq N$, that is, all together $[(n-2) \cdot 2^{n-1} + 1]$ independent equations of this form; and finally n equations of form (9.4). Thus we are still short of $(2^n - 2)$ equations, i.e. of one equation for each of the $(2^n - 2)$ sectional syndicates $S \subset N$ of the game. These additional equations we shall obtain by deriving conditions for the optimal threat strategies ϑ^S and $\vartheta^{\bar{S}}$ of each pair of complementary syndicates S and $\bar{S}$.

Let t_i^S ($S \subseteq N$; $i \in S$) be the quantities defined by (7.8) and (7.9). By (7.10), (7.11) and (8.1), for any syndicate S having two or more members, we can write

$$(9.5) \qquad t_i^S = \sum_{\substack{R \ni i \\ R \subset S}} (-1)^{s-r+1} u_i^R , \qquad s > 1 ,\ i \in S ,\ S \subseteq N .$$

Equation (9.3) can also be written as

$$(9.6) \qquad a_i \cdot (u_i^S - t_i^S) = a_j \cdot (u_j^S - t_j^S) \qquad i, j \in S ,\ S \subseteq N .$$

We have already pointed out that at a given point u the derivatives $H_1, \cdots, H_n$ cannot be all zero. Hence $\sum H_i$ will always be nonzero. On the other hand, equations (9.3) and (9.6) remain true if we divide all weights $a_1, \cdots, a_n$ by the same nonzero proportionality factor. Hence we can normalize the weights of the game by means of replacing equation (9.4) with the equation

$$(9.7) \qquad a_i = H_i(u_1^N, \cdots, u_n^N) / \sum_{j \in N} H_j(u_1^N, \cdots, u_n^N) ,$$

which implies

$$(9.8) \qquad \sum_{i \in N} a_i = 1 .$$

We shall assume that threat strategy ϑ^S will be chosen by some particular member i of syndicate S, whereas threat strategy $\vartheta^{\bar{S}}$ will be chosen by some particular member j of syndicate $\bar{S}$. The choice of ϑ^S by i and the choice of $\vartheta^{\bar{S}}$ by j will be regarded as a threat subgame $\tilde{\Gamma}_{ij}^{S\bar{S}}$ between these two players i and j. (We shall see that the outcome of this threat subgame will be the same irrespective of which member i is assumed to play for syndicate S and which member j is assumed to play for syndicate $\bar{S}$.)

We shall assume that subgame $\tilde{\Gamma}_{ij}^{S\bar{S}}$ is played under the following conditions. Player i, in choosing ϑ^S from the set $\tilde{\theta}^S$, will be concerned only with maximizing his own final payoff $u_i = u_i^N$. Likewise, player j, in choosing $\vartheta^{\bar{S}}$ from the set $\tilde{\theta}^{\bar{S}}$, will be concerned only with maximizing his own final payoff $u_j = u_j^N$. The threat strategies ϑ^R of all other syndicates $R \neq S, \bar{S}$ are regarded as given. Therefore, in view of (6.3), the same is true about all conflict payoffs from any syndicate $R \neq S, \bar{S}$ and $\subset N$. Hence, by (9.5), the quantities t_k^S and $t_m^{\bar{S}}$ are also given for all $k \in S$ (both $k = i$ and $k \neq i$) and for all $m \in \bar{S}$ (both $m = j$ and $m \neq j$). Now players i and j have to choose ϑ^S and $\vartheta^{\bar{S}}$ in such a way that requirement (9.6) is satisfied for these given quantities t_k^S and $t_m^{\bar{S}}$.

Finally it will be assumed that, for the purposes of subgame $\tilde{\Gamma}_{ij}^{S\bar{S}}$, the weights of the game, $a_1, \cdots, a_n$, are also regarded as given. For, the ratio a_i/a_k for any $k \neq i, j$ (which determines the distribution of dividends between players i and k in any syndicate R of which both i and k are members) is decided by bargaining between players i and k within their bargaining subgame Γ_{ik}, and so should be regarded as given for the purposes of the threat subgame $\tilde{\Gamma}_{ij}^{S\bar{S}}$ between players i and j. Likewise, the ratio a_j/a_k is decided by bargaining between players j and k within their bargaining subgame Γ_{jk}, and so should be regarded as given for the purposes of the threat subgame $\tilde{\Gamma}_{ij}^{S\bar{S}}$ between players i and j. But if the ratios a_i/a_k and a_j/a_k are given then the ratio a_i/a_j is also given. Hence all ratios a_k/a_m $(k, m \in N)$ are given and so, in view of (9.8), the weights $a_1, \cdots, a_n$ are fully determined.

As we are assuming that equations (9.5) and (9.6) are satisfied, all players' final payoffs will be defined by equation (8.7). Moreover, as the threat strategies ϑ^R of all sectional syndicates $R \neq S, \bar{S}$ are given, the quantities Z^R for all these syndicates are also given. Therefore the final payoff u_k^N of any member k of syndicate S can be written in the form

$$u_k^N = \frac{1}{a_k} \cdot \left[bZ^N + \frac{c}{s} \cdot (Z^S - Z^{\bar{S}}) \right] + \frac{d_k}{a_k}, \tag{9.9}$$

where s is the number of players in S; a_k, b and c are positive constants; and d_k is also a constant (which may have either sign). On the other hand, the final payoff u_m of any member m of syndicate $\bar{S}$ can be written in the general form

$$(9.10) \qquad u_m^N = \frac{1}{a_m} \cdot \left[bZ^N + \frac{c}{n-s} \cdot (Z^{\bar{S}} - Z^S) \right] + \frac{d_m}{a_m} .$$

Now suppose that the threat strategy $\vartheta^{\bar{S}}$ chosen by player j is kept constant. Then the quantities Z^S and $Z^{\bar{S}}$ become functions of the threat strategy ϑ^S alone, chosen by player i. Moreover, equations (9.1), (9.5) and (9.6) [as applied to $S = N$] make Z^N also a function of ϑ^S.

In view of (9.9), as ϑ^S varies, the final payoffs u_k of the various members k of syndicate S will vary all in the same direction. Again, the final payoffs u_m of the various members m of syndicate $\bar{S}$ will likewise vary all in the same direction. In contrast, in view of (5.2), the final payoffs u_k of the players k in S and the final payoffs u_m of the players m in $\bar{S}$ will always vary in opposite directions.

Therefore, player i will maximize his own final payoff u_i if he maximizes the quantity

$$\frac{1}{s} \cdot \sum_{k \in S} a_k u_k^N - \frac{1}{n-s} \cdot \sum_{m \in \bar{S}} a_m u_m^N$$
$$= \frac{nc}{s \cdot (n-s)} \cdot (Z^S - Z^{\bar{S}}) + \frac{1}{s} \cdot \sum_{k \in S} d_k - \frac{1}{n-s} \cdot \sum_{m \in \bar{S}} d_m ,$$

which is equivalent to maximizing the difference

$$(9.11) \qquad \Delta^S = Z^S - Z^{\bar{S}} = \sum_{k \in S} a_k u_k^S - \sum_{m \in \bar{S}} a_m u_m^{\bar{S}} .$$

The maximization of Δ^S is of course, according to our assumptions, subject to equation (9.6), that is, to the constraints

$$(9.12) \qquad \begin{aligned} a_i \cdot (u_i^S - t_i^S) &= a_k \cdot (u_k^S - t_k^S) , \qquad & i, k \in S ; \\ a_j \cdot (u_j^{\bar{S}} - t_j^{\bar{S}}) &= a_m \cdot (u_m^{\bar{S}} - t_m^{\bar{S}}) , \qquad & j, m \in \bar{S} . \end{aligned}$$

By similar reasoning, if the threat strategy ϑ^S chosen by player i is kept constant, player j will maximize his own final payoff u_j^N if he maximizes $-\Delta^S = Z^{\bar{S}} - Z^S$, that is, if he minimizes $\Delta^S = Z^S - Z^{\bar{S}}$ itself, subject to (9.12). Hence the threat subgame $\Gamma_{ij}^{S\bar{S}}$ can be regarded as a two-person zero-sum game, in which player i chooses ϑ^S so as to maximize Δ^S, whereas player j chooses $\vartheta^{\bar{S}}$ so as to minimize Δ^S, in both cases subject to constraint (9.12).

To show that this two-person zero-sum game possesses a solution, we must show that the two players will have strategy pairs ϑ^S and

$\vartheta^{\bar{S}}$ such that satisfy constraint (9.12). But this follows from the fact that, by our assumptions, the members of each syndicate S and $\bar{S}$ are free to reduce their own conflict payoffs by any amounts.

This assumption implies that constraint (9.12) can be omitted if we redefine the conflict payoff of each player k in S and of each player m in $\bar{S}$ as

$$\tilde{u}_k^S = t_k^S + \frac{1}{a_k} \min_{h \in S} (a_h u_h^S - a_h t_h^S) ,$$

(9.13)

$$\tilde{u}_m^{\bar{S}} = t_m^{\bar{S}} + \frac{1}{a_m} \min_{h \in \bar{S}} (a_h u_h^{\bar{S}} - a_h t_h^{\bar{S}}) .$$

We then can regard the threat subgame $\tilde{\Gamma}_{ij}^{S\bar{S}}$ as a two-person zero-sum game in which player i tries to maximize, and player j tries to minimize the quantity

(9.14) $$\tilde{\Delta}^S = \sum_{k \in S} a_k \tilde{u}_k^S - \sum_{m \in \bar{S}} a_m \tilde{u}_m^{\bar{S}}$$

subject to no constraint.

The solution of this two-person zero-sum game will be the maximin (or equivalently minimax) value of $\tilde{\Delta}^S$, subject to no constraint, or, what is the same thing, the maximin (minimax) value of Δ^S, subject to constraint (9.12). Hence we can write

(9.15)

$$\Delta^S = Z^S - Z^{\bar{S}} = \sum_{k \in S} a_k X_k(\vartheta_0^S, \vartheta_0^{\bar{S}}) - \sum_{m \in \bar{S}} a_m X_m(\vartheta_0^S, \vartheta_0^{\bar{S}})$$

$$= \max_{\vartheta^S \in \tilde{\theta}^S} \min_{\vartheta^{\bar{S}} \in \tilde{\theta}^{\bar{S}}} [\sum_{k \in S} a_k X_k(\vartheta^S, \vartheta^{\bar{S}}) - \sum_{m \in \bar{S}} a_m X_m(\vartheta^S, \vartheta^{\bar{S}})] ,$$

where the maximin operator is subject to (9.12).

Clearly, equation (9.15) makes the solution of the threat subgame $\tilde{\Gamma}_{ij}^{S\bar{S}}$ independent of which member i of syndicate S and which member j of syndicate $\bar{S}$ represents his own syndicate S or $\bar{S}$ within this subgame.

Hence we can regard a pair of threat strategies ϑ_0^S and $\vartheta_0^{\bar{S}}$ satisfying (9.15) as being optimal threat strategies from the standpoint of the relevant syndicate S or $\bar{S}$ as a whole.

10. DEFINITION OF THE SOLUTION

Consequently we can define our solution for a general n-person cooperative game Γ as follows. The payoff vector $u = (u_1, \cdots, u_n) = (u_1^N, \cdots, u_n^N)$ is called a *solution* if

(10.1) $$H(u_1^N, \cdots, u_n^N) = 0 ;$$

(10.2) $$a_i = H_i(u_1^N, \cdots, u_n^N) , \qquad i \in N ;$$

(10.3) $$u_i^S = X_i(\vartheta_0^S, \vartheta_0^{\bar{S}}) , \qquad i \in S , \ S \subset N ;$$

(10.4) $$t_i^S = \sum_{\substack{R \ni i \\ R \subset S}} (-1)^{s-r+1} u_i^R , \qquad s > 1 , \ i \in S , \ S \subseteq N ;$$

(10.5) $$a_i \cdot (u_i^N - t_i^N) = a_j \cdot (u_j^N - t_j^N) , \qquad i, j \in N ;$$

(10.6) $$\begin{aligned} \sum_{i \in S} a_i u_i^S - \sum_{j \in \bar{S}} a_j u_j^{\bar{S}} &= \sum_{i \in S} a_i X_i(\vartheta_0^S, \vartheta_0^{\bar{S}}) - \sum_{j \in \bar{S}} a_j X_j(\vartheta_0^S, \vartheta_0^{\bar{S}}) \\ &= \max_{\vartheta^S \in \bar{\theta}^S} \min_{\vartheta^{\bar{S}} \in \bar{\theta}^{\bar{S}}} [\sum_{i \in S} a_i X_i(\vartheta^S, \vartheta^{\bar{S}}) - \sum_{j \in \bar{S}} a_j X_j(\vartheta^S, \vartheta^{\bar{S}})] , \qquad S, \bar{S} \subset N , \end{aligned}$$

subject to:

(10.7) $$\begin{aligned} a_i \cdot (u_i^S - t_i^S) &= a_k \cdot (u_k^S - t_k^S) , \qquad i, k \in S , \\ a_j \cdot (u_j^{\bar{S}} - t_j^{\bar{S}}) &= a_m \cdot (u_m^{\bar{S}} - t_m^{\bar{S}}) , \qquad j, m \in \bar{S} , \end{aligned}$$

where for the purposes of the maximin operation the quantities a_i, a_k, a_j, a_m and $t_i^S, t_k^S, t_j^{\bar{S}}, t_m^{\bar{S}}$ are regarded as constants.

Of course, instead of (10.2) we could have also used (9.7), which subjects the a_i's to the normalizing requirement $\sum a_i = 1$.

In order to obtain conditions that do not assume the existence of the derivatives $H_1, \cdots, H_n$, we can replace equations (10.1) and (10.2) by the requirements

(10.1a) $$a_i \geqq 0 , \qquad i \in N ;$$

(10.2a) $$\sum_{i \in N} a_i u_i^N = \max_{\substack{x \in P \\ a_1, \cdots, a_n = \text{const.}}} \sum_{i \in N} a_i x_i .$$

The new conditions so obtained are equivalent to conditions (10.1) to (10.7) in all cases where the derivatives $H_1, \cdots, H_n$ exist at the solution point u.[9] In cases where these derivatives do not exist, the new conditions can be justified by the method outlined in footnote 6 above.

It is easy to see that these conditions contain the conditions defining the Nash solution (Section 3) and those defining the modified Shapley value (Section 4) as special cases.

In the case of a two-person game, we have only three syndicates within the game, viz. (1), (2) and (12). We must set $u_1^{(1)} = t_1^{(12)} = t_1$; $u_2^{(2)} = t_2^{(12)} = t_2$; $u_1^{(12)} = u_1$; $u_2^{(12)} = u_2$. Then our conditions directly yield the Nash solution.

Our solution, like the Nash solution itself, is invariant with respect to order-preserving linear transformations of any player's utility function. If we disregard the degenerate case where one or more of the a_i's

[9] In John C. Harsanyi, "Measurement of Social Power in *n*-Person Power Situations," *Behavioral Science*, VII (January, 1962), 87, equations (10.1*a*), (10.2*a*), and (10.3) to (10.7) of this paper have been referred to as equations ($\alpha\alpha$), ($\beta\beta$), and (γ) to (η).

are zero, by appropriate linear transformations of the players' utility functions we can always achieve $a_1 = \cdots = a_n = 1$. After these transformations the close relationship between our solution and the modified Shapley values becomes quite obvious.

Let Γ be a general cooperative game for which $a_1 = \cdots = a_n$. We can approximate Γ by a game with transferable utility $\bar{\Gamma}$ as follows. Let X_i be the payoff function of the original game Γ. In the new game $\bar{\Gamma}$, if all n players cooperate in using strategy ϑ^N then their joint payoff $\bar{X}^N$ will be defined as

$$\bar{X}^N(\vartheta^N) = \sum_{i \in N} X_i(\vartheta^N) . \tag{10.8}$$

If only the members of a given sectional syndicate S cooperate in using the joint strategy ϑ^S, whereas the other players are using the joint strategy $\vartheta^{\bar{S}}$, then the joint payoff $\bar{X}^S$ of syndicate S will be defined as

$$\bar{X}^S(\vartheta^S, \vartheta^{\bar{S}}) = \sum_{i \in S} t_i^S + s \cdot \min_{i \in S} [X_i(\vartheta^S, \vartheta^{\bar{S}}) - t_i^S] , \tag{10.9}$$

where t_i^S is the quantity defined by equation (10.4) for the original game Γ, while s is the number of players in S. (Of course in the new game $\bar{\Gamma}$ it is assumed that these joint payoffs $\bar{X}^N$ and $\bar{X}^S$ can be freely redistributed among the members of syndicates N and S respectively, since $\bar{\Gamma}$ is a game with transferable utility.)

Then, the modified Shapley values $u_1, \cdots, u_n$ for $\bar{\Gamma}$ will be equal to the final payoffs $u_1, \cdots, u_n$ that our solution defines for the original game Γ. Clearly, if the original game Γ already has transferable utility, then $\bar{\Gamma} = \Gamma$, and the final payoffs defined by our solution will simply be the modified Shapley values of Γ itself.

11. THE EXISTENCE THEOREM

We shall call a game Γ *regular* if its prospect space and its strategy spaces satisfy the assumptions of Section 5 above.

THEOREM 2 (Existence Theorem). *For every regular n-person cooperative game Γ there exists a solution $u = (u_1, \cdots, u_n)$ satisfying conditions* (10.1a), (10.2a), *as well as* (10.3) *to* (10.7).

PROOF. We shall restrict ourselves to the case where the partial derivatives $H_1, \cdots, H_n$ exist at all points of the extended upper boundary $\tilde{H}$. Our proof can be extended to the general case by the method outlined in footnote 6.

Suppose that each of the $(2^n - 1)$ syndicates of the n-person cooperative game Γ appoints a trustee and that these trustees play a $(2^n - 1)$-

person non-cooperative game Γ^* among themselves. The trustee of the n-player syndicate N will be called "player N" while the trustee of any given sectional syndicate $S \subset N$ will be called "player S".

Player N has the task of selecting a vector $a = (a_1, \cdots, a_n)$, subject to the conditions

$$\sum_{i \in N} a_i = 1 \,, \tag{11.1}$$

$$a_i \geqq 0 \,, \qquad \text{for all } i \in N \,. \tag{11.2}$$

Each player S representing a sectional syndicate S has the task of selecting a joint strategy ϑ_0^S for his syndicate S, subject to

$$\vartheta_0^S \in \theta^S \,. \tag{11.3}$$

We now define

$$\bar{u}_i^S = X_i(\vartheta_0^S, \vartheta_0^{\bar{S}}) \qquad \text{for all } S \subset N \text{ and all } i \in S \,, \tag{11.4}$$

where $\bar{S} = N - S$.

If $S = (i)$ has only one member then we write

$$u_i^S = \bar{u}_i^S \,; \tag{11.5}$$

whereas if $S \subset N$ has more than one member, we define

$$u_i^S = \bar{u}_i^S \qquad \text{if } a_i = 0 \,; \tag{11.6}$$

$$u_i^S = t_i^S + \frac{1}{a_i} \min_{j \in S} (a_j \bar{u}_j^S - a_j t_j^S) \qquad \text{if } a_i > 0 \,, \tag{11.7}$$

where

$$t_j^S = \sum_{\substack{R \ni j \\ R \subset S}} (-1)^{s-r+1} u_j^R \qquad \text{for all } S \subseteq N \text{ and all } j \in S \,. \tag{11.8}$$

Clearly, equations (11.4) to (11.8) provide a recursive definition of the variables u_i^S in terms of the quantities $a_1, \cdots, a_n$ as well as the strategies ϑ_0^S, for all syndicates $S \subset N$. This is so because equations (11.4) and (11.5) define the u_i^S's for all one-member syndicates, and the remaining equations allow computation of the u_i^S's first for all two-member syndicates, then for all three-member syndicates, etc.

We also write

$$Z^S = \sum_{i \in S} a_i u_i^S \qquad \text{for all } S \subset N \,. \tag{11.9}$$

The variables $u_1 = u_1^N, \cdots, u_n = u_n^N$ will be defined as follows. Let L be the straight line which is the locus of all points $x = (x_1, \cdots, x_n)$ satisfying $(n-1)$ independent equations of the form:

(11.10) $$a_i \cdot (x_i - t_i^N) = a_j \cdot (x_j - t_j^N) , \qquad i, j \in N .$$

Let K be the intersection of this line L and of the extended upper boundary $\tilde{H}$ of the prospect space. The set K is always nonempty. If all a_i's are positive then the nonemptiness of K directly follows from the definition of $\tilde{H}$ irrespective of whether the point $t^N = (t_1^N, \cdots, t_n^N)$ lies in $\tilde{P}$ or not. On the other hand, if some of the a_i's are zero, let M be the set of all i for which $a_i = 0$. In the latter case again the set K will be nonempty provided that $t^N \in \tilde{P}$. But this condition is always satisfied because by (11.5), (11.6) and (11.7) for any $S \subset N$, we can write

$$u_i^S = u_i^M = t_i^N ,$$
$$t_i^N \leqq \bar{u}_i^M , \qquad \text{if } i \in M ;$$

whereas

$$u_i^S = u_i^{\bar{M}} = t_i^N ,$$
$$t_i^N \leqq \bar{u}_i^{\bar{M}} , \qquad \text{if } i \in \bar{M} = N - M .$$

Hence for all $i \in N$ we have

$$t_i^N \leqq X_i(\vartheta_0^M, \vartheta_0^{\bar{M}}) ;$$

but

$$X(\vartheta_0^M, \vartheta_0^{\bar{M}}) \in P .$$

Hence

$$t^N \in \tilde{P} .$$

The set K may consist of a unique point $\tilde{x} = (\tilde{x}_1, \cdots, \tilde{x}_n)$ or may consist of infinitely many points. (The latter possibility can arise only if at least one of the a_i's is zero.) In the latter case we define $\tilde{x}$ as that point in K which (weakly) dominates all other points in K. Moreover, we write

(11.11) $$u_i = u_i^N = \tilde{x}_i \qquad \text{for all } i \in N .$$

Finally, we define

(11.12) $$a_i^* = \frac{H_i(u_1, \cdots, u_n)}{\sum_{j \in N} H_j(u_1, \cdots, u_n)} .$$

We shall assume that in the non-cooperative game Γ^* each player S representing a sectional syndicate $S \subset N$ will try to maximize the quantity

(11.13) $$\Delta^S = Z^S - Z^{\bar{S}} .$$

On the other hand, player N will try to maximize the quantity

$$A = -\sum_{i \in N} (a_i - a_i^*)^2 \,. \tag{11.14}$$

By Debreu's equilibrium-point existence theorem [1], this non-cooperative game Γ^* will always have at least one equilibrium point. It is easy to see that Γ^* satisfies the continuity and contractibility requirements of Debreu's theorem.

Yet, for any equilibrium point of Γ^* we have

$$a_i = a_i^* \qquad \text{for all } i \in N\,; \tag{11.15}$$

and so any equilibrium point of Γ^* defines a solution u for our original cooperative game Γ, satisfying equations (10.1a), (10.2a), as well as (10.3) to (10.7). This completes the proof of our theorem.

12. THE QUESTION OF UNIQUENESS

It is known that the solution is always unique at least in the following three special cases: (i) when there are only two players (in which case our solution is identical with the Nash solution of the game); (ii) when the disagreement payoffs the players will receive if they cannot reach a unanimous agreement are uniquely determined in advance by the rules of the game [Cf. 3, (§ 3)]; (iii) when the upper boundary H of the prospect space is a hyperplane. (Case (iii) contains all games with transferable utility, but is more general than that.)

By continuity, it is reasonable to assume that this uniqueness property also holds for games which at least "come close" to satisfying either (ii) or (iii). But it is known from counter-examples that in general the solution is not unique.

As I have argued in an earlier paper [3, (347–48)], if we assume perfect rationality on the part of the players then, in case a given game Γ has several solutions $u^1, \cdots, u^k$, none of these will be stable. Each player will try to press for that particular solution which would yield him the highest payoff, and different players will have opposite interests in this matter. But, as I have shown (loc. cit.), one can define a unique "stable solution" $\overline{\overline{u}}$ for the game by assuming an additional final round of bargaining among the players.

For empirical applications, however, it now seems to me the various solutions $u^1, \cdots, u^k$ satisfying the conditions of Section 10 are much more interesting concepts than the "stable solution" $\overline{\overline{u}}$ we can derive from them. It is more realistic to assume that the players will usually know the prospect space of the game only in the immediate vicinity of their actual position. Hence any one of the solutions $u^1, \cdots, u^k$

will be stable in practice because it is locally stable (since it satisfies the conditions of the Nash solution for each of the subgames into which the main game can be divided) and because the players will be usually unaware of the existence of alternative solutions. Thus games possessing several solutions should be regarded as social systems with multiple equilibria.

Wayne State University, U.S.A.

REFERENCES

[1] DEBREU, G., "A Social Equilibrium Existence Theorem," *Proceedings of the National Academy of Sciences*, XXXVIII (August, 1952), 886–93.

[2] HARSANYI, JOHN C., "Approaches to the Bargaining Problem before and after the Theory of Games: A Critical Discussion of Zeuthen's, Hicks', and Nash's Theories," *Econometrica*, XXIV (April, 1956), 144–57.

[3] ———, "A Bargaining Model for the Cooperative *n*-Person Game," in *Contributions to the Theory of Games*, IV, ed. A. W. Tucker and R. D. Luce (Princeton: Princeton University Press, 1959), 325–55.

[4] ISBELL, J. R., "A Modification of Harsanyi's Bargaining Model," *Bulletin of the American Mathematical Society*, LXVI (March, 1960), 70–73.

[5] LUCE R. D. and H. RAIFFA, *Games and Decisions* (New York: John Wiley and Sons, Inc., 1957).

[6] NASH, J. F., "The Bargaining Problem," *Econometrica*, XVIII (April, 1950), 155–62.

[7] ———, "Two-person Cooperative Games," *Econometrica*, XXI (January, 1953), 128–40.

[8] SHAPLEY, L. S., "A Value for *n*-Person Games," in *Contributions to the Theory of Games*, II, ed. H. W. Kuhn and A. W. Tucker (Princeton: Princeton University Press, 1953) 307–17.

SECTION B

PROPERTIES OF EQUILIBRIUM POINTS

CHAPTER 4

GAMES WITH RANDOMLY DISTURBED PAYOFFS: A NEW RATIONALE FOR MIXED-STRATEGY EQUILIBRIUM POINTS[1, 2]

Abstract: Equilibrium points in mixed strategies seem to be unstable, because any player can deviate without penalty from his equilibrium strategy even if he expects all other players to stick to theirs. This paper proposes a model under which most mixed-strategy equilibrium points have full stability. It is argued that for any game Γ the players' uncertainty about the other players' exact payoffs can be modeled as a disturbed game Γ^*, i.e., as a game with small random fluctuations in the payoffs. Any equilibrium point in Γ, whether it is in pure or in mixed strategies, can "almost always" be obtained as a limit of a pure-strategy equilibrium point in the corresponding disturbed game Γ^* when all disturbances go to zero. Accordingly, mixed-strategy equilibrium points are stable – even though the players may make no deliberate effort to use their pure strategies with the probability weights prescribed by their mixed equilibrium strategies – because the random fluctuations in their payoffs will *make* them use their pure strategies approximately with the prescribed probabilities.

1. Introduction

On the face of it, equilibrium points in mixed strategies are unstable because any player can deviate without penalty from his equilibrium strategy even if all other players stick to theirs. (He can shift to any pure strategy to which his mixed equilibrium strategy assigns a positive probability; he can also shift to any arbitrary probability mixture of these pure strategies.) This instability seems to pose a serious problem because many games have only mixed-strategy equilibrium points.

However, as we shall see, the stability of these equilibrium points will appear in a different light if we take due account of the uncertainty in which every player finds himself about the precise payoffs that other players associate with alternative strategy combinations. Classical game theory assumes that in any game Γ every

---

[1] Prof. JOHN C. HARSANYI, University of California, Berkeley School of Business Administration Berkeley, Cal. 94720, U.S.A.

[2] The author is grateful to the National Science Foundation for supporting this research by Grant GS-3222 to the University of California, Berkeley, through the University's Center for Research in Management Science. Special thanks are due to Professor REINHARD SELTEN, of the University of Bielefeld, West Germany, for suggesting Theorem 1 and for many valuable discussions. The author is indebted for helpful comments also to Professor ROBERT J. AUMANN of the Hebrew University, Jerusalem; to Professors GERARD DEBREU and ROY RADNER of the University of California, Berkeley; and to Dr. LLOYD S. SHAPLEY of RAND Corporation, Santa Monica.

player has precise knowledge of the payoff function of every other player (as well as of his own). But it is more realistic to assume that – even if each player i does have exact knowledge of his own payoff function U_i – he can have at best only somewhat inexact information about the other players' payoff functions U_j, $j \neq i$. This assumption, of course, formally transforms the game into a game with incomplete information. However, it can be shown [HARSANYI, 1967–68] that such games can also be modeled, more conveniently, as games with complete information involving appropriate random variables (chance moves), where the players' ignorance about any aspect of the game situation is represented as ignorance about the actual values of these random variables.

Accordingly, we shall propose the following model, to be called a disturbed game $\Gamma^* = \Gamma^*(\varepsilon)$. The payoff function U_i of every player i is subject to small random disturbances within a given range $[-\varepsilon, +\varepsilon]$, due to small stochastic fluctuations in his subjective and objective conditions (e.g., in his mood, taste, resources, social situation, etc.). The probability laws governing these disturbances are known to all players, but their precise effects on any given player's payoff function U_i at the time of his strategy choice are known only to this player i himself. As we shall see, under our assumptions (see below), all equilibrium points of any disturbed game $\Gamma^*(\varepsilon)$ will be in pure strategies. Moreover, any mixed-strategy equilibrium point s in an ordinary game Γ can "almost always" be obtained as the limit of a pure-strategy equilibrium point $s(\varepsilon)$ in the corresponding disturbed game $\Gamma^*(\varepsilon)$ when all random disturbances go to zero (i.e., when the parameter ε goes to zero). Therefore, such a mixed-strategy equilibrium point s can be interpreted as a somewhat unprecise description of this pure-strategy equilibrium point $s(\varepsilon)$ of the disturbed game $\Gamma^*(\varepsilon)$, and can be regarded as having the same stability as $s(\varepsilon)$ has.

More specifically, our results imply the following – somewhat surprising – conclusion. The players may make no deliberate effort to use their pure strategies with the probability weights prescribed by their mixed equilibrium strategies $s_1, \ldots, s_n$. Nevertheless, as a result of the random fluctuations in their payoffs, they will in fact use their pure strategies approximately with the prescribed probabilities.

2. Definitions and Notations

Let Γ be a finite n-person noncooperative game. The k-th pure strategy of player i will be called a_i^k while the set of all his K_i pure strategies will be called A_i. Let

$$K = \prod_i K_i . \tag{1}$$

We shall assume that the K possible n-tuples of pure strategies will be numbered consecutively (e.g., in a lexicographical order) as $a^1, \ldots, a^m, \ldots, a^K$ – but see Convention (92) below). Let

$$a^m = (a_1^{k_1}, \ldots, a_i^{k_i}, \ldots, a_n^{k_n}). \tag{2}$$

Then we shall write

$$a^m(i) = a_i^{k_i} \tag{3}$$

to denote the pure strategy used by player i in the strategy n-tuple a^m. The set of all possible pure-strategy n-tuples a^m will be called A. We can write $A = A_1 \times \cdots \times A_n$.

Any mixed strategy s_i of player i will be of the form

$$s_i = \sum_k p_i^k a_i^k, \tag{4}$$

where p_i^k is the probability that this mixed strategy s_i assigns to the pure strategy a_i^k. Of course,

$$p_i^k \geqq 0 \quad \text{for} \quad k = 1, \ldots, K_i \quad \text{and} \quad \sum_k p_i^k = 1. \tag{5}$$

The set of all mixed strategies available to player i will be called S_i. The set of all possible n-tuples $s = (s_1, \ldots, s_n)$ will be called S. We have $S = S_1 \times \cdots \times S_n$. We shall write $s = (s_i, \bar{s}_i)$, where $\bar{s}_i = (s_1, \ldots, s_{i-1}, s_{i+1}, \ldots, s_n)$ is the strategy $(n-1)$-tuple representing the strategies of all $(n-1)$ players other than player i.

The set $C(s_i)$ of pure strategies to which s_i assigns positive probabilities is called the *carrier* of s_i. The set $C^*(s_i)$ of all mixed strategies t_i that distribute all probability over the pure strategies in $C(s_i)$ is called the *extended carrier* of s_i. Thus $C^*(s_i)$ is the convex hull of $C(s_i)$.

A mixed strategy s_i whose carrier $C(s_i)$ contains only one pure strategy a_i^k will be identified with this latter so that we shall write $s_i = a_i^k$. On the other hand, if $C(s_i)$ contains two or more pure strategies, then s_i will be called a *proper* mixed strategy.

Suppose that the i-th component of the pure-strategy n-tuple a^m is $a^m(i) = a_i^k$, and that the mixed strategy s_i of player i assigns the probability p_i^k to this pure strategy a_i^k. Then we shall write

$$q_i^m(s_i) = p_i^k. \tag{6}$$

Of course, if $s_i = a_i^k$ is a pure strategy, then we have

$$q_i^m(a_i^k) = 1 \quad \text{when} \quad a^m(i) = a_i^k, \tag{7}$$

whereas

$$q_i^m(a_i^k) = 0 \quad \text{when} \quad a^m(i) \neq a_i^k. \tag{8}$$

If the n players use a pure-strategy n-tuple a^m, then player i $(i = 1, \ldots, n)$ will obtain the payoff

$$U_i(a^m) = v_i^m, \tag{9}$$

while if they use a mixed-strategy n-tuple $s = (s_1, \ldots, s_n)$, then his payoff will be

$$U_i(s) = \sum_m [\prod_i q_i^m(s_i)] v_i^m . \qquad (10)$$

A given strategy s_i of player i is a *best reply* to some strategy combination $\bar{s}_i$ used by the other $(n-1)$ players if

$$U_i(s_i, \bar{s}_i) \geqq U_i(t_i, \bar{s}_i) \quad \text{for all} \quad t_i \in S_i . \qquad (11)$$

A given strategy n-tuple $s = (s_1, \ldots, s_n)$ is an *equilibrium point* [NASH, 1951] if every component s_i of s is a best reply to the corresponding strategy combination $\bar{s}_i$ of the $(n-1)$ other players.

An equilibrium point s is called *strong* [3]) if all n components s_i of s satisfy (11) with the strong inequality sign $>$ for all $t_i \neq s_i$. That is, s is a strong equilibrium point if every player's equilibrium strategy s_i is his *only* best reply to the other players' strategy combination $\bar{s}_i$. An equilibrium point is called *weak* if it is not strong. Any equilibrium point s is always weak if at least one player's equilibrium strategy s_i is a proper mixed strategy because for this player i every strategy t_i in the extended carrier $C^*(s_i)$ of his mixed equilibrium strategy s_i is a best reply to $\bar{s}_i$, and there are infinitely many such strategies t_i in $C^*(s_i)$.

An equilibrium point s is *quasi-strong* if no player i has best replies to $\bar{s}_i$ other than the strategies t_i belonging to the extended carrier $C^*(s_i)$ of his equilibrium strategy s_i. Though an equilibrium point in proper mixed strategies is never strong, it may be (and mostly is) quasi-strong. An equilibrium point that is not even quasi-strong is called *extra-weak*. If an equilibrium point in pure strategies is weak then it is always extra-weak.[4])

3. Disturbed Games

In a disturbed game, $\Gamma^* = \Gamma^*(\varepsilon)$, the payoff of any player i for any given n-tuple a^m of pure strategies can be written as

$$U_i(a^m) = V_i(a^m) + \varepsilon \xi_i^m = v_i^m + \varepsilon \xi_i^m , \quad i = 1, \ldots, n; \quad m = 1, \ldots, K , \qquad (12)$$

where $V_i(a^m) = v_i^m$ and ε are constants with $\varepsilon > 0$, whereas ξ_i^m is a random variable. As V_i is the nonrandom part of player i's payoff function U_i, V_i will be called player i's *basic payoff function*. We shall write

$$\xi_i = (\xi_i^1, \ldots, \xi_i^K), \quad i = 1, \ldots, n . \qquad (13)$$

Thus ξ_i is a random vector consisting of all random variables ξ_i^m occurring in player i's payoff function U_i.

---

3) I am using the term "strong equilibrium point" in a sense different from AUMANN'S [1960a, p. 363].

4) It can be shown that games possessing extra-weak equilibrium points, whether in pure or in mixed strategies, are very "exceptional": they correspond to a closed set of measure zero in the space of all games of a given size.

We shall assume that the random vectors ξ_i associated with different players i are statistically independent. The probability distribution of each random vector ξ_i will be written as

$$F_i = F_i(\xi_i) = F_i(\xi_i^1, \ldots, \xi_i^K). \tag{14}$$

We shall assume that the nonrandom basic payoff functions $V_1, \ldots, V_n$, the parameter ε, and the probability distributions $F_1, \ldots, F_n$ are known to all n players. But the actual values of the variables $\xi_i^1, \ldots, \xi_i^K$ are known only to player i himself ($i = 1, \ldots, n$).

We shall make the following assumptions about the probability distributions F_i:

The range of each random variable $\xi_i^m (i = 1, \ldots, n; m = 1, \ldots, K)$ is a subset (or possibly the whole) of the closed interval $I = [-1, +1]$. (15)

All probability distributions F_i are absolutely continuous. Moreover, (16)

For each F_i, the corresponding probability density function (17)

$$f_i(\xi_i) = f_i(\xi_i^1, \ldots, \xi_i^K) = \frac{\partial^K F_i}{\partial \xi_i^1, \ldots, \partial \xi_i^K}$$

is continuously differentiable for all possible values of ξ_i.

Suppose that player i uses the pure strategy a_i^k while the other $(n-1)$ players use the mixed-strategy combination $\bar{s}_i = (s_1, \ldots, s_{i-1}, s_{i+1}, \ldots, s_n)$. Then, in view of (10) and (12), player i will receive the payoff

$$U_i(a_i^k, \bar{s}_i) = \sum_m [q_i^m(a_i^k) \prod_{j \neq i} q_j^m(s_j)] v_i^m + \varepsilon \sum_m [q_i^m(a_i^k) \prod_{j \neq i} q_j^m(s_j)] \xi_i^m. \tag{18}$$

For convenience, we shall write

$$V_i(a_i^k, \bar{s}_i) = \sum_m [q_i^m(a_i^k) \prod_{j \neq i} q_j^m(s_j)] v_i^m \tag{19}$$

and

$$\zeta_i^k = \sum_m [q_i^m(a_i^k) \prod_{j \neq i} q_j^m(s_j)] \xi_i^m, \tag{20}$$

so that (18) can now be written as

$$U_i(a_i^k, \bar{s}_i) = V_i(a_i^k, \bar{s}_i) + \varepsilon \zeta_i^k. \tag{21}$$

The quantity ζ_i^k is obviously a random variable whose probability distribution depends on the strategy combination $\bar{s}_i$ used by the other $(n-1)$ players. Corresponding to the K_i pure strategies $a_i^1, \ldots, a_i^{K_i}$ available to any given player i, Eq. (20) defines K_i random variables $\zeta_i^1, \ldots, \zeta_i^{K_i}$ for him. Let ζ_i be the random vector

$$\zeta_i = (\zeta_i^1, \ldots, \zeta_i^{K_i}), \quad i = 1, \ldots, n. \tag{22}$$

Clearly, for each player $i (i = 1, \ldots, n)$, the probability distribution $G_i = G_i(\zeta_i | \bar{s}_i)$ of this random vector will once more depend on $\bar{s}_i$. More particularly, let $\Omega_i(\bar{s}_i)$ be the set of all points $\xi_i = (\xi_i^1, \ldots, \xi_i^K)$ satisfying the K_i inequalities of the form

$$\sum_m [q_i^m(a_i^k) \prod_{j \neq i} q_j^m(s_j)] \xi_i^m \leqq \zeta_i^k, \quad \text{for} \quad k = 1, \ldots, K_i. \tag{23}$$

Then we can define

$$G_i(\zeta_i|\bar{s}_i) = G_i(\zeta_i^1, \ldots, \zeta_i^{K_i}|\bar{s}_i) = \underset{\xi_i \in \Omega_i(\bar{s}_i)}{\int \mathrm{d}\xi_i^1 \cdots \int \mathrm{d}\xi_i^K} f_i(\xi_i). \tag{24}$$

For any random variable or random vector, we define its *range space* as the set of all possible values it can take in game Γ^*. Thus, we define the range spaces $\Xi_i^m = \{\xi_i^m\}$, $Z_i^k = \{\zeta_i^k\}$, $\Xi_i = \{\xi_i\}$, and $Z_i = \{\zeta_i\}$, with $i = 1, \ldots, n$; $m = 1, \ldots, K$; $k = 1, \ldots, K_i$. In view of (15), we have

$$\Xi_i^m \subseteq I = [-1, +1] \quad \text{for all } i \text{ and } m, \tag{25}$$

which, by (20), in turn implies

$$Z_i^k \subseteq I = [-1, +1] \quad \text{for all } i \text{ and } k. \tag{26}$$

Moreover, $\Xi_i = \Xi_i^1 \times \cdots \times \Xi_i^K$ and $Z_i = Z_i^1 \times \cdots \times Z_i^{K_i}$.

4. Strategies and Equilibrium Points in Disturbed Games

The disturbed game $\Gamma^*(\varepsilon)$ as defined in Section 3 is not in normal form. One reason for this is that our definitions have not made explicit the dependence of each player's strategy choice on his random payoff-disturbance vector ξ_i. We shall now define *normalized* strategies (i.e., strategies appropriate to the normal form of $\Gamma^*(\varepsilon)$).[5] In contrast, the pure strategies a_i^k and the mixed strategies s_i introduced in Section 2 will be called *ordinary* strategies. When we speak of a pure or mixed strategy without further specification we shall always mean an ordinary pure or mixed strategy a_i^k or s_i.

A *normalized pure strategy*, or *n-pure strategy*, a_i^* of any player i ($i = 1, \ldots, n$) will be defined as a measurable function from the range space Ξ_i of player i's random vector ξ_i to the set A_i of his ordinary pure strategies a_i^k. For any specific value of the random vector ξ_i, this n-pure strategy a_i^* will select some specific ordinary pure strategy

$$a_i^k = a_i^*(\xi_i) \tag{27}$$

as the strategy to be used by player i whenever ξ_i takes this particular value. Let

$$X_i^k = X_i^k(a_i^*) = \{\xi_i | \xi_i \in \Xi_i \quad \text{and} \quad a_i(\xi_i) = a_i^k\}. \tag{28}$$

Thus, X_i^k is the set of all possible ξ_i values that will make player i choose the ordinary pure strategy a_i^k if he follows the n-pure strategy a_i^*. Clearly, any n-pure strategy a_i^* can also be defined as a partitioning of the range space Ξ_i into K_i disjoint measurable subsets $X_i^1, \ldots, X_i^{K_i}$. (Of these K_i sets, as many as $(K_i - 1)$ sets may be empty.) The set of all n-pure strategies a_i^* will be called A_i^*.

An n-pure strategy a_i^* is called a *constant* n-pure strategy and will be denoted as $a_i^* = [a_i^k]$ if it assigns the *same* ordinary pure strategy $a_i^k = a_i^*(\xi_i)$ to all possible ξ_i values, $\xi_i \in \Xi_i$.

---

[5]) However, for convenience, we shall keep the payoff functions U_i in their present form and shall not introduce normalized payoff functions (which would have to be defined as the expected values of the U_i's). That is, we shall use the semi-normal form of $\Gamma^*(\varepsilon)$ [HARSANYI, 1967–68, p. 182].

Let $D(a_i^*, a_i^{*\prime})$ be the set of all ξ_i values where the two n-pure strategies a_i^* and $a_i^{*\prime}$ disagree, i.e., where $a_i^*(\xi_i) \neq a_i^{*\prime}(\xi_i)$. Two n-pure strategies a_i^* and $a_i^{*\prime}$ will be called essentially identical if $D(a_i^*, a_i^{*\prime})$ is a set of measure zero. In view of (16), if two n-pure strategies are essentially identical, then with probability one they will prescribe choosing the same ordinary pure strategies. Two n-tuples of n-pure strategies, $a^* = (a_1^*, \ldots, a_n^*)$ and $a^{*\prime} = (a_1^{*\prime}, \ldots, a_n^{*\prime})$, are called *essentially identical* if their corresponding components a_i^* and $a_i^{*\prime}$ are essentially identical (for $i = 1, \ldots, n$). If two n-pure strategies, or two n-tuples of such strategies, are *not* essentially identical, then they are called *essentially different*.

A *normalized mixed strategy*, or *n-mixed strategy*, s_i^* of any player i is a probability distribution over the set A_i^* of his n-pure strategies. As A_i^* is a function space, proper definition of such probability distributions poses certain technical difficulties, which can be overcome in various ways [Aumann, 1960b, 1961]. We shall not go into these problems because we are introducing n-mixed strategies only in order to show that there is no need for them in disturbed games $\Gamma^*(\varepsilon)$. We shall simply assume that any n-mixed strategy s_i^* to be discussed has been defined in a mathematically satisfactory way.

If an n-mixed strategy s_i^* is a probability mixture of two or more essentially identical strategies $a_i^*, a_i^{*\prime}, \ldots$, then s_i^* will be called an *essentially pure* normalized strategy, and will be regarded as being essentially identical with each one of these n-pure strategies $a_i^*, a_i^{*\prime}, \ldots$.

Lemma 1:

If player i follows the n-pure strategy a_i^* then he will use any ordinary pure strategy a_i^k of his with the probability

$$p_i^k = \pi_i^k(a_i^*) = \int_{\xi_i \in X_i^k(a_i^*)} \mathrm{d}\xi_i^1 \ldots \int \mathrm{d}\xi_i^K f_i(\xi_i), \quad k = 1, \ldots, K_i, \tag{29}$$

where, for each particular value of k, $X_i^k(a_i^*)$ is the set defined by (28). Therefore, his behavior will be such as if he followed the ordinary mixed strategy

$$s_i = \pi(a_i^*) = \sum_k p_i^k a_i^k = \sum_k \pi_i^k(a_i^*) a_i^k . \tag{30}$$

This strategy $s_i = \pi(a_i^*)$ will be called the mixed strategy *induced* by the n-pure strategy a_i^*.

Proof:

In view of (27), if player i follows the n-pure strategy a_i^*, then the ordinary pure strategy he will be using will become a function of the random vector ξ_i. He will use a_i^k whenever $\xi_i \in X_i^k(a_i^*)$. But the probability of this event is given by (29).

Lemma 1 directly implies:

Lemma 2:

Let s_i^* be an n-mixed strategy of player i, representing a probability mixture of two or more n-pure strategies $a_i^*, a_i^{*\prime}, \ldots$. Let $s_i = \pi(a_i^*)$, $s_i' = \pi(a_i^{*\prime})$,... be the ordinary mixed strategies induced by these component strategies $a_i^*, a_i^{*\prime}, \ldots$. Let

t_i be the ordinary mixed strategy representing a probability mixture of these induced mixed strategies $s_i, s_i', \ldots$ such that assigns the same probability weights (or the same probability densities) to $s_i, s_i', \ldots$ as s_i^* itself assigns to the corresponding n-pure strategies $a_i^*, a_i^{*\prime}, \ldots$. Then, whenever player i follows this n-mixed strategy s_i^*, he will be using all his ordinary pure strategies a_i^k with probabilities p_i^k corresponding to the ordinary mixed strategy t_i.

This strategy t_i will be called the ordinary mixed strategy *induced* by s_i^*, and we shall write $t_i = \pi(s_i^*)$.

Let $a^* = (a_1^*, \ldots, a_n^*)$ and let $s_i = \pi(a_i^*)$ be the mixed strategy induced by a_i^*, for $i = 1, \ldots, n$. Let $s = (s_1, \ldots, s_n)$. Then we shall write $s = \pi(a^*)$. Moreover, let $\overline{a_i^*} = (a_1^*, \ldots, a_{i-1}^*, a_{i+1}^*, \ldots, a_n^*)$ and $\bar{s}_i = (s_1, \ldots, s_{i-1}, s_{i+1}, \ldots, s_n)$. Then we shall write $\bar{s}_i = \pi(\overline{a_i^*})$.

Likewise, let $s^* = (s_1^*, \ldots, s_n^*)$ and let $t_i = \pi(s_i^*)$ for $i = 1, \ldots, n$. Let $t = (t_1, \ldots, t_n)$. Then we shall write $t = \pi(s^*)$. Finally, let $\overline{s_i^*} = (s_1^*, \ldots, s_{i-1}^*, s_{i+1}^*, \ldots, s_n^*)$ and $\bar{t}_i = (t_1, \ldots, t_{i-1}, t_{i+1}, \ldots, t_n)$. Then we shall write $\bar{t}_i = \pi(\overline{s_i^*})$.

Lemma 3:

Suppose that player i uses the n-pure strategy a_i^*, and therefore chooses the ordinary pure strategy $a_i^k = a_i^*(\xi_i)$ at the point ξ_i. If at the same time the other $(n-1)$ players use the n-pure-strategy combination $\overline{a_i^*}$ then player i will obtain the (expected) payoff

$$U_i(a_i^*, \overline{a_i^*}) = U_i(a_i^k, \bar{s}_i) = V_i(a_i^k, \bar{s}_i) + \varepsilon \zeta_i^k(\xi_i, \bar{s}_i), \tag{31}$$

where

$$\bar{s}_i = \pi(\overline{a_i^*}) \tag{32}$$

and where $\zeta_i^k = \zeta_i^k(\xi_i, \bar{s}_i)$ is the random variable defined by (20). On the other hand, if the other $(n-1)$ players use the n-mixed-strategy combination $\overline{s_i^*}$ then he will obtain the payoff

$$U_i(a_i^*, \overline{s_i^*}) = U_i(a_i^k, \bar{t}_i) = V_i(a_i^k, \bar{t}_i) + \varepsilon \zeta_i^k(\xi_i, \bar{t}_i), \tag{33}$$

where

$$\bar{t}_i = \pi(\overline{s_i^*}). \tag{34}$$

Proof:

By Lemmas 1 and 2, each player $j \neq i$ will use his ordinary pure strategies with probabilities corresponding to the mixed strategy $s_j = \pi(a_j^*)$, or to the mixed strategy $t_j = \pi(s_j^*)$. On the other hand, player i himself will use the ordinary pure strategy $a_i^k = a_i^*(\xi_i)$. Therefore, in view of (21), Eqs. (31) and (33) follow.

A given n-pure strategy a_i^* of player i will be called a *(uniformly) best reply* to some combination $\overline{a_i^*}$ of n-pure strategies (or to some combination $\overline{s_i^*}$ of n-mixed strategies) used by the other $(n-1)$ players if, for *every* possible value of the random vector ξ_i, this strategy a_i^* maximizes player i's payoff U_i as specified by (31) (or by (33)) when the strategy combination $\overline{a_i^*}$ (or $\overline{s_i^*}$) is kept constant. In view of (31) and (33), in this case we shall also say that a_i^* is a best reply to the combination $\bar{s}_i = \pi(\overline{a_i^*})$ (or to the combination $\bar{t}_i = \pi(\overline{s_i^*})$) of ordinary mixed strategies.

Equilibrium points in a disturbed game $\Gamma^*(\varepsilon)$ will be defined in terms of this concept of uniformly best replies.

Theorem 1:

The best reply of any player i to some combination $\overline{a_i^*}$ of n-pure strategies, or to some combination $\overline{s_i^*}$ of n-mixed strategies, used by the other $(n-1)$ players is always an *essentially unique* n-pure strategy a_i^*. (That is, if *both* a_i^* and $a_i^{*\prime}$ are best replies to $\overline{a_i^*}$ or to $\overline{s_i^*}$ then a_i^* and $a_i^{*\prime}$ must be essentially identical.)

Proof:

We shall discuss only the case where the other $(n-1)$ players use some combination $\overline{a_i^*}$ of n-pure strategies. The proof is basically the same in the case where they use some combination $\overline{s_i^*}$ of n-mixed strategies.

In view of (31), for any specific value of the random vector ξ_i, a given ordinary pure strategy a_i^k will maximize player i's payoff U_i if and only if

$$V_i(a_i^k, \overline{s_i}) - V_i(a_i^{k'}, \overline{s_i}) \geqq \varepsilon \cdot (\zeta_i^{k'} - \zeta_i^k), \quad \text{for all} \quad k' \neq k. \tag{35}$$

This maximizing strategy a_i^k will be unique only if

$$V_i(a_i^k, \overline{s_i}) - V_i(a_i^{k'}, \overline{s_i}) > \varepsilon \cdot (\zeta_i^{k'} - \zeta_i^k), \quad \text{for all} \quad k' \neq k. \tag{36}$$

On the other hand, two different strategies a_i^k and $a_i^{k'}$ can *both* yield a maximal payoff only if

$$V_i(a_i^k, \overline{s_i}) - V_i(a_i^{k'}, \overline{s_i}) = \varepsilon \cdot (\zeta_i^{k'} - \zeta_i^k). \tag{37}$$

But in view of (20), for any given $\overline{s_i}$, the variables ζ_i^k and $\zeta_i^{k'}$ are convex combinations of the variables $\xi_i^1, \ldots, \xi_i^K$. Consequently, for any given choice of $\overline{s_i}$, and of a_i^k and $a_i^{k'}$, the locus $L^{kk'}$ of all points $\xi_i = (\xi_i^1, \ldots, \xi_i^K)$ satisfying (20) and (37) is a $(K-1)$-dimensional hyperplane. Let $\hat{L}^{kk'}$ be the intersection of $L^{kk'}$ with the K-dimensional range space Ξ_i of the random vector ξ_i. Clearly, $\hat{L}^{kk'}$ will be a set of measure zero in this range space Ξ_i. Let L^* be the union of the $K_i(K_i-1)/2$ sets $\hat{L}^{kk'}$ defined for all possible pairs (k,k') with $k,k' = 1, \ldots, K_i$ and with $k' \neq k$. Obviously, L^* will be again a set of measure zero in Ξ_i.

Let $Y_i^k = Y_i^k(\overline{s_i})$ be the set of all points $\xi_i = (\xi_i^1, \ldots, \xi_i^K)$ satisfying (20) and (36) for a given strategy a_i^k of player i. Y_i^k will be a relatively open polyhedral subset of Ξ_i, bounded by the various hyperplanes $L^{kk'}$, $L^{kk''}$, etc. (For some values of k, Y_i^k may be empty.)

Any n-pure strategy a_i^* can be a best reply to $\overline{a_i^*}$ only if it makes player i choose his ordinary pure strategy in such a way that

$$a_i^*(\xi_i) = a_i^k, \quad \text{for all} \quad \xi_i \in Y_i^k(\overline{s_i}), \quad \text{for all} \quad k = 1, \ldots, K. \tag{38}$$

This means that if two n-pure strategies a_i^* and $a_i^{*\prime}$ are both best replies to $\overline{a_i^*}$ then they must agree over all sets $Y_i^k (k = 1, \ldots, K_i)$ and can disagree at most only over the set L^* of measure zero (or over some subset of L^*). Consequently, a_i^* and $a_i^{*\prime}$ will be essentially identical.

Corollary to Theorem 1:

An n-mixed strategy s_i^* can be a best reply only if it is an *essentially pure* strategy (i.e., it must be a probability mixture of essentially identical n-pure strategies).

This corollary and Lemma 3 in turn imply:

Theorem 2:

In a disturbed game $\Gamma^*(\varepsilon)$ every equilibrium point must be in pure – or at least in essentially pure – normalized strategies. Moreover, for every equilibrium point $s^* = (s_1^*, \ldots, s_n^*)$ partly or wholly in mixed strategies (as we have seen, the latter must be "essentially pure" mixed strategies), there exists an essentially identical equilibrium point $a^* = (a_1^*, \ldots, a_n^*)$ wholly in pure strategies, yielding every player i the same payoff $U_i(a^*) = U_i(s^*)$.

Theorem 2 and the Corollary to Theorem 1 can be regarded as extensions of the results obtained by BELLMAN and BLACKWELL [1949] and by BELLMAN [1952]: if the game itself already contains enough "randomness" (random variables) then the players themselves need not – indeed, should not – introduce any additional randomness by a use of mixed strategies.

Theorem 1 also implies:

Theorem 3:

Every equilibrium point $a^* = (a_1^*, \ldots, a_n^*)$ of a disturbed game is *essentially strong*, in the sense that no player i can counter the strategy combination $\overline{a_i^*}$ of the other $(n-1)$ players by a best reply essentially different from his equilibrium strategy a_i^*.

Note:

If a given player i does shift from his equilibrium strategy a_i^* to another best-reply strategy $a_i^{*\prime}$ essentially identical to a_i^*, then this will not destabilize the situation. This is so because, by Lemma 3, this shift will leave his own and all other players' payoffs unchanged, and the resulting new strategy n-tuple $a^{*\prime} = (a_1^*, \ldots, a_{i-1}^*, a_i^{*\prime}, a_{i+1}^*, \ldots, a_n^*)$ will again be an equilibrium point. Therefore this shift will not give the other players any incentive to change their own strategies.

Theorem 4:

Every disturbed game $\Gamma^*(\varepsilon)$ has at least one equilibrium point.

Proof:

Let $s = (s_1, \ldots, s_n)$ be an n-tuple of ordinary mixed strategies. For each player i, let a_i^* be his best reply to $\overline{s_i} = (s_1, \ldots, s_{i-1}, s_{i+1}, \ldots, s_n)$ in the sense that, for every possible value of the random vector ξ_i, the ordinary pure strategy $a_i^k = (a_i^*(\xi_i)$ prescribed by a_i^* will satisfy (35) with respect to all $a_i^{k'} \neq a_i^k$. Let a^* be the n-tuple $a^* = (a_1^*, \ldots, a_n^*)$. Finally, let $s' = (s_1', \ldots, s_n')$ be the n-tuple of ordinary mixed strategies induced by a^*, i.e., let $s' = \pi(a^*)$.

Now consider the mapping M: $s \to s'$. Clearly, M is a mapping of the set S of all possible n-tuples of ordinary mixed strategies into itself. This set S is a finite-dimensional convex and compact set.

Moreover, the mapping M is single-valued. Even though in general there will be many n-pure strategies a_i^* that are best replies to a given $(n-1)$-tuple $\overline{s_i}$, all of these strategies a_i^* will be essentially identical (see the proof of Theorem 1). Therefore, all of them will induce exactly the same ordinary mixed strategy $s_i' = \pi(a_i^*)$.

Finally, the mapping M is continuous. For any given n-tuple s, the n-tuple $s' = M(s)$ is defined by conditions (29) and (30). But in (29) the integrand is the continuous function f_i whereas the region of integration, the set $X_i^k(a_i^*)$, can be written as

$$X_i^k(a_i^*) = Y_i^k(\overline{s}_i) \cup D_i^k(\overline{s}_i), \tag{39}$$

where Y_i^k is defined by conditions (20) and (36), whereas D_i^k is a set of measure zero, and therefore does not affect the result of the integration in (29). On the other hand, the set Y_i^k itself depends continuously on the n-tuple s in terms of the Hausdorff metric. This continuous dependence of Y_i^k on s and the continuity of f_i establish the continuity of the mapping M.

Consequently, by Brouwer's Fixed-point Theorem [see, e.g., LEFSCHETZ, 1949, p. 117], the mapping M will have at least one fixed point where $s' = M(s) = s$. Yet, any n-tuple a^* corresponding to such a fixed point $s' = s$ will be an equilibrium point in game $\Gamma^*(\varepsilon)$. First of all, by the definition of M, each component a_i^* of a^* will have the property that $a_i^k = a_i^*(\xi_i)$ will satisfy (35) with respect to all $a_i^{k'} \neq a_i^k$ for all values of ξ_i if we set $\overline{s}_i = \overline{s}_i' = \pi(\overline{a}_i^*)$. But, by Lemma 3, this very property establishes the fact that a_i^* is a best reply to $\overline{a}_i^*$. As this is true for all players i, a^* will be an equilibrium point in $\Gamma^*(\varepsilon)$.

Let $a^* = (a_1^*, \ldots, a_n^*)$ be an equilibrium point in some disturbed game $\Gamma^*(\varepsilon)$, and let $s = (s_1, \ldots, s_n = \pi(a^*)$ be the n-tuple of ordinary mixed strategies induced by a^*. Then s will be called the *s-representation* of the equilibrium point a^*. We shall also say that s is an *s-equilibrium point* in game $\Gamma^*(\varepsilon)$.

5. Equilibrium Points in Ordinary Games as Limits of *s*-Equilibrium Points in Disturbed Games

Let $\{\Gamma^*(\varepsilon)\}$ be a one-parameter family of disturbed games, each game $\Gamma^*(\varepsilon)$ being characterized by a different value of the parameter ε, but all of them having the same n-tuple of basic payoff functions $V_1, \ldots, V_n$ and the same n-tuple of probability distributions $F_1, \ldots, F_n$. If we set $\varepsilon = 0$ then, in view of (12), we obtain an ordinary (undisturbed) game $\Gamma = \Gamma^*(0)$ with the payoff functions $U_1 = V_1, \ldots, U_n = V_n$.

Theorem 5:

Let $\{s(\varepsilon)\}$ be a one-parameter family of n-tuples of ordinary mixed strategies such that, for all ε values in some open interval $(0, \varepsilon^*)$ with $\varepsilon^* > 0$, each n-tuple $s(\varepsilon) = (s_1(\varepsilon), \ldots, s_n(\varepsilon))$ is an s-equilibrium point in the corresponding game $\Gamma^*(\varepsilon)$. Suppose that

$$\lim_{\varepsilon \to 0} s(\varepsilon) = s = (s_1, \ldots, s_n). \tag{40}$$

Then s will be an equilibrium point in game $\Gamma = \Gamma^*(0)$.

Proof:

In view of (40), for any sufficiently small ε we must have

$$C(s_i) \subseteq C(s_i(\varepsilon)), \quad \text{for} \quad i = 1, \ldots, n; \tag{41}$$

where $C(s_i)$ and $C(s_i(\varepsilon))$ are the carriers of strategies s_i and $s_i(\varepsilon)$, respectively. Since $s(\varepsilon)$ is an s-equilibrium point in $\Gamma^*(\varepsilon)$, any ordinary pure strategy a_i^k in the carrier $C(s_i(\varepsilon))$ must satisfy requirement (34) with respect to all other pure strategies $a_i^{k'}$ of player i over some nonempty set Y_i^k of possible ξ_i values. When ε goes to zero, requirement (35) will take the form

$$V_i(a_i^k, \bar{s}_i) - V_i(a_i^{k'}, \bar{s}_i) \geqq 0, \quad \text{for all} \quad k' \neq k. \tag{42}$$

In view of (41), this requirement will be satisfied for all pure strategies a_i^k in $C(s_i)$. But this is precisely the condition that must be satisfied if s is to be an equilibrium point in game $\Gamma = \Gamma^*(0)$. This establishes the theorem.

Let s be an equilibrium point in game $\Gamma = \Gamma^*(0)$. We shall say that s is *approachable* by some s-equilibrium point $s(\varepsilon)$ of a disturbed game $\Gamma^*(\varepsilon)$ if there is some s-equilibrium point $s(\varepsilon)$ (or, more exactly, if there is some family $\{s(\varepsilon)\}$ of s-equilibrium points) satisfying (40) with respect to this equilibrium point s.

Theorems 4 and 5 imply that every ordinary game $\Gamma = \Gamma^*(0)$ has at least *one* approachable equilibrium point s. We shall now show that every *strong* equilibrium point of any game Γ is always approachable, and that a *quasi-strong* equilibrium point is "almost always" approachable (in the sense that this property can fail only for a small class of games, corresponding to a closed set of measure zero in the space of all games of a given size).

Theorem 6:

Let $a = (a_1^{k_1}, \ldots, a_n^{k_n})$ be a *strong* equilibrium point [6] in game $\Gamma = \Gamma^*(0)$. Let $a^* = (a_1^*, \ldots, a_n^*)$ be an n-tuple of *constant* n-pure strategies such that $a_i^* = [a_i^{k_i}]$ for $i = 1, \ldots, n$. Then, for any small enough ε, a^* will be an equilibrium point in game $\Gamma^*(\varepsilon)$.

Proof:

Let $\bar{a}_i$ be the $(n-1)$-tuple we obtain if we omit the i-th component, the strategy $a_i^{k_i}$, from the n-tuple a. Since a is a strong equilibrium point in game Γ, we must have

$$V_i(a_i^{k_i}, \bar{a}_i) - V_i(a_i^{k'}, \bar{a}_i) > 0, \quad \text{for all} \quad k' \neq k_i, \quad \text{and for} \quad i = 1, \ldots, n. \tag{43}$$

Consequently, for any small enough ε, we can write

$$V_i(a_i^{k_i}, \bar{a}_i) - V_i(a_i^{k'}, \bar{a}_i) \geqq \varepsilon \cdot (\zeta_i^{k_i} - \zeta_i^{k'}) \tag{44}$$

[6] A strong equilibrium point is always in pure strategies.

for all possible values of $\zeta_i^{k_i}$ and $\zeta_i^{k'}$. Hence, $a_i^{k_i}$ will satisfy (35), and therefore $a_i^{k_i}$ will be a best reply to the strategy combination $\bar{a}_i$ used by the other $(n-1)$ players. As this will be true for all players i, this means that a will be an equilibrium point in any game $\Gamma^*(\varepsilon)$ with a small enough ε.

Consider the constant n-pure strategy $a_i^* = [a_i^{k_i}]$. Clearly, if player i follows this constant strategy then he will choose the corresponding ordinary pure strategy $a_i^{k_i}$ with probability one. Consequently, the "mixed" strategy induced by this n-pure strategy $a_i^* = [a_i^{k_i}]$ is simply the pure strategy $a_i^{k_i}$ itself, so that we can write $\pi(a_i^*) = a_i^{k_i}$. Accordingly, the s-equilibrium point corresponding to the equilibrium point a^* mentioned in Theorem 6 will be simply the n-tuple $a = \pi(a^*)$ of ordinary pure strategies.

Lemma 4:

Let $s = (s_1, \ldots, s_n)$ be an n-tuple of ordinary mixed strategies, and suppose that the probabilities p_i^k characterizing these mixed strategies s_i satisfy the equation

$$p_i^k = \int_{\zeta_i \in \Psi_i^k} \mathrm{d}\zeta_i^1 \cdots \int \mathrm{d}\zeta_i^{K_i} g_i(\zeta_i|\bar{s}_i), \quad \text{for} \quad i = 1, \ldots, n; \quad k = 1, \ldots, K_i; \tag{45}$$

where

$$g_i(\zeta_i|\bar{s}_i) = g_i(\zeta_i^1, \ldots, \zeta_i^{K_i}|\bar{s}_i) = \frac{\partial^{K_i} G_i(\zeta_i|\bar{s}_i)}{\partial\zeta_i^1, \ldots, \partial\zeta_i^{K_i}} \tag{46}$$

is the probability density function corresponding to the probability distribution $G_i(\zeta_i|\bar{s}_i)$ defined by (24); whereas, Ψ_i^k is the set of all points $\zeta_i = (\zeta_i^1, \ldots, \zeta_i^{K_i})$ satisfying Condition (36) for a given positive value of ε, and for a given pure strategy a_i^k, with respect to all other strategies $a_i^{k'}$, $k' \neq k$, of the same player i.

Then s will be an s-equilibrium point in game $\Gamma^*(\varepsilon)$.

Proof:

For each player i, let a_i^* be his best reply to $\bar{s}_i = (s_1, \ldots, s_{i-1}, s_{i+1}, \ldots, s_n)$, in the sense specified in the proof of Theorem 4. Let $a^* = (a_1^*, \ldots, a_n^*)$. We shall show that a^* is an equilibrium point in game $\Gamma^*(\varepsilon)$, and that s is the corresponding s-equilibrium point. As a_i^* is a best reply to $\bar{s}_i$, it will select any given pure strategy a_i^k whenever a_i^k satisfies Condition (36). Therefore a_i^* will choose each pure strategy a_i^k with the probability p_i^k defined by Eq. (45). (To be sure, a_i^* may make player i choose this pure strategy a_i^k also at some points ξ_i where (36) is not satisfied, though the weaker Condition (35) is. But the total probability mass associated with such points ξ_i will always be zero.) Consequently, for each player i, a_i^* will induce the mixed strategy s_i corresponding to the probabilities p_i^k, so that $s_i = \pi(a_i^*)$. Therefore, the strategy n-tuple s will be a fixed point $s = M(s)$ of the mapping M defined in the proof of Theorem 4, which implies, as we have seen, that a^* is an equilibrium point in $\Gamma^*(\varepsilon)$ whereas $s = \pi(a^*)$ is the corresponding s-equilibrium point.

Let $s = (s_1, \ldots, s_n)$ be an n-tuple of ordinary mixed strategies. Suppose that the carriers $C(s_1), \ldots, C(s_n)$ of the strategies $s_1, \ldots, s_n$ contain $\gamma_1, \ldots, \gamma_n$ pure strategies, respectively. Then, without loss of generality, we can introduce the following notational convention:

The pure strategies a_i^k of each player i are numbered in such a way that the carrier $C(s_i)$ of his mixed strategy s_i will contain his *first* γ_i pure strategies $a_i^1, \ldots, a_i^{\gamma_i}$. (47)

Clearly, the $(\gamma_i - 1)$ probabilities $p_i^1, \ldots, p_i^{\gamma_i - 1}$ fully characterize s_i since

$$p_i^{\gamma_i} = 1 - \sum_{k=1}^{\gamma_i - 1} p_i^k$$

and

$$p_i^k = 0, \quad \text{for} \quad k = \gamma_i + 1, \ldots, K_i .$$

Let p_i denote the vector

$$p_i = (p_i^1, \ldots, p_i^{\gamma_i - 1}) . \tag{48}$$

Let $\overline{p_i}$ denote the composite vector

$$\overline{p_i} = (p_1, \ldots, p_{i-1}, p_{i+1}, \ldots, p_n) , \tag{49}$$

where $p_1, \ldots, p_n$ are the vectors defined by (48). In order to make the dependence of the functions V_i and g_i on the probabilities p_i^k more explicit, we shall now write

$$V_i(a_i^k, \overline{s_i}) = V_i(a_i^k, \overline{p_i}) \tag{50}$$

and

$$g_i(\zeta_i | \overline{s_i}) = g_i(\zeta_i | \overline{p_i}) . \tag{51}$$

For each player i ($i = 1, \ldots, n$), we introduce the random variables

$$\delta_i^{kk'} = \zeta_i^{k'} - \zeta_i^k, \quad \text{for} \quad k, k' = 1, \ldots, K_i; \; k' \neq k; \tag{52}$$

as well as the random vectors

$$\delta_i^k = (\delta_i^{k1}, \delta_i^{k2}, \ldots, \delta_i^{k(k-1)}, \delta_i^{k(k+1)}, \delta_i^{k(k+2)}, \ldots, \delta_i^{kK_i}), \quad \text{for} \quad k = 1, \ldots, \gamma_i - 1 . \tag{53}$$

For convenience, we shall also write

$$\delta_i^{kk} = \zeta_i^k - \zeta_i^k = 0 . \tag{54}$$

Lemma 5:

Let $s = (s_1, \ldots, s_n)$ be an n-tuple of ordinary mixed strategies. Suppose that for each player i ($i = 1, \ldots, n$) the probabilities p_i^k characterizing his mixed strategy s_i, together with some appropriately chosen quantities y_i^k and $z_i^{kk'}$, satisfy the following conditions with respect to a specific positive value of ε:

$$\varepsilon y_i^k = V_i(a_i^k, \overline{p_i}) - V_i(a_i^{k+1}, \overline{p_i}), \quad \text{for} \quad k = 1, \ldots, \gamma_i - 1; \tag{55}$$

$$z_i^{kk'} = \sum_{j=k}^{k'-1} y_i^j, \quad \text{for} \quad k = 1, \ldots, \gamma_i - 1; \quad k' = k+1, \ldots, \gamma_i; \tag{56}$$

$$z_i^{kk'} = - \sum_{j=k'}^{k-1} y_i^j, \quad \text{for} \quad k = 2, \ldots, \gamma_i; \quad k' = 1, \ldots, k-1; \tag{57}$$

$$p_i^k = H_i^k(z_i^{k1}, \ldots, z_i^{k(k-1)}, z_i^{k(k+1)}, \ldots, z_i^{k\gamma_i}; \overline{p_i}) = \tag{58}$$

$$= \int_{-\infty}^{\sigma_1} \mathrm{d}\,\delta_i^{k1} \ldots \int_{-\infty}^{\sigma_{k-1}} \mathrm{d}\,\delta_i^{k(k-1)} \int_{-\infty}^{\sigma_{k+1}} \mathrm{d}\,\delta_i^{k(k+1)} \ldots \int_{-\infty}^{\sigma_{K_i}} \mathrm{d}\,\delta_i^{kK_i} \int_{-\infty}^{+\infty} \mathrm{d}\zeta_i^k h_i^k(\delta_i^k;\zeta_i^k),$$

where

$$\sigma_j = z_i^{kj}, \quad \text{for} \quad j = 1,\ldots,k-1,k+1,\ldots,\gamma_i; \tag{59}$$

but

$$\sigma_j = +\infty, \quad \text{for} \quad j = \gamma_i + 1,\ldots,K_i; \tag{60}$$

whereas

$$\begin{aligned} h_i^k(\delta_i^k;\zeta_i^k) &= h_i^k(\delta_i^{k1},\ldots,\delta_i^{k(k-1)},\delta_i^{k(k+1)},\ldots,\delta_i^{kK_i};\zeta_i^k) = \\ &= g_i(\zeta_i^k + \delta_i^{k1},\ldots,\zeta_i^k + \delta_i^{kj},\ldots,\zeta_i^k + \delta_i^{kK_i}|\bar{p}_i); \end{aligned} \tag{61}$$

and finally

$$V_i(a_i^k,\bar{p}_i) - V_i(a_i^{k'},\bar{p}_i) > \varepsilon \cdot (\zeta_i^{k'} - \zeta_i^k) \tag{62}$$

for all possible values of $\zeta_i^{k'}$ and $\zeta_i^k(\zeta_i^{k'} \in Z_i^{k'}$ and $\zeta_i^k \in Z_i^k)$; and for $k = 1,\ldots,\gamma_i$; $k' = \gamma_i + 1,\ldots,K_i$.

Then s will be an s-equilibrium point in game $\Gamma^*(\varepsilon)$. Moreover, the quantities $z_i^{kk'}$ will satisfy

$$-2 < z_i^{kk'} < +2. \tag{63}$$

Proof:

As $\varepsilon > 0$, (55), (56), and (57) imply that

$$z_i^{kk'} = [V_i(a_i^k,\bar{p}_i) - V_i(a_i^{k'},\bar{p}_i)]/\varepsilon. \tag{64}$$

Hence, (58) can be obtained from (45) by a change of variables. In view of (36), (52), and (64), for each variable $\delta_i^{kk'}$ with $k,k' = 1,\ldots,\gamma_i$ and with $k' \neq k$, the upper limit of integration is $\sigma_{k'} = z_i^{kk'}$. Moreover, since in this case both a_i^k and $a_i^{k'}$ are in $C(s_i)$, each upper limit $z_i^{kk'}$ will have the property that the two inequalities $\delta_i^{kk'} < z_i^{kk'}$ and $\delta_i^{kk'} > z_i^{kk'}$ are *both* satisfied with positive probabilities. Therefore, each quantity $z_i^{kk'}$ must lie in the interior of the range of the corresponding random variable $\delta_i^{kk'}$. By (26) and (52), this range is a subset of $[-2, +2]$, which implies (63).

In contrast, for the variables $\delta_i^{kk'}$ with $k \leqq \gamma_i < k'$, the upper limit of integration can be taken to be $\sigma_{k'} = +\infty$. This is so because now the inequality $\delta_i^{kk'} < \sigma_{k'}$ must be satisfied with probability one, since in this case, $a_i^k \in C(s_i)$, whereas $a_i^{k'} \notin C(s_i)$. This completes the proof.

Lemma 6:

Equation (58) can also be written as

$$p_i^k = E_i^k(y_i^1,\ldots,y_i^{\gamma_i-1};\bar{p}_i), \quad \text{for} \quad i = 1,\ldots,n; \quad k = 1,\ldots,\gamma_i - 1. \tag{65}$$

These functions E_i^k are continuously differentiable.

Proof:

(56) and (57) make the quantities $z_i^{kk'}$ functions of the variables y_i^k. Continuous differentiability follows from (17) and from Leibnitz's Rule [Olmsted, 1959, p. 417].

Lemma 7:

Let $s = (s_1, \ldots, s_n)$ be an n-tuple of ordinary mixed strategies, characterized by the probabilities p_i^k. Then, we can always find some quantities y_i^k and $z_i^{kk'}$ which, together with these probabilities p_i^k, will satisfy (56) to (61).

Proof:

First, we shall show that we can find quantities y_i^k satisfying (65) for $i = 1, \ldots, n$ and for $k = 1, \ldots, \gamma_i - 1$. Let I^* be the closed interval $I^* = [-2, +2]$. For each player i, we define a set $Y_i^* = I_1 \times \cdots \times I_{\gamma_i}$ where $I_1 = \cdots = I_{\gamma_i} = I^*$. For each vector $y_i^* = (y_i^1, \ldots, y_i^{\gamma_i})$ in Y_i^* we define a vector $\omega_i = (\omega_i^1, \ldots, \omega_i^{\gamma_i})$ as follows. Let

$$\pi_i^k = E_i^k(y_i^1, \ldots, y_i^{\gamma_i - 1}), \quad \text{for} \quad k = 1, \ldots, \gamma_i - 1; \tag{66}$$

$$\pi_i^{\gamma_i} = 1 - \sum_{k=1}^{\gamma_i - 1} \pi_i^k, \tag{67}$$

and

$$p_i^{\gamma_i} = 1 - \sum_{k=1}^{\gamma_i - 1} p_i^k. \tag{68}$$

Finally, we set

$$\omega_i^k = y_i^k[1 - \max(0, \pi_i^k - p_i^k)], \quad \text{for} \quad k = 1, \ldots, \gamma_i. \tag{69}$$

For each player i, Eqs. (66) to (69) establish a mapping $M_i : y_i^* \to \omega_i$. Clearly M_i is a mapping of the finite-dimensional convex and compact set Y_i^* into itself. Moreover, by Lemma 6, M_i is continuously differentiable and therefore continuous. Consequently, by Brouwer's Fixed-Point Theorem, M_i must have a fixed point where

$$\omega_i = M_i(y_i^*) = y_i^*. \tag{70}$$

For each player i, the first $(\gamma_i - 1)$ components $y_i^1, \ldots, y_i^{\gamma_i - 1}$ of any vector y_i^* satisfying (70) will satisfy (65). Moreover, if we substitute these variables y_i^k in (56) and (57), then we shall obtain quantities $z_i^{kk'}$ satisfying (58). This completes the proof.

Any vector y_i^* satisfying (70) must lie in the interier of Y_i^* [cf. the proof of Lemma 5, regarding Condition (63)]. Thus we can write

$$-2 < y_i^k < +2, \quad \text{for } i = 1, \ldots, n; \quad k = 1, \ldots, \gamma_i - 1. \tag{71}$$

(This inequality is, of course, true also for $k = \gamma_i$, but now we do not need this fact.)

We now define

$$\gamma^* = \sum_{i=1}^{n} (\gamma_i - 1) = \sum_{i=1}^{n} \gamma_i - n, \tag{72}$$

and

$$y_i = (y_i^1, \ldots, y_i^{\gamma_i - 1}), \quad \text{for} \quad i = 1, \ldots, n. \tag{73}$$

Let p and y denote the composite vectors

$$p = (p_1, \ldots, p_n) \tag{74}$$

and

$$y = (y_1, \ldots, y_n), \tag{75}$$

where $p_1, \ldots, p_n$ and $y_1, \ldots, y_n$ are the vectors defined by (48) and (73). Clearly, p has γ^* components of form p_i^k, and y has γ^* components of form y_i^k.

Let P be the set of all vectors p satisfying

$$p_i^k > 0, \quad \text{for} \quad i = 1, \ldots, n; \quad \text{and for } k = 1, \ldots, \gamma_i - 1; \tag{76}$$

and

$$\sum_{k=1}^{\gamma_i - 1} p_i^k < 1, \quad \text{for} \quad i = 1, \ldots, n. \tag{77}$$

Let Y be the set of all vectors y satisfying (71).

In vector notation, we can now write Eq. (65) as

$$p_i = E_i(y_i; \bar{p}_i), \quad \text{for} \quad i = 1, \ldots, n. \tag{78}$$

Lemma 8:

There exists a mapping $\tau: p \to y$ from set P to set Y with the following properties:

(a) For every p in P, the vectors p and $y = \tau(p)$ together satisfy (65).

(b) The set P^0 of all points p in P where τ fails to be continuously differentiable is closed and has measure zero as a subset of P.

Proof:

Let J be the Jacobian

$$J = \frac{\partial(E_1^1, \ldots, E_1^{\gamma_1 - 1}; \ldots; E_i^1, \ldots, E^{\gamma_i - 1}; \ldots; E_n^1, \ldots, E_n^{\gamma_n - 1})}{\partial(y_1^1, \ldots, y^{\gamma_1 - 1}; \ldots; y_i^1, \ldots, y_i^{\gamma_i - 1}; \ldots; y_n^1, \ldots, y_n^{\gamma_n - 1})}. \tag{79}$$

It is easy to verify that

$$J = \prod_{i=1}^{n} J_i, \tag{80}$$

where J_i is the smaller Jacobian

$$J_i = \frac{\partial(E_i^1, \ldots, E_i^{\gamma_i - 1})}{\partial(y_i^1, \ldots, y_i^{\gamma_i - 1})}, \quad \text{for} \quad i = 1, \ldots, n. \tag{81}$$

By Lemma 7, a mapping τ always exists. By Lemma 6 and the Implicit Function Theorem, τ can be chosen so as to be continuously differentiable at every point $(p, \tau(p))$, where the Jacobian J is nonzero; i.e., where all n Jacobians $J_1, \ldots, J_n$ are nonzero. We shall now show that this condition will be met everywhere except on a closed set P^0 of measure zero.

Let 0_i be the set of all points $(y_i, \bar{p}_i)$ where the Jacobian $J_i = J_i(y_i, \bar{p}_i)$ vanishes. Since J_i depends continuously on $(y_i, \bar{p}_i)$, and since both y_i and $\bar{p}_i$ are bounded, 0_i is a compact set.

Now, consider the mapping τ_i^*: $(y_i, \bar{p}_i) \to (p_i, \bar{p}_i) = p$, where $p_i = E_i(y_i; \bar{p}_i)$. By Lemma 6, this mapping τ_i^* is continuously differentiable. Therefore, by Sard's Theorem [SARD, 1942], the set $P_i^0 = \tau_i^*(0_i)$ of all points p associated with zeroes of J_i is a set of measure zero in P. Moreover, as P_i^0 is the image of 0_i under the continuous mapping τ_i^*, P_i^0 itself is also a compact set, and therefore closed.

As each set P_i^0 is a closed set of measure zero, their union, the set

$$P^0 = \bigcup_{i=1}^{n} P_i^0 \tag{82}$$

will have the same properties. But continuous differentiability of τ can fail only on this set P^0. This establishes the lemma.

We now introduce the (nK)-vector

$$v = (v_1^1, \ldots, v_1^K; \ldots; v_i^1, \ldots, v_i^K; \ldots; v_n^1, \ldots, v_n^K). \tag{83}$$

Let us write Eq. (55) in the form

$$\begin{aligned} \varphi_i^k(p,y,v,\varepsilon) &= \varepsilon y_i^k - V_i(a_i^k, \overline{p}_i) + V_i(a_i^{k+1}, \overline{p}_i) = \\ &= \varepsilon y_i^k - \sum_{m=1}^{K} \left[q_i^m(a_i^k) \prod_{j \neq i} q_j^m(s_j) \right] v_i^m \\ &+ \sum_{m=1}^{K} \left[q_i^m(a_i^{k+1}) \prod_{j \neq i} q_j^m(s_j) \right] v_i^m = 0, \quad \text{for } i = 1, \ldots, n; \\ &\qquad\qquad\qquad\qquad\qquad\qquad\qquad k = 1, \ldots, \gamma_i - 1. \end{aligned} \tag{84}$$

In view of Lemma 8, this can also be written as

$$\varphi_i^k(p, \tau(p), v, \varepsilon) = \psi_i^k(p, v, \varepsilon) = 0. \tag{85}$$

Lemma 9:

Let $s = (s_1, \ldots, s_n)$ be a *quasi-strong* equilibrium point, characterized by the probability vector p, in game $\Gamma = \Gamma^*(0)$. Suppose that

(i) The mapping τ from P to Y is continuously differentiable on some neighborhood N of p; and that

(ii) The Jacobian

$$J^* = \frac{\partial(\psi_1^1, \ldots, \psi_1^{\gamma_1 - 1}; \ldots; \psi_i^1, \ldots, \psi_i^{\gamma_i - 1}; \ldots; \psi_n^1, \ldots, \psi_n^{\gamma_n - 1})}{\partial(p_1^1, \ldots, p_1^{\gamma_1 - 1}; \ldots; p_i^1, \ldots, p_i^{\gamma_i - 1}; \ldots; p_n^1, \ldots, p_n^{\gamma_n - 1})} \neq 0 \tag{86}$$

at the point p if we set $\varepsilon = 0$.

Then, there exists a family $\{s(\varepsilon)\}$ of n-tuples of ordinary strategies such that, for any sufficiently small positive ε, $s(\varepsilon)$ is an s-equilibrium point in game $\Gamma^*(\varepsilon)$, and such that $s(\varepsilon)$ and s satisfy (40).

Proof:

First, we propose to show that p satisfies (55) if we set $\varepsilon = 0$. The left-hand side of (55) now will be zero, and the same will be true about the right-hand side since s is an equilibrium point in Γ. Therefore, in view of (86) and the Implicit Function Theorem, for any small enough positive ε, there exists a vector $p(\varepsilon)$ satisfying (55) in conjunction with the vector $y(\varepsilon) = \tau(p(\varepsilon))$. This vector $p(\varepsilon)$ will depend continuously on ε. In view of Lemmas 7 and 8, these vectors $p(\varepsilon)$ and $y(\varepsilon)$, together with some appropriately chosen quantities $z_i^{kk'}(\varepsilon)$, will also satisfy Conditions (56) to (61). Finally, as s is a quasi-strong equilibrium point in Γ, we have

$$V_i(a_i^k, \overline{p}_i) - V_i(a_i^{k'}, \overline{p}_i) > 0, \quad \text{for } k = 1, \ldots, \gamma_i; \ k' = \gamma_i + 1, \ldots, K_i. \tag{87}$$

This is so because $a_i^k \in C(s_i)$ whereas $a_i^{k'} \notin C(s_i)$. Consequently, for any small enough positive ε, the strategy n-tuple $s(\varepsilon)$ defined by $p(\varepsilon)$ will satisfy (62). Thus, by Lemma 5, $s(\varepsilon)$ will be an s-equilibrium point in $\Gamma^*(\varepsilon)$. Finally, (40) follows from the continuous dependence of $s(\varepsilon)$ on ε.

We now define

$$m(i,1) = 1, \quad \text{for} \quad i = 1, \ldots, n; \tag{88}$$

$$m(1,k) = k, \quad \text{for} \quad k = 2, \ldots, \gamma_1 - 1; \tag{89}$$

and

$$m(i,k) = 1 + \sum_{j=1}^{i-1} (\gamma_j - 2) + k = \sum_{j=1}^{i-1} \gamma_j - 2i + k + 3, \tag{90}$$
$$\text{for} \quad i = 2, \ldots, n; \; k = 2, \ldots, \gamma_i - 1.$$

Thus, $m(i,k)$ ranges over all positive integers from $m(i,1) = 1$ to

$$m(n, \gamma_n - 1) = \sum_{j=1}^{n} \gamma_j - 2n + 3 = \gamma^* - n + 3 = \gamma^{**}. \tag{91}$$

In addition to (47), we now introduce the following notational convention, which again involves no loss of generality:

> The first γ^{**} pure-strategy n-tuples $a^1, \ldots, a^{\gamma^{**}}$ are numbered in such a way that for any $m^* = m(i,k)$ with $i = 1, \ldots, n$ and with $k = 1, \ldots, \gamma_i - 1$, we have
> $$a^{m^*} = (a_1^1, \ldots, a_{i-1}^1, a_i^k, a_{i+1}^1, \ldots, a_n^1). \tag{92}$$

We now define

$$v(i,k) = v_i^{m^*} = v_i^{m(i,k)} = V_i(a_1^1, \ldots, a_{i-1}^1, a_i^k, a_{i+1}^1, \ldots, a_n^1), \tag{93}$$
$$\text{for} \quad i = 1, \ldots, n; \; k = 1, \ldots, \gamma_i - 1.$$

Let v^* be the vector formed of those γ^* components of vector v which can be written in form (93). Let v^{**} be the vector formed of the remaining $(nK - \gamma^*)$ components of v. Thus

$$v = (v^*, v^{**}). \tag{94}$$

Let $\mathscr{I} = \mathscr{I}(n; K_1, \ldots, K_n)$ be the set of all ordinary (undisturbed) n-person games in which players $1, \ldots, n$ have exactly $K_1, \ldots, K_n$ strategies, respectively. Thus $\mathscr{I}$ is the set of all games of a given size. Each game Γ in $\mathscr{I}$ can be characterized by its vector $v = v(\Gamma)$ of possible payoffs for pure-strategy combinations. As v is an (nK)-vector, where $K = \prod_i K_i$, $\mathscr{I}$ can be regarded as an (nK)-dimensional Euclidean space, and we can write $\mathscr{I} = \{v\}$. Let $\bar{\mathscr{I}}$ be the set of all games Γ in $\mathscr{I}$ for which a given statement $\mathscr{S}$ is false. We shall say that $\mathscr{S}$ is true for *almost all* games Γ if, for every possible set $\mathscr{I} = \mathscr{I}(n; K_1, \ldots, K_n)$, this set $\bar{\mathscr{I}}$ is closed and has measure zero in $\mathscr{I}$. [As to the requirement of closure for $\bar{\mathscr{I}}$, cf. DEBREU, 1970, p 387.]

Theorem 7:

For *all* probability distributions $F_1, \ldots, F_n$ satisfying Conditions (15) to (17), and for *almost all* games Γ, the following statement is true. Let $s = (s_1, \ldots, s_n)$ be a *quasi-strong* equilibrium point in Γ. Let $\{\Gamma^*(\varepsilon)\}$ be a family of disturbed games characterized by these probability distributions $F_1, \ldots, F_n$ with $\Gamma^*(0) = \Gamma$. Then there exists a family $\{s(\varepsilon)\}$ of n-tuples of ordinary mixed strategies with the property that, for any small enough positive ε, $s(\varepsilon)$ is an s-equilibrium point in the disturbed game $\Gamma^*(\varepsilon)$, with

$$\lim_{\varepsilon \to 0} s(\varepsilon) = s. \tag{95}$$

Proof:

Each equation of form (55) (or, equivalently, of form (84) and (85)) can be written as

$$v^{m^*} = \frac{\varepsilon y_i^k - \sum\limits_{\substack{1 \leq m \leq K \\ m \neq m^*}} \left[q_i^m(a_i^k) \prod\limits_{j \neq i} q_j^m(s_j) \right] v_i^m + \sum\limits_{\substack{1 \leq m \leq K \\ m \neq m^*}} \left[q_i^m(a_i^{k+1}) \prod\limits_{j \neq i} q_j^m(s_j) \right] v_i^m}{\prod\limits_{j \neq i} q_j^{m^*}(s_j)}, \tag{96}$$

where $m^* = m(i,k)$. Each equation of from (55) can be written in this way because $q_j^{m^*}(s_j) = p_j^1 > 0$ for all $j \neq i$ since $a_i^1 \in C(s_j)$. Moreover, $q_i^{m^*}(a_i^k) = 1$, whereas $q_i^{m^*}(a_i^{k+1}) = 0$ since the i-th component of the strategy combination a^{m^*} is a_i^k. Note that each quantity $v_i^{m^*}$ occurs only in *one* equation of form (96) with a nonzero coefficient. Consequently, the γ^* equations of form (96) define a continuously differentiable mapping ρ: $(p, y) \to v^*$, which is a mapping from $(P \times Y)$ to the γ^*-dimensional Euclidean space $\mathscr{I}^* = \{v^*\}$.

Let P^0 be the set defined by (82), and let $Y^0 = \tau(P^0)$. By the proof of Lemma 8, $(P^0 \times Y^0)$ is a compact set of measure zero in $(P \times Y)$. Therefore, its image $\mathscr{I}^{*0} = \rho(P^0 \times Y^0)$ under the continuously differentiable mapping ρ will again be a compact set with measure zero in $\mathscr{I}^*$. Let $\mathscr{I}^{**} = \{v^{**}\}$ be the set of all possible vectors v^{**}. $\mathscr{I}^{**}$ will be an $(nK - \gamma^*)$-dimensional Euclidean space. We have $\mathscr{I} = \mathscr{I}^* \times \mathscr{I}^{**}$. Let $\mathscr{I}^0 = \mathscr{I}^{*0} \times \mathscr{I}^{**}$. Clearly, $\mathscr{I}^0$ will be a closed set of measure zero in $\mathscr{I}$.

We can use the mapping ρ to define another mapping $\rho^*: (p, v^{**}) \to (v^*, v^{**}) = v$, by setting $v^* = \rho(p, \tau(p))$. This mapping ρ^* is from set $(P \times \mathscr{I}^{**})$ to $\mathscr{I}$, and is continuously differentiable whenever $v \in (\mathscr{I} - \mathscr{I}^0)$ because in this case, τ will be continuously differentiable. Therefore, by Sard's Theorem, the set $\mathscr{I}^{00}$ of all points v in $(\mathscr{I} - \mathscr{I}^0)$ corresponding to zeroes of the Jacobian J^* of (86) will have measure zero in $(\mathscr{I} - \mathscr{I}^0)$ and therefore also in $\mathscr{I}$.

We shall now show that $\mathscr{I}^{00}$ will be a closed set. We set

$$u_i = \min_m v_i^m \quad \text{and} \quad w_i = \max_m v_i^m, \quad \text{for} \quad i = 1, \ldots, n. \tag{97}$$

We define

$$\dot{v}_i^m = \mu(v_i^m) = \frac{v_i^m - u_i}{w_i - u_i}, \quad \text{if} \quad u_i < w_i, \tag{98}$$

and

$$\dot{v}_i^m = \mu(v_i^m) = 0, \quad \text{if} \quad u_i = w_i, \quad \text{for} \quad i = 1, \ldots, n; \; m = 1, \ldots, K. \tag{99}$$

We shall write

$$\dot{v} = \mu(v) = (\dot{v}_1^1, \ldots, \dot{v}_1^K; \ldots; \dot{v}_i^1, \ldots, \dot{v}_i^K; \ldots; \dot{v}_n^1, \ldots, \dot{v}_n^K), \tag{100}$$

and also

$$\dot{v}^* = \mu(v^*) \quad \text{and} \quad \dot{v}^{**} = \mu(v^{**}). \tag{101}$$

We also define

$$\dot{\mathscr{I}} = \{\dot{v}\}, \; \dot{\mathscr{I}}^* = \{\dot{v}^*\}, \quad \text{and} \quad \dot{\mathscr{I}}^{**} = \{\dot{v}^{**}\}. \tag{102}$$

Any component $\dot{v}_i^m$ of any vector $\dot{v}$ in $\dot{\mathscr{I}}$ will satisfy

$$0 \leqq \dot{v}_i^m \leqq 1, \quad \text{for} \quad i = 1, \ldots, n; \; m = 1, \ldots, K. \tag{103}$$

Clearly, $\mu \colon v \to \dot{v}$ is a linear mapping from $\mathscr{I}$ to $\dot{\mathscr{I}}$, which normalizes each player's payoff so as to satisfy Condition (103).

Let $\mathscr{F}$ be the set of all points $(p, \dot{v}^{**})$ where the Jacobian J^* vanishes. As J^* is a continuous function of $(p, \dot{v}^{**})$, $\mathscr{F}$ will be a closed set. Indeed, $\mathscr{F}$ will be a compact set since both $P = \{p\}$ and $\dot{\mathscr{I}}^{**} = \{\dot{v}^{**}\}$ are bounded. Consequently, $\dot{\mathscr{I}}^{00} = \rho^*(\mathscr{F})$ will also be compact since ρ^* is continuous. But $\dot{\mathscr{I}}^{00} = \mu(\mathscr{I}^{00})$, and μ is continuous. Therefore, $\mathscr{I}^{00}$ will be a closed set.

Thus, we can conclude that the assumptions of Lemma 9 are everywhere satisfied on $\mathscr{I}$, except on a closed set $\bar{\mathscr{I}} = \mathscr{I}^0 \cup \mathscr{I}^{00}$ with measure zero in $\mathscr{I}$. This completes the proof.

6. Numerical Examples

Consider the following three games:

Game Γ_1

	a_2^1	a_2^2
a_1^1	(3,4)	(2,2)
a_1^2	(1,1)	(2,1)

Game Γ_2

	a_2^1	a_2^2
a_1^1	(1,5)	(4,1)
a_1^2	(2,1)	(0,3)

Game Γ_3

	a_2^1	a_2^2
a_1^1	(1,1)	(1,1)
a_1^2	(1,1)	(1,1)

In all three games player i's payoff funtion will be called V_i $(i = 1, 2)$. For each game Γ_t we shall define a disturbed game $\Gamma_t^*(\varepsilon)$ by assuming that each player's payoff function U_i will be of form (12). For the sake of simplicity, we shall also assume that in each game Γ_t^* the probability functions F_1 and F_2 are such that the random variables $\delta_i^{kk'} = \delta_i^{12}$ defined by (52) are uniformly distributed and that this distribution is independent of the other player's strategy $s_j, j \neq i$. More particularly, δ_1^{12} is uniformly distributed over the interval $[-\alpha, +\alpha]$ whereas δ_2^{12} is uniformly distributed over the interval $[-\beta, +\beta]$, with $\alpha \leqq 1$ and $\beta \leqq 1$. We shall write $\alpha^* = \varepsilon\alpha$ and $\beta^* = \varepsilon\beta$.

Let $s = (s_1, s_2)$ be a pair of proper ordinary mixed strategies with $s_1 = qa_1^1 + (1-q)a_1^2$ and $s_2 = ra_2^1 + (1-r)a_2^2$. In view of (58), s will be an s-equilibrium point in $\Gamma_t{}^*$ if we have

$$[rv_1^{11} + (1-r)v_1^{12}] - [rv_1^{21} + (1-r)v_1^{22}] = \alpha^*\left(q - \frac{1}{2}\right) \tag{104}$$

and

$$[qv_2^{11} + (1-q)v_2^{21}] - [qv_2^{21} + (1-q)v_2^{22}] = \beta^*\left(r - \frac{1}{2}\right), \tag{105}$$

where

$$v_i^{km} = V_i(a_1^k, a_2^m), \quad \text{for} \quad i, k, m = 1, 2. \tag{106}$$

More particularly, Γ_1 has two equilibrium points in pure strategies. Of these, $a^1 = (a_1^1, a_2^1)$ is *strong* so that the pair $a^* = ([a_1^1], [a_2^1])$ of constant n-pure strategies is an equilibrium point in $\Gamma_1^*(\varepsilon)$ with a small ε. In contrast, $a^2 = (a_1^2, a_2^2)$ is an *extra weak* equilibrium point. Eqs. (104) and (105) yield:

$$q(\alpha^*, \beta^*) = -\beta^*\left(1 + \frac{\alpha^*}{2}\right)/(4 - \alpha^*\beta^*)$$

and

$$r(\alpha^*, \beta^*) = -\alpha^*\left(1 + \frac{\beta^*}{2}\right)/(4 - \alpha^*\beta^*).$$

These quantities q and r are close to the probabilities $q = 0$ and $r = 0$ characterizing a^2 but they are negative and so cannot be probabilities. Therefore a^2 is not approachable by any s-equilibrium point of game $\Gamma_1^*(\varepsilon)$.

Γ_2 has only one *quasi-strong* equilibrium point in mixed strategies, viz. $s = (s_1, s_2)$ with $s_1 = (\frac{1}{3}a_1^1 + \frac{2}{3}a_1^2)$ and $s_2 = (\frac{4}{5}a_2^1 + \frac{1}{5}a_2^2)$. By (104) and (105), s is approachable by the s-equilibrium point $s(\alpha^*, \beta^*)$, characterized by the probabilities

$$q(\alpha^*, \beta^*) = \frac{1}{3}\left(1 + \frac{3}{20}\beta^* + \frac{1}{20}\alpha^*\beta^*\right)\bigg/\left(1 + \frac{1}{30}\alpha^*\beta^*\right)$$

and

$$r(\alpha^*, \beta^*) = \frac{4}{5}\left(1 - \frac{1}{24}\alpha^* + \frac{1}{48}\alpha^*\beta^*\right)\bigg/\left(1 + \frac{1}{30}\alpha^*\beta^*\right).$$

Clearly, for small values of α^* and β^*, these probabilities will be very close to the probabilities $q = \frac{1}{3}$ and $r = \frac{4}{5}$ characterizing s itself.

Finally, Γ_3 has infinitely many equilibrium points. In fact, *any* possible pair $s = (s_1, s_2)$ of pure or mixed strategies is an equilibrium point. Those equilibrium points where one or both players use pure strategies are *extra weak*. Those where both use proper mixed strategies are *quasi-strong*. However, the Jacobian J^* of (86) vanishes at all these equilibrium points. (Actually, it can be shown that J^* will *always* vanish at all equilibrium points belonging to any infinite connected family of equilibrium points. This is not inconsistent with Theorem 7 since all games containing such families of equilibrium points form a closed set of measure zero in any $\mathscr{I}$.) Eqs. (104) and (105) now yield the probabilities

$$q(\alpha^*, \beta^*) = r(\alpha^*, \beta^*) = \frac{1}{2},$$

which happen to be independent of α^* and β^*. This means that, out of the infinitely many equilibrium points in game Γ_3, only *one* is approachable, and this is the strategy pair $s = (s_1, s_2)$ with $s_i = \frac{1}{2}a_i^1 + \frac{1}{2}a_i^2$ for $i = 1, 2$. But it is easy to verify that, in general, different choices of probability distributions for the random variables δ_1^{12} and δ_2^{12} would have made other equilibrium points of Γ_3 approachable. (Once more, it is a common occurrence that, out of any infinite connected family of equilibrium points, only a finite subset is approachable, and that this subset itself strongly depends on the choice of the probability distributions $F_1, \ldots, F_n$ used in defining the disturbed game $\Gamma^*(\varepsilon)$.)

7. Conclusion

We have found that the players' uncertainty about the exact payoffs that the other players will associate with various strategy combinations can be modeled as a game $\Gamma^*(\varepsilon)$ with randomly disturbed payoffs. Under this model, every ordinary game Γ will have at least one stable (i.e., approachable) equilibrium point. Indeed, all strong and "almost all" quasi-strong equilibrium points will be stable, including "almost all" equilibrium points in mixed strategies. Moreover, under this model, these equilibrium points will be stable, not because the players will make any deliberate effort to use their pure strategies with the probabilities prescribed by their mixed equilibrium strategies, but rather because the random fluctuations in their payoffs will *make* them use their pure strategies approximately with the prescribed probabilities.

References

AUMANN, R. J.: Acceptable Points in Games of Perfect Information. Pacific Journal of Mathematics **10**, 381–417, 1960a.

—: Spaces of Measurable Transformations. Bulletin of the American Mathematical Society **66**, 301–304, 1960b.

—: Borel Structures for Function Spaces. Illinois Journal of Mathematics **5**, 614–630, 1961.

BELLMAN, R.: On Games Involving Bluffing. Rendiconti del Circulo Mathematico di Palermo, Ser. 2 **1**, 139–156, 1952.

BELLMAN, R., and D. BLACKWELL: Some Two-person Games Involving Bluffing. Proceedings of the National Academy of Sciences **35**, 600–605, 1949.

DEBREU, G.: Economies with a Finite Set of Equilibria. Econometrica **38**, 387–392, 1970.

DIEUDONNÉ, J.: Foundations of Modern Analysis. Academic Press, New York 1969.

HARSANYI, J. C.: Games with Incomplete Information Played by 'Bayesian' Players. Management Science **14**, 159–182, 320–334, and 486–502, Parts I–III, 1967–68.

LEFSCHETZ, S.: Introduction to Topology. Princeton University Press, Princeton, N. J. 1949.

NASH, J. F.: Noncooperative Games. Annals of Mathematics **54**, 286–295, 1951.

OLMSTED, J. M. H.: Real Variables. Appleton-Century-Crofts, New York 1959.

SARD, A.: The Measure of the Critical Values of Differentiable Maps. Bulletin of the American Mathematical Society **48**, 883–890, 1942.

Received December, 1972.

CHAPTER 5

ODDNESS OF THE NUMBER OF EQUILIBRIUM POINTS: A NEW PROOF[1]

Abstract: A new proof is offered for the theorem that, in "almost all" finite games, the number of equilibrium points is *finite* and *odd*. The proof is based on constructing a one-parameter family of games with logarithmic payoff functions, and studying the topological properties of the graph of a certain algebraic function, related to the graph of the set of equilibrium points for the games belonging to this family. In the last section of the paper, it is shown that, in the space of all games of a given size, those "exceptional" games which fail to satisfy the theorem (by having an even number or an infinity of equilibrium points) is a closed set of measure zero.

1. Introduction

WILSON [1971, Theorem 1 on p. 85] has shown that, apart from certain degenerate cases, in any finite game, the number of equilibrium points is *finite* and *odd*. The purpose of this paper is to offer a new proof for WILSON's theorem.

Let Γ be a finite noncooperative game. The k-th pure strategy of player $i\,(i = 1,\ldots,n)$ will be called a_i^k, whereas the set of all his K_i pure strategies will be called A_i. Let

$$K = \prod_{i=1}^{n} K_i\,. \tag{1}$$

We shall assume that the K possible n-tuples of pure strategies are numbered consecutively as $a^1,\ldots,a^m,\ldots,a^K$. Let

$$a^m = (a_1^{k_1},\ldots,a_i^{k_i},\ldots,a_n^{k_n})\,. \tag{2}$$

Then we shall write

$$a^m(i) = a_i^{k_i}\,, \tag{3}$$

to denote the pure strategy used by player i in the strategy n-tuple a^m. The set of all K possible pure-strategy n-tuples will be called A. We have $A = A_1 \times \cdots \times A_n$.

Any mixed strategy of a given player $i\,(i = 1,\ldots,n)$ can be identified with a probability vector p_i of the form

$$p_i = (p_i^1,\ldots,p_i^k,\ldots,p_i^{K_i})\,, \tag{4}$$

where $p_i^1,\ldots,p_i^{K_i}$ are the probabilities that this mixed strategy assigns to his pure strategies $a_i^1,\ldots,a_i^{K_i}$. The set P_i of all mixed strategies available to player i is a

[1] Prof. JOHN C. HARSANYI, University of California, Berkeley, Cal. 94720. This research has been supported by Grant GS-3222 of the National Science Foundation, through the Center for Research in Management Science, University of California, Berkeley.

simplex, consisting of all K_i-vectors satisfying the conditions

$$p_i^k \geqq 0, \quad \text{for} \quad k = 1, \ldots, K_i, \tag{5}$$

and

$$\sum_{k=1}^{K_i} p_i^k = 1. \tag{6}$$

The set $P = P_1 \times \cdots \times P_n$ of all possible n-tuples $p = (p_1, \ldots, p_n)$ of mixed strategies is a compact and convex polyhedron, and will be called the *strategy space* of game Γ. We shall write $p = (p_i, \overline{p_i})$, where $\overline{p_i} = (p_1, \ldots, p_{i-1}, p_{i+1}, \ldots, p_n)$ is the strategy $(n-1)$-tuple representing the mixed strategies of the $(n-1)$ players other than player i.

The set $C(p_i)$ of all pure strategies a_i^k to which the mixed strategy p_i assigns positive probabilities $p_i^k > 0$ is called the *carrier* of p_i. If the carrier $C(p_i)$ of a given strategy p_i contains only *one* pure strategy a_i^k, then p_i will be identified with this pure strategy $a_i^k = p_i$. On the other hand, if $C(p_i)$ contains *all* K_i pure strategies of player i, then p_i will be called a *complete(ly) mixed* strategy. Finally, if p_i is neither pure nor complete, then it will be called an *incomplete(ly) mixed* strategy.

For any strategy n-tuple $p = (p_1, \ldots, p_n)$, the *carrier* $C(p)$ of p will be defined as the union of the carriers of its component strategies, that is, as

$$C(p) = \bigcup_{i=1}^{n} C(p_i). \tag{7}$$

Suppose that the i-th component of the pure-strategy n-tuple a^m is $a^m(i) = a_i^k$, and that a given mixed strategy p_i of player i assigns the probability p_i^k to this pure strategy a_i^k. Then, we shall write

$$q_i^m(p_i) = p_i^k. \tag{8}$$

Of course, if $p_i = a_i^k$ is a pure strategy, then we have

$$q_i^m(a_i^k) = 1, \quad \text{when} \quad a^m(i) = a_i^k, \tag{9}$$

but

$$q_i^m(a_i^k) = 0 \quad \text{when} \quad a^m(i) \neq a_i^k. \tag{10}$$

When the n players use the pure-strategy n-tuple a^m, then player i $(i = 1, \ldots, n)$ will receive the payoff

$$U_i(a^m) = u_i^m, \tag{11}$$

whereas if they use the mixed-strategy n-tuple $p = (p_1, \ldots, p_n)$, then his payoff will be

$$U_i(p) = \sum_{m=1}^{K} \left[\prod_{i=1}^{n} q_i^m(p_i) \right] u_i^m. \tag{12}$$

Let $\mathscr{I} = \mathscr{I}(n; K_1, \ldots, K_n)$ be the set of all n-person games in which players $1, \ldots, n$ have exactly $K_1, \ldots, K_n$ pure strategies, respectively. Thus, $\mathscr{I}$ is the set of all games of a given size. Each specific game Γ in $\mathscr{I}$ can be characterized by the (nK)-vector

$$u = (u_1^1, \ldots, u_1^{K_1}; \ldots; u_i^1, \ldots, u_i^{K_i}; \ldots; u_n^1, \ldots, u_n^{K_n}), \tag{13}$$

whose components $u_i^m = U_i(a^m)$ are the payoff to various players i for the different pure-strategy combinations a^m. We can identify each game Γ with its vector $u = u(\Gamma)$ of possible payoffs for pure-strategy combinations, and can regard the set $\mathscr{I}$ as an (nK)-dimensional Euclidean space $\mathscr{I} = \{u\}$.

Let $\bar{\mathscr{S}}(\mathscr{I})$ be the set of all games Γ in $\mathscr{I}$ for which a given mathematical statement $\mathscr{S}$ is false. We shall say that statement $\mathscr{S}$ is true for *almost all* games if, for every possible set $\mathscr{I}$ of games of a particular size, this set $\bar{\mathscr{S}}(\mathscr{I})$ is a *closed* set of *measure zero* within the relevant set $\mathscr{I}$, regarded as an (nK)-dimensional Euclidean space. (Concerning the closure requirement for $\bar{\mathscr{S}}(\mathscr{I})$, see DEBREU [1970, p. 387].)

2. Logarithmic Games

Let Λ be an n-person noncooperative game, where the n players have the same simplexes $P_1, \ldots, P_n$ they have in game Γ as strategy spaces, but where the payoff function L_i of each player $i (i = 1, \ldots, n)$ is of the form[2])

$$L_i(p) = L_i(p_i) = \sum_{k=1}^{K_i} \log p_i \,. \tag{14}$$

Thus, Λ is a "degenerate" game, in which each player's payoff L_i depends only on his own strategy p_i, and does not depend on the other players' strategies $p_j, j \neq i$.

Finally, we define a one-parameter family of games $\{\Lambda^*(t)\}$, with $0 \leqq t \leqq 1$. In any particular game $\Lambda^*(t)$ with a specific value of the parameter t, the payoff function of player $i (i = 1, \ldots, n)$ is

$$L^*(p,t) = (1-t)\,U_i(p) + t\,L_i(p_i)\,. \tag{15}$$

Obviously, $\Lambda^*(0) = \Gamma$, whereas $\Lambda^*(1) = \Lambda$. All games $\Lambda^*(t)$ with $0 < t \leqq 1$ will be called *logarithmic* games. Γ will sometimes be called the *original* game, while Λ will be called the *pure* logarithmic game.

---

[2]) Since the payoff function L_i are logarithmic functions (instead of being multilinear functions in the probabilities p_i^k as is the case in ordinary finite games), this game Λ – as well as the games $\Lambda^*(t)$ to be defined below – are best regarded as being *infinite* games in which the *pure* strategy of every player i consists in choosing a specific point p_i from the simplex P_i, which makes each p_i a pure strategy, rather than a mixed strategy, in game Λ (or $\Lambda^*(t)$). But, for convenience, we shall go on calling any given strategy p_i a (complete or incomplete) *mixed* strategy – even in discussing the infinite games Λ and $\Lambda^*(t)$ – if p_i would represent a (complete or incomplete) mixed strategy in the finite game Γ. This terminology will not give rise to any confusion because, in analyzing these infinite games, we shall never consider mixed strategies having the nature of probability mixtures of two or more strategies p_i.

3. Equilibrium Points

A given strategy p_i of player i will be called a *best reply* to a strategy combination $\bar{p}_i$ used by the other $(n-1)$ players in game $\Lambda^*(t)$ if

$$L^*(p_i,\bar{p}_i,t) \geqq L^*(p_i',\bar{p}_i,t) \quad \text{for all} \quad p_i' \in P_i\,. \tag{16}$$

A given strategy n-tuple $p = (p_1,\ldots,p_n)$ is an *equilibrium point* [NASH, 1951] in game $\Lambda^*(t)$ if every component p_i of p is a best reply to the corresponding strategy combination $\bar{p}_i$ of the other $(n-1)$ players.

An equilibrium point p is called *strong* [3]) if all n components p_i of p satisfy (16) with the strong inequality sign $>$ for all $p_i' \neq p_i$. That is, p is a strong equilibrium point if every player's equilibrium strategy p_i is his *only* best reply to the other players' strategy combination $\bar{p}_i$. An equilibrium point is called *weak* if it is not strong.

An equilibrium point p is quasi-strong [4]) if no player i has pure-strategy best replies to $\bar{p}_i$ other than the pure strategies belonging to the carrier $C(p_i)$ of his equilibrium strategy p_i. An equilibrium point that is not even quasi-strong is called *extra-weak.*

A given game Γ itself will be called *quasi-strong* if *all* its equilibrium points are quasi-strong; and it will be called *extra-weak* if at least *one* of its equilibrium points is extra-weak.

In the original game $\Gamma = \Lambda^*(0)$, a best reply p_i to any given strategy combination $\bar{p}_i$ may be a pure strategy or may be a mixed strategy. (It can be a mixed strategy only if *all* pure strategies a_i^k in its carrier $C(p_i)$ are themselves best replies to $\bar{p}_i$.) In contrast, in a logarithmic game $\Lambda^*(t)$ with $t > 0$, only a *complete* mixed strategy can be a best reply. This is so because, in view of (12), (14), and (15), any player i will obtain an infinite negative payoff $L_i^* = -\infty$ if he uses a pure or an incompletely mixed strategy, but will always obtain a finite payoff $L_i^* > -\infty$ if he uses a completely mixed strategy. Consequently, in these logarithmic games, all equilibrium points will be in completely mixed strategies.

In the original game Γ, in general, the mathematical conditions characterizing an equilibrium point $p = (p_1,\ldots,p_n)$ will be partly equations and partly inequalities. The former will be of the form:

$$U_i(a_i^k,\bar{p}_i) = U_i(a_i^{k'},\bar{p}_i) \quad \text{if} \quad a_i^k, a_i^{k'} \in C(p_i)\,; \tag{17}$$

whereas the latter will be of the form:

$$U_i(a_i^k,\bar{p}_i) \geqq U_i(a_i^{k'},\bar{p}_i) \quad \text{if} \quad a_i^k \in C(p_i) \quad \text{while} \quad a_i^{k'} \notin C(p_i)\,. \tag{18}$$

Only in the special case where all n equilibrium strategies $p_1,\ldots,p_n$ are *pure*

[3]) I am using the term "strong equilibrium point" in a different sense from AUMANN's [1959, p. 300].
[4]) Many of the concepts used in this paper were first introduced in HARSANYI [1973].

strategies, will all conditions characterizing p be inequalities of form (18); and only in the special case where all n equilibrium strategies are *complete* mixed strategies, will all these conditions be equations of form (17).

An equilibrium point p of game Γ will be *quasi-strong* if and only if, for every player i, and for every strategy a_i^k in the carrier $C(p_i)$ of i's equilibrium strategy, and for every strategy $a_i^{k'}$ *not* in this carrier, condition (18) is satisfied with the *strong* inequality sign $>$.

In contrast to equilibrium points in the finite game Γ, every equilibrium point p in any logarithmic game $\Lambda^*(t)$ with $t > 0$, is always characterized by *equations* of the following form:

$$\left(\frac{\partial L_i^*(p,t)}{\partial p_i^k}\right)_{\sum_i p_i^k = 1} = 0, \qquad \text{for} \quad k = 1,\ldots,K_i - 1 \quad \text{and for} \quad i = 1,\ldots,n. \tag{19}$$

Here each partial derivative $\partial L_i^*/\partial p_i^k$ must be evaluated at the equilibrium point p itself. Of course, these eqs. (19) express only the first-order conditions for maximizing the payoff function L_i^* with respect to the vector p_i. But, since each function L_i^* is strictly concave in p_i, the second-order conditions are always satisfied, so that the eqs. (19) are both necessary and sufficient conditions for maximization.

The functions L_i^* can also be written as

$$L_i^*(p,t) = (1 - t)\sum_{k=1}^{K_i} p_i^k U_i(a_i^k,\bar{p}_i) + \sum_{k=1}^{K_i} \log p_i^k. \tag{20}$$

Therefore, using the fact that

$$p_i^1 = 1 - \sum_{k=2}^{K_i} p_i^k, \tag{21}$$

we can write (19) in the form

$$(1 - t)[U_i(a_i^k,\bar{p}_i) - U_i(a_i^1,\bar{p}_i)] + \frac{t}{p_i^k} - \frac{t}{p_i^1} = 0, \tag{22}$$

or, equivalently, in the form

$$(1 - t)p_i^1 p_i^k[U_i(a_i^k,\bar{p}_i) - U_i(a_i^1,\bar{p}_i)] + t(p_i^1 - p_i^k) = 0, \quad \text{for} \quad i = 1,\ldots,n; \quad \text{and for} \quad k = 2,\ldots,K_i. \tag{23}$$

The number of equations of form (23) is:

$$K^* = \sum_{i=1}^{n}(K_i - 1) = \sum_{i=1}^{n} K_i - n. \tag{24}$$

Thus, together with the n equations of form (6), we have all together

$$K^{**} = K^* + n = \sum_{i=1}^{n} K_i \tag{25}$$

independent equations for characterizing each equilibrium point p, which is the same as the number of the variables p_i^k determined by these equations.

In view of (12), all equations of form (23) are algebraic equations in the variables p_i^k and in the parameter t. All equations of form (6) are likewise algebraic. Let S be the set of all $(K^{**} + 1)$-vectors (t, p) satisfying the K^{**} equation of forms (6) and (23). Clearly, S will be typically a one-dimensional algebraic variety, i.e., an algebraic curve. (In degenerate cases, however, S may also contain zero-dimensional subsets, i.e., isolated points, and/or subsets of more than one dimension, i.e., algebraic surfaces of various dimensionality.)

Let T be the set of all vectors (t, p) satisfying, not only the K^{**} equations of forms (6) and (23), but also the K^{**} inequalities of form (5). Clearly, T is simply that part of the algebraic variety S which lies within the compact and convex cylinder (polyhedron) $R = P \times I$, where $I = [0,1]$ is the closed unit interval. Since T is the locus of all solutions (t, p) to the simultaneous equations and inequalities (5), (6), and (23), T will be called the *solution graph* for the latter.

For any point (t, p), t will be called its *first coordinate.* Within the cylinder R, the strategy space P of any specific game $\Lambda^*(t)$ is represented by the set R^t of all points (t, p) in R whose first coordinate is the relevant t value.

For any game $\Lambda^*(t)$, let E^t be the set of all points (t, p) in R^t such that p is an equilibrium point of $\Lambda^*(t)$. Finally, let T^t be the intersection of R^t with the solution graph T. We can now state:

Lemma 1:

For all t with $0 < t \leqq 1$, $E^t = T^t$. In contrast, for $t = 0$, in general, we have only $E^0 \subsetneq T^0$.

Proof:

For all t with $0 < t \leqq 1$, conditions (5), (6), and (23) are sufficient and necessary conditions for any given point p to be an equilibrium point of game $\Lambda^*(t)$. On the other hand, for $t = 0$, it is easy to verify that all equilibrium points p of the game $\Lambda^*(0) \mp \Gamma$ satisfy all these conditions but, in general, so will also some strategy combinations p that are *not* equilibrium points of Γ. For example, all these conditions will be satisfied by any pure-strategy n-tuple $p = a^m$, whether it is an equilibrium point of game Γ or not.

4. Some Topological Properties of the Solution Graph T

Consider the mapping $\mu: t \to T^t$. The Jacobian of this mapping, as evaluated at any given point (p, t) of T^t, can be written as

$$J(t,p) = \frac{\partial(F_1^1,\ldots,F_i^k,\ldots,F_n^{K_n})}{\partial(p_1^1,\ldots,p_i^k,\ldots,p_n^{K_n})}, \qquad (26)$$
$$i = 1,\ldots,n;$$
$$\text{and, for each } i, \quad k = 1,\ldots,K_i.$$

Here

$$F_i^1(t,p) = \sum_{i=1}^{n} p_i^k - 1, \qquad i = 1,\ldots,n; \qquad (27)$$

whereas

$$F_i^k(t,p) = (1-t)p_i^1 p_i^k[U_i(a_i^k,\bar{p}_i) - U_i(a_i^1,\bar{p}_i)] + t(p_i^1 - p_i^k), \qquad (28)$$
$$i = 1,\ldots,n;$$
$$\text{and, for each } i, \quad k = 2,\ldots,K_i.$$

For points of the form $(t,p) = (0,p)$ in set T^0, the functions $F_i^k (k \neq 1)$ take the following simpler form:

$$F_i^k(0,p) = p_i^1 p_i^k[U_i(a_i^k,\bar{p}_i) - U_i(a_i^1,\bar{p}_i)], \qquad i = 1,\ldots,n; \qquad (29)$$
$$\text{and, for each } i, \quad k = 2,\ldots,K_i.$$

Any equilibrium point p of the original game Γ will be called regular if $J(0,p) \neq 0$; and will be called *irregular* if $J(0,p) = 0$. A given game Γ itself will be called *regular* if *all* of its equilibrium points are regular; and it will be called *irregular* if at least *one* of its equilibrium points is irregular.

We shall now state two lemmas, based on well-known facts in algebraic geometry.

Lemma 2:

Let (x^0,x^*) be an arc of an algebraic curve S in a ν-dimensional Euclidean space X^ν, connecting the two points $x^0 = (x_1^0,\ldots,x_\nu^0)$ and $x^* = (x_1^*,\ldots,x_\nu^*) \neq x^0$. Then, this arc (x^0,x^*) can be uniquely continued analytically beyond point x^* (and beyond point x^0).

Proof:

If x^* is not a singular point, then the possibility of analytic continuation follows from the Implicit Function Theorem. On the other hand, if x^* is a singular point, then this possibility follows from PUISEUX's Theorem [VAN DER WAERDEN, 1939, Theorem 14]. By this theorem, if x^* is a point of some branch S^* of a given algebraic curve S, then, whether x^* is a singular point or not, in some neighborhood $N(x^*)$ of x^*, the coordinates x_i of any point $x = (x_1,\ldots,x_\nu)$ of this branch S^* can be represented by ν convergent power series $\pi_i(y)$ in an auxiliary parameter y, so that we can write $x_i = \pi_i(y)$ for a suitably chosen value of $y (i = 1,\ldots,\nu)$. Moreover, we can select the ν functions π_i in such a way that x^* itself will correspond to $y = 0$ (so that $x_i^* = \pi_i(0)$, for $i = 1,\ldots,\nu$), and in such a way that all other points

x of the arc (x^0, x^*) will correspond to negative values of y. Then, by assigning positive values to y, we can analytically continue the arc (x^0, x^*) beyond x^*. Even though we can choose the v functions π_i in many different ways, all choices will yield the same curve as the analytic continuation of (x^0, x^*).

Corollary:

Let S be an algebraic curve, and x be an arbitrary point. Then, the number of arcs belonging to S and originating from x is always even (possibly zero). These arcs always uniquely partition themselves into pairs, so that the two arcs belonging to the same pair are analytic continuations of each other, and are not analytic continuations of any other arc originating from x.

Lemma 3:

Let (x^0, x^*) be an arc of an algebraic curve S. Suppose that (x^0, x^*) lies wholly within a given compact and convex set R with a nonempty interior, and that x^0 is a boundary point of R whereas x^* is an interior point of R. Then, by analytically continuing (x^0, x^*) far enough beyond x^*, we shall once more eventually reach a boundary point x^{00} of R.

Proof:

Let S^* be the curve we obtain if we continue (x^0, x^*) beyond x^* as far as possible without leaving set R. For each coordinate x_i, let

$$m_i = \inf_{x \in S^*} x_i \quad \text{and} \quad m^i = \sup_{x \in S^*} x_i \,. \tag{30}$$

Since S^* is not an isolated point, at least for one coordinate x_i, its variation on S^*, $\Delta_i = m^i - m_i$, must be positive. On the other hand, since S^* is an arc of an algebraic curve, it can be divided up into a *finite* number of segments $\alpha^1, \ldots, \alpha^\mu, \ldots, \alpha^M$, such that, as we move away from x^0 along any given segment α^μ, this coordinate x_i is either strictly increasing or is strictly decreasing. Let us assume that, starting from x^0 and moving along S^*, we reach these segments in the order they have been listed. Now, first suppose that, along the *last* segment α^M, x_i *increases*. Then, since R is a compact set, x_i must reach a local maximum at some point x^{00} of α^M. Obviously, this point x^{00} can only be the endpoint of α^M furthest away from x^0. Moreover, it can only be a *boundary* maximum point for x_i because, if it were an interior maximum point, then α^M could not be the last segment of S^*. Therefore, this point x^{00} must be a boundary point of R. By the same token, if x_i *decreases* along α^M, then the endpoint x^{00} of α^M must be a local boundary minimum point for x_i and, therefore, it must be boundary point of R. Thus, in either case, S^* will eventually reach a boundary point x^{00} of R.

Let $\bar{P}$ be the boundary of the strategy space P. Thus, $\bar{P}$ is the set of all strategy n-tuples $p = (p_1, \ldots, p_n)$ having at least one pure or incompletely mixed strategy p_i as a component. Let $I^0 = (0,1)$ be the open unit interval. Let B be the set

$B = \bar{P} \times I^0$. Clearly, the boundary hypersurface $\bar{R}$ of cylinder R is made up of the three disjoint sets P^0, B, and P^1.

Lemma 4:

Let $\bar{T}$ be the intersection of the solution graph T and of the boundary hypersurface $\bar{R}$ of cylinder R. Let (t,p) be any nonisolated point of $\bar{T}$. Then, *either* $(t,p) = (1,\tilde{p})$, where $\tilde{p}$ is the unique equilibrium point of the pure logarithmic game $\Lambda^*(1)$; *or* $(t,p) = (0,p^*)$, where p^* is an equilibrium point of the original game $\Lambda^*(0) = \Gamma$.

Proof:

For any t with $0 < t \leqq 1$, the vector p characterizing any given point (t,p) of $\bar{T}$ must be an equilibrium point of the game $\Lambda^*(t)$, because (t,p) is a point of the solution graph T. Therefore, (t,p) cannot belong to set B, since the logarithmic games $\Lambda^*(t)$ with $0 < t \leqq 1$ have no equilibrium point using a pure or an incompletely mixed strategy p_i as equilibrium strategy. Hence, if $t > 0$, then (t,p) can only be a point belonging to set P^1, which is possible only if $(t,p) = (1,\tilde{p})$.

On the other hand, if $t = 0$, then $(t,p) = (0,p)$ is a point belonging to set P^0. As (t,p) is a nonisolated point of T, it is a limit point of some convergent point sequence $(t^1,p^1),\ldots,(t^j,p^j),\ldots$, where each p^j is an equilibrium point in game $\Lambda^*(t^j)$, with $t^j > 0$. Consequently, p itself is an equilibrium point in game $\Lambda^*(0) = \Gamma$, because the correspondence $\mu^*: t \to E^t$ is upper semi-continuous (where E^t is the set mentioned in Lemma 1). This completes the proof.

Lemma 5:

The point $(t,p) = (1,p)$, corresponding to the unique equilibrium point $\tilde{p}$ of the pure logarithmic game $\Lambda^*(1) = \Lambda$ is always a nonsingular point of the graph T, and is the endpoint of exactly one branch $\beta(\tilde{p})$ of T.

Proof:

As is easy to verify, $J(1,\tilde{p}) \neq 0$. Consequently, $(1,\tilde{p})$ is nonsingular and, by the Implicit Function Theorem, it lies on exactly one branch $\alpha(\tilde{p})$ of T.

Lemma 6:

Let Γ be a *regular* and *quasi-strong* game. Then, any point $(t,p) = (0,p)$ corresponding to an equilibrium point p of game $\Lambda^*(0) = \Gamma$ is always a nonsingular point of the graph T, and is the endpoint of exactly one branch $\beta(p)$ of T.

Proof:

Since Γ is regular, we have $J(0,p) \neq 0$. Hence, if p is an interior point of the strategy space P, then the present lemma can be established by the same reasoning as was used in the proof of Lemma 5. However, if p is a boundary point of P, then this reasoning shows only that $(0,p)$ lies on exactly one branch $\beta(p)$ of the algebraic variety S. In order to prove the lemma, we have to show also that $\beta(p)$ belongs to the graph T, i.e., that it lies within cylinder R. In other words, we have to show that $\beta(p)$ goes from $(0,p)$ towards the interior of R, which is equivalent to showing that, for any zero component $p_i^k = 0$ of the vector p, the total derivative $\mathrm{d}p_i^k/\mathrm{d}t$ is positive at the point $(0,p)$. Now, by differentiating eq. (23) with respect to t, and then setting $t = p_i^k = 0$, we obtain

$$p_i^1[U_i(a_i^k,\bar{p}_i) - U_i(a_i^1,\bar{p}_i)] + p_i^1 = 0\,. \tag{31}$$

Since the numbering of player i's pure strategies is arbitrary, without loss of generality we can assume that

$$p_i^1 > 0\,. \tag{32}$$

On the other hand, since $p_i^1 > 0$ and $p_i^k = 0$, we have $p_i^1 \in C(p_i)$ but $p_i^k \notin C(p_i)$. Since p is a quasi-strong equilibrium point, condition (18) must be satisfied by a *strong* inequality sign if we set $a_i^{k'} = a_i^1$. Therefore,

$$U_i(a_i^k,\bar{p}_i) - U_i(a_i^1,\bar{p}_i) < 0\,. \tag{33}$$

But (31), (32), and (33) together imply that $\mathrm{d}p_i^k/\mathrm{d}t > 0$, as desired.

In what follows, when we say that two points are "connected", we shall mean that they are connected by some branch α of the solution graph T.

Theorem 1:

Let Γ be a *regular* and *quasi-strong* finite game. Then, the number of equilibrium points in Γ is finite. Moreover, there exists exactly one distinguished equilibrium point p^* in Γ such that the corresponding point $(0,p^*)$ is connected with the point $(1,\tilde{p})$, associated with the unique equilibrium point $\tilde{p}$ of the pure logarithmic game $\Lambda^*(1) = \Lambda$. All other equilibrium points of Γ form pairs, such that the two equilibrium points belonging to the same pair are connected with each other and with no other equilibrium point. Therefore, the number of equilibrium points in Γ is odd.

Proof:

By Lemma 6, every equilibrium point p of Γ lies on some branch $\beta(p)$ of T. But T, being the intersection of an algebraic variety S and of a compact and convex set R, can have only a finite number of branches. Moreover, on any given branch β, there can be at most two equilibrium points, corresponding to the two endpoints of β. Therefore, the number of equilibrium points in Γ is finite.

By Lemma 5, there exists a unique branch $\alpha(p)$ of T, originating from the point $(1,\tilde{p})$. By Lemmas 1 and 2, this branch $\alpha(\tilde{p})$ must lead to a boundary point x^{00} of R. As $J(1,\tilde{p}) \neq 0$, we must have $x^{00} \neq (1,\tilde{p})$, because otherwise T would have two local branches originating from $(1,\tilde{p})$, contrary to the Implicit Function Theorem. Consequently, by Lemma 4, $x^{00} = (0,p^*)$, where p^* is an equilibrium point – called the *distinguished* equilibrium point – of game Γ.

Finally, let $p \neq p^*$ be any equilibrium point of Γ, other than the distinguished equilibrium point p^*. By Lemma 6, there exists a unique branch $\beta(p)$ of T, originating from the point $(0,p)$. By an argument similar to the one used in the last paragraph, it can be shown that $\beta(p)$ must lead to another boundary point x^{00} of R, with $x^{00} = (0,p')$, where $p' \neq p$ and $\neq p^*$ is another equilibrium point of Γ. Hence, all equilibrium points of Γ, other than the distinguished equilibrium point $\tilde{p}$, are pairwise connected. But this means that the number of these latter equilibrium points is *even*, which makes the total number of equilibrium points in Γ *odd*.

Note:

The proof of Theorem 1 shows that, for any game $\Lambda^*(t)$ with $0 \leqq t \leqq 1$, the set Q^t of all equilibrium points in $\Lambda^*(t)$ is *nonempty*. This is so because branch $\alpha(\tilde{p})$ of graph T connects the two points $(1,\tilde{p})$ and $(0,p^*)$. Therefore, $\alpha(\tilde{p})$ intersects every set R^t with $0 \leqq t \leqq 1$ at some point (t,p^t). As is easy to verify, the strategy n-tuple p^t defining this point must be an equilibrium point of game $\Lambda^*(t)$.

5. Three "Almost All" Theorems

Within a given set $\mathscr{I} = \mathscr{I}(n; K_1, \ldots, K_n)$ of games of a particular size, let $\mathscr{F}(C^*)$ be the set of all games Γ that have at least one equilibrium point p with a specified set $C^* = C(p)$ as its carrier. There are only a *finite* number of different sets $\mathscr{F}(C^*)$ in $\mathscr{I}$ because, for all games Γ in $\mathscr{I}$, the number of possible carrier sets C^* is finite. This is so because any set C^* is a subset of the finite set

$$A^* = \bigcup_{i=1}^{n} A_i\,, \tag{34}$$

consisting of the set of all K^{**} pure strategies a_i^k for the n players in each game Γ, where

$$K^{**} = \sum_{i=1}^{n} K_i\,. \tag{35}$$

(Of course, two sets $\mathscr{F}(C^*)$ corresponding to different carrier sets C^* will in general overlap.) We can now state the following theorem.

Theorem 2:

Almost all finite games are quasi-strong.

Proof:

Let $\bar{\mathscr{F}}(C^*)$ be the set of all games Γ in $\mathscr{I}$ that have at least one extra-weak equilibrium point p with the set $C^* = C(p)$ as its carrier. Obviously, $\bar{\mathscr{F}}(C^*) \subset \mathscr{F}(C^*)$. Let $\bar{\bar{\mathscr{F}}}(C^*) = \mathscr{F}(C^*) - \bar{\mathscr{F}}(C^*)$. Thus, all games Γ in $\bar{\bar{\mathscr{F}}}(C^*)$ have the property that they contain one or more equilibrium points p with the set $C^* = C(p)$ as their carrier set, and all these equilibrium points p are quasi-strong.

All games Γ in a given set $\mathscr{F}(C^*)$ are characterized by the fact that their defining vector $u = u(\Gamma)$ satisfies a finite number of algebraic equations and algebraic weak inequalities, of forms (17) and (18), in which the functions U_i are defined by (12). Thus, if we regard the set $\mathscr{I}$ as an (nK)-dimensional Euclidean space $\mathscr{I} = \{u\}$, then each set $\mathscr{F}(C^*)$ will correspond to a subset of $\mathscr{I}$, bounded by pieces of a finite number of algebraic hypersurfaces. In view of (12) and (18), these bounding hypersurfaces are multilinear, i.e., they are hyperboloids. Within each set $\mathscr{F}(C^*)$, all games Γ belonging to $\bar{\mathscr{F}}(C^*)$ are characterized by the fact that their defining vectors $u = u(\Gamma)$ satisfy *all* the inequalities of form (18) used in defining this set $\mathscr{F}(C^*)$, with a *strong in*equality sign $>$. In contrast, all games Γ belonging to $\bar{\bar{\mathscr{F}}}(C^*)$ have a defining vector $u = u(\Gamma)$ satisfying one or more of these weak inequalities with an *equality* sign $=$. Therefore, all games in $\bar{\mathscr{F}}(C^*)$ correspond to interior points u of $\mathscr{F}(C^*)$, whereas all games in $\bar{\bar{\mathscr{F}}}(C^*)$ correspond to boundary points of $\mathscr{F}(C^*)$. Hence, as a subset of the (nK)-dimensional Euclidean space $\mathscr{I}$, $\bar{\bar{\mathscr{F}}}(C^*)$ consists of pieces of a finite number of hyperboloids of at most $(nK - 1)$ dimensions. Consequently, each set $\bar{\bar{\mathscr{F}}}(C^*)$ is a set of measure zero in $\mathscr{I}$.

Let $\bar{\bar{\mathscr{F}}}^*$ be the set of all extra-weak games in $\mathscr{I}$. $\bar{\bar{\mathscr{F}}}$ is the union of all sets $\bar{\bar{\mathscr{F}}}(C^*)$, corresponding to various possible carrier sets C^*. Thus, $\bar{\bar{\mathscr{F}}}^*$ is a union of a finite number of sets of measure zero in $\mathscr{I}$. Therefore, $\bar{\bar{\mathscr{F}}}^*$ itself is also a set of *measure zero* in $\mathscr{I}$.

Next, we shall show that $\bar{\bar{\mathscr{F}}}^*$ is a *closed* set. Let $\Gamma^1, \Gamma^2, \ldots$ be a sequence of extra-weak games in $\mathscr{I}$, with the defining vectors $u^1 = u(\Gamma^1), u^2 = u(\Gamma^2), \ldots$. Suppose that the sequence $u^1, u^2, \ldots$ converges to a given vector u^0. Let Γ^0 be the game corresponding to this vector $u^0 = u(\Gamma^0)$. We have to show that Γ^0 is likewise an extra-weak game.

Since the games $\Gamma^j (j = 1, 2, \ldots)$ are extra-weak, each vector u^j satisfies one or more inequalities of form (18), with an equality sign. Yet, there are only a finite number of inequalities of this form. Therefore, at least one of these inequalities – let us call it inequality (18)* – will be satisfied by infinitely many vectors u^j, with an equality sign. As the sequence of these latter vectors, being a subsequence of the original sequence $\{u^j\}$, converges to u^0, this vector u^0 itself will also satisfy (18)* with an equality sign, which makes the corresponding game Γ^0 extra-weak, as desired. This completes the proof of Theorem 2.

Let p be an equilibrium point in game Γ belonging to set $\mathscr{I}$, with the carrier $C^* = C(p) = \bigcup_i C(p_i)$. Thus $\Gamma \in \mathscr{F}(C^*)$. Suppose the carriers $C(p_1), \ldots, C(p_n)$ of the equilibrium strategies $p_1, \ldots, p_n$ consist of exactly $\gamma_1, \ldots, \gamma_n$ pure strategies, respectively. In studying games Γ in set $\mathscr{F}(C^*)$, we shall adopt the following notational convention, which, of course, involves no loss of generality:

> The pure strategies a_i^k of each player $i (i = 1, \ldots, n)$ have been renumbered in such a way that the carrier $C(p_i)$ of his equilibrium strategy p_i now contains his *first* γ_i pure strategies $a_i^1, \ldots, a_i^{\gamma_i}$. (36)

We can fully characterize each equilibrium strategy p_i by the $(\gamma_i - 1)$ probability numbers $p_i^2, p_i^3, \ldots, p_i^{\gamma_i}$, since we have

$$p_i^1 = 1 - \sum_{j=2}^{\gamma_i} p_i^j, \tag{37}$$

and

$$p_i^k = 0, \quad \text{for} \quad k = \gamma_i + 1, \ldots, K_i. \tag{38}$$

Let π_i be the probability vector

$$\pi_i = (p_i^2, \ldots, p_i^{\gamma_i}), \quad \text{for} \quad i = 1, \ldots, n. \tag{39}$$

Thus, π_i is a subvector of the probability vector p_i.

Let Π_i $(i = 1, \ldots, n)$ be the set of all $(\gamma_i - 1)$-vectors satisfying the two conditions

$$p_i^k > 0, \quad k = 2, \ldots, \gamma_i; \tag{40}$$

and

$$\sum_{j=2}^{\gamma_i} < 1. \tag{41}$$

Let π be the composite vector

$$\pi = (\pi_1, \ldots, \pi_n). \tag{42}$$

Thus, π is a vector consisting of γ^* probability numbers p_i^k, where

$$\Gamma^* = \sum_{i=1}^{n} (\gamma_i - 1) = \sum_{i=1}^{n} \gamma_i - n. \tag{43}$$

Clearly, π is a subvector of the probability vector p.

Let Π be the set of all γ^*-vectors π whose subvectors $\pi_1, \ldots, \pi_n$ satisfy conditions (40) and (41). Clearly, $\Pi = \Pi_1 \times \cdots \times \Pi_n$.

We now define

$$m^*(1,k) = k - 1, \quad \text{for} \quad k = 2, \ldots, \gamma_1; \tag{44}$$

and

$$m^*(i,k) = \sum_{j=1}^{i-1} (\gamma_j - 1) + (k - 1) = \sum_{j=1}^{i-1} \gamma_j - i + k, \quad \text{for} \quad i = 2, \ldots, n;\ k = 2, \ldots, \gamma_i. \tag{45}$$

In addition to (36), we now introduce the following further notational convention, which again involves no loss of generality:

> The pure-strategy n-tuples a^m of the game have been re-numbered in such a way that the first γ^* pure-strategy n-tuples $a^1, \ldots, a^{\gamma^*}$ will now have the following form. For any m with $1 \leqq m \leqq \gamma^*$, let i and k be the unique pair of numbers satisfying $m^*(i,k) = m$.

Then

$$a^m = (a_1^1, \ldots, a_{i-1}^1, a_i^k, a_{i+1}^1, \ldots, a_m^1). \tag{46}$$

Thus, we can write

$$u_i^m = u^{m^*(i,k)} = U_i(a_1^1,\ldots,a_{i-1}^1,a_i^k,a_{i+1}^1,\ldots,a_n^1), \text{ for } i = 1,\ldots,n;\ k = 2,\ldots,\gamma_i. \tag{47}$$

Let u^* be the vector formed of those γ^* components u_i^m of vector u which can be written in form (47). Let u^{**} be the vector formed of the remaining $(nK - \gamma^*)$ components of u. Hence

$$u = (u^*,u^{**}). \tag{48}$$

The set of all possible vectors u^* is a γ^*-dimensional Euclidean space, to be denoted as $\mathscr{I}^* = \{u^*\}$; whereas the set of all possible vectors u^{**} is an $(nK - \gamma^*)$-dimensional Euclidean space, to be denoted as $\mathscr{I}^{**} = \{u^{**}\}$. Clearly, $\mathscr{I}^* \times \mathscr{I}^{**} = \mathscr{I}$.

Since p is an equilibrium point in game Γ, it must satisfy condition (17). This condition can also be written as

$$p_i^1 p_i^k[U_i(a_i^k,\bar{p}_i) - U_i(a_i^1,\bar{p}_i)] = 0, \quad \text{for} \quad i = 1,\ldots,n;\ k = 2,\ldots,\gamma_i. \tag{49}$$

Since p_i^1 and $p_i^k > 0$, (49) is equivalent to (17). In view of (12), this condition can also be written as

$$u^{m^*} = \sum_{\substack{m \neq m^* \\ m \in M}} \frac{[q_i^m(a_i^k)\prod_{j \neq i} q_j^m(\bar{p}_j)]u_i^m}{\prod_{j \neq i} q_j^{m^*}(p_j)} \tag{50}$$

$$- \sum_{m \in M} \frac{[q_i^m(a_i^1)\prod_{j \neq i} q_j^m(\bar{p}_j)]u_i^m}{\prod_{j \neq i} q_j^{m^*}(p_j)}, \quad \text{for} \quad i = 1,\ldots,n;\ k = 2,\ldots,\gamma_i.$$

Here $M = \{1,2,\ldots,K\}$ and $m^* = m^*(i,k)$. It is permissible to write eq. (17) [or (49)] in form (50) because, by (46), we have $q_j^{m^*}(p_j) = p_j^1$ for all $j \neq i$, and $p_j^1 > 0$ since $a_j^1 \in C(p_j)$.

Note that each quantity $u_i^{m^*}$ for a specific value of $m^* = m^*(i,k)$ occurs, with a nonzero coefficient, only in *one* equation of form (50) (where it occurs on the left-hand side). This is so because, by (10) and (46), for any $k' \neq k$, we have $q_i^{m^*}(a_i^{k'}) = 0$. Therefore, if we know the γ^* components p_i^k of vector π, and know the $(nK - \gamma^*)$ components u_i^m of vector u^{**}, then we can compute each one of the γ^* components $u_i^{m^*}$ of vector u^* separately, from the relevant equation of form (50). Consequently, the γ^* equations of form (50) define a mapping $\rho: (\pi,u^{**}) \to u^*$ from set $\Pi \times \mathscr{I}^{**}$ to set $\mathscr{I}^*$. This mapping ρ is continously differentiable because, by (40), for each point π in Π we have $p_i^k > 0$ for $k = 2,\ldots,\gamma_i$; so that the denominators on the right-hand side of (50) never vanish within Π.

We can use this mapping ρ to define another mapping $\rho^*: (\pi,u^{**}) \to (u^*,u^{**}) = u$, where $u^* = \rho(\pi,u^{**})$. This mapping ρ^* is from set $\Pi \times \mathscr{I}^*$ to set $\mathscr{I}^* \times \mathscr{I}^{***} = \mathscr{I}$; and it is continuously differentiable since ρ is.

We can now state the following theorem:

Theorem 3:

Almost all finite games are regular.

Proof:

Instead of using the γ^* equations of form (50), we can also use the equivalent γ^* equations of form (49), in order to define the mappings ρ and ρ^*. But if we do so, then the Jacobian of mapping ρ^* can be written as

$$J^*(\pi,u^{**}) = \frac{\partial(F_i^2,\ldots,F_i^k,\ldots,F_n^{\gamma_n})}{\partial(p_i^2,\ldots,p_i^k,\ldots,p_n^{\gamma_n})}, \quad i = 1,\ldots,n; \quad \text{and, for each} \quad k = 2,\ldots,\gamma_n, \tag{51}$$

where the F_i^k's are the functions $F_i^k = F_i^k(0,p)$ defined by (29). This means that $J^*(\pi,u^{**})$ is a subdeterminant of the Jacobian determinant $J(0,p)$ defined by (26, (27), and (29); it is that particular subdeterminant that we obtain if, for each player i, we cross out the rows and the columns corresponding to $k = 1$, and to $k = \gamma_i + 1,\ldots,K_i$. It is easy to verify that, owing to the special form of the functions F_i^1 $(i = 1,\ldots,n)$ as defined by (27), and owing to the fact that $p_i^{\gamma_i+1} = \cdots = p_i^{K_i} = 0$, this crossing out of these rows and columns does not change the value of the original determinant $J(0,p)$. Hence, $J^*(\pi,u^{**}) = J(0,p)$ if π is the subvector of p defined by (39) and (42).

Let $\mathscr{E}(C^*)$ be the set of all games Γ in $\mathscr{I}$ having at least one *irregular* equilibrium point p with set $C^* = C(p)$ as its carrier set. Equivalently, $\mathscr{E}(C^*)$ can also be defined as the set of all vectors $u = \rho^*(\pi,u^{**})$ corresponding to those points (π,u^{**}) in set $(\Pi \times \mathscr{I}^{**})$ at which the Jacobian $J^*(\pi,u^{**}) = J(0,p)$ vanishes. By SARD's Theorem [SARD, 1942], this set $\mathscr{E}(C^*)$ is a set of measure zero in the (nK)-dimensional Euclidean space $\mathscr{I}$.

Let $\mathscr{E}^*$ be the set of all games Γ in $\mathscr{I}$ having at least one irregular equilibrium point p, regardless of what its carrier $C^* = C(p)$ is. Thus, $\mathscr{E}^*$ is simply the set of all irregular games in $\mathscr{I}$. $\mathscr{E}^*$ is the union of a finite number of sets $\mathscr{E}(C^*)$, corresponding to different carrier sets C^*. Since each set $\mathscr{E}(C^*)$ is a set of *measure zero* in $\mathscr{I}$, their union $\mathscr{E}^*$ will also have this property.

Next, we shall show that $\mathscr{E}^*$ is a *closed* set. Let $\Gamma^1,\Gamma^2,\ldots$ be a sequence of irregular games, with the defining vectors $u^1 = u(\Gamma^1), u^2 = u(\Gamma^2),\ldots$. Suppose that the sequence $u^1,u^2,\ldots$ converges to a given vector u^0. Let Γ^0 be the game corresponding to $u^0 = u(\Gamma^0)$. We have to show that Γ^0 is likewise an irregular game.

Let $p^1,p^2,\ldots$ be a sequence of strategy n-tuples, such that p^j $(j = 1,2,\ldots)$ is an irregular equilibrium point in game Γ^j. All these points p^j lie in the compact set P. Consequently, the sequence $\{p^j\}$ must contain a convergent subsequence. Suppose the latter consists of the points $p^{j_1},p^{j_2},\ldots$, and that it converges to some point p^0 in P. Then:

(1) This point p^0 will be an *equilibrium point* of game Γ^0. This is so because the set $Q(\Gamma)$ of all equilibrium points in any given game is an upper semi-continuous set function of the defining vectors $u = u(\Gamma)$ of Γ, i.e., of the payoffs u_i^m of Γ.

(2) This point p^0 will be an *irregular* equilibrium point of game Γ^0. This is so because $J(0, p^{j_1}) = J(0, p^{j_2}) = \cdots = 0$ since $p^{j_1}, p^{j_2}, \ldots$ are irregular equilibrium points. Consequently, $J(0, p^0) = 0$ since p^0 is the limit of the sequence $p^{j_1}, p^{j_2}, \ldots$, and since $J(0, p)$ is a continuous function of p.

Consequently, p^0 is an irregular equilibrium point in Γ^0 and, therefore, Γ^0 itself is an irregular game, as desired. This completes the proof of Theorem 3.

Theorems 1, 2, and 3 directly imply:

Theorem 4:

In almost all finite games, the number of equilibrium points is finite and odd.

References

AUMANN, R. J.: Acceptable Points in General Cooperative n-person Games. Contributions to the Theory of Games, IV (edited by A. W. Tucker and R. D. Luce). Princeton, N. J., pp. 287–324, 1959.

DEBREU, G.: Economies with a Finite Number of Equilibria. Econometrica, **38**, 387–392, 1970.

HARSANYI, J. C.: Games with Randomly Disturbed Payoffs. International Journal of Game Theory, **2**, 1–23, 1973.

NASH, J. F.: Noncooperative Games. Annals of Mathematics, **54**, 286–295, 1951.

SARD, A.: A Measure of Critical Values of Differentiable Maps. Bulletin of the Mathematical Society, **48**, 883–890, 1942.

VAN DER WAERDEN, B. L.: Einführung in die algebraische Geometrie. Berlin, 1939.

WILSON, R.: Computing Equilibria in N-person Games. SIAM Journal of Applied Mathematics, **21**, 80–87, 1971.

Received August, 1973

SECTION C

PROBABILISTIC MODELS

CHAPTER 6

GAMES WITH INCOMPLETE INFORMATION PLAYED BY "BAYESIAN" PLAYERS, I–III

PART I. THE BASIC MODEL*†[1]

The paper develops a new theory for the analysis of games with incomplete information where the players are uncertain about some important parameters of the game situation, such as the payoff functions, the strategies available to various players, the information other players have about the game, etc. However, each player has a subjective probability distribution over the alternative possibilities.

In most of the paper it is assumed that these probability distributions entertained by the different players are mutually "consistent", in the sense that they can be regarded as conditional probability distributions derived from a certain "basic probability distribution" over the parameters unknown to the various players. But later the theory is extended also to cases where the different players' subjective probability distributions fail to satisfy this consistency assumption.

In cases where the consistency assumption holds, the original game can be replaced by a game where nature first conducts a lottery in accordance with the basic probablity distribution, and the outcome of this lottery will decide which particular subgame will be played, i.e., what the actual values of the relevant parameters will be in the game. Yet, each player will receive only partial information about the outcome of the lottery, and about the values of these parameters. However, every player will know the "basic probability distribution" governing the lottery. Thus, technically, the resulting game will be a game with complete information. It is called the Bayes-equivalent of the original game. Part I of the paper describes the basic model and discusses various intuitive interpretations for the latter. Part II shows that the Nash equilibrium points of the Bayes-equivalent game yield "Bayesian equilibrium points" for the original game. Finally, Part III considers the main properties of the "basic probablity distribution".

---

* Received June 1965, revised June 1966, accepted August 1966, and revised June 1967.

† Parts II and III of "Games with Incomplete Information Played by 'Bayesian' Players" will appear in subsequent issues of *Management Science: Theory*.

[1] The original version of this paper was read at the Fifth Princeton Conference on Game Theory, in April, 1965. The present revised version has greatly benefitted from personal discussions with Professors Michael Maschler and Robert J. Aumann, of the Hebrew University, Jerusalem; with Dr. Reinhard Selten, of the Johann Wolfgang Goethe University, Frankfurt am Main; and with the other participants of the International Game Theory Workshop held at the Hebrew University in Jerusalem, in October–November 1965. I am indebted to Dr. Maschler also for very helpful detailed comments on my manuscript.

This research was supported by Grant No. GS-722 of the National Science Foundation as well as by a grant from the Ford Foundation to the Graduate School of Business Administration, University of California. Both of these grants were administered through the Center for Research in Management Science, University of California, Berkeley. Further support has been received from the Center for Advanced Study in the Behavioral Sciences, Stanford.

TABLE OF CONTENTS

Glossary of Mathematical Notation

I-game ··· A game with *incomplete* information.
C-game ··· A game with *complete* information.
G ··· The I-game originally given to us.
G^* ··· The Bayesian game equivalent to G. (G^* is a C-game.)
G^{**} ··· The Selten game equivalent to G and to G^*. (G^{**} is likewise a C-game.)
$\mathfrak{N}(G)$, $\mathfrak{N}(G^*)$, $\mathfrak{N}(G^{**})$ ··· The normal form of G, G^* and G^{**} respectively.
$\mathcal{S}(G)$, $\mathcal{S}(G^*)$ ··· The semi-normal form of G and G^* respectively.
s_i ··· Some strategy (pure or mixed) of player i, with $i = 1, \cdots, n$.
$S_i = \{s_i\}$ ··· Player i's strategy space.
c_i ··· Player i's attribute vector (or information vector).
$C_i = \{c_i\}$ ··· The range space of vector c_i.
$c = (c_1, \cdots, c_n)$ ··· The vector obtained by combining the n vectors $c_1, \cdots, c_n$ into one vector.

$C = \{c\}$ ··· The range space of vector c.

$c^i = (c_1, \cdots, c_{i-1}, c_{i+1}, \cdots, c_n)$ ··· The vector obtained from vector c by omitting subvector c_i.

$C^i = \{c^i\}$ ··· The range space of vector c^i.

x_i ··· Player i's payoff (expressed in utility units).

$x_i = U_i(s_1, \cdots, s_n) = V_i(s_1, \cdots, s_n; c_1, \cdots, c_n)$ ··· Player i's payoff function.

$P_i(c_1, \cdots, c_{i-1}, c_{i+1}, \cdots, c_n) = P_i(c^i) = R_i(c^i \mid c_i)$ ··· The subjective probability distribution entertained by player i.

$R^* = R^*(c_1, \cdots, c_n) = R^*(c)$ ··· The basic probability distribution of the game.

$R_i^* = R^*(c_1, \cdots, c_{i-1}, c_{i+1}, \cdots, c_n \mid c_i) = R^*(c^i \mid c_i)$ ··· The conditional probability distribution obtained from R^* for a given value of vector c_i.

k_i ··· The number of different values that player i's attribute vector c_i can take in the game (in cases where this number is finite).

$K = \sum_{i=1}^{n} k_i$ ··· The number of players in the Selten game G^{**} (when this number is finite).

s_i^* ··· A normalized strategy of player i. (It is a function from the range space C_i of player i's attribute vector c_i, to his strategy space S_i.)

$S_i^* = \{s_i^*\}$ ··· The set of all normalized strategies s_i^* available to player i.

$\mathcal{E}$ ··· The expected-value operator.

$\mathcal{E}(x_i) = W_i(s_1^*, \cdots, s_n^*)$ ··· Player i's normalized payoff function, stating his *unconditional* payoff expectation.

$\mathcal{E}(x_i \mid c_i) = Z_i(s_1^*, \cdots, s_n^* \mid c_i)$ ··· Player i's semi-normalized payoff function, stating his *conditional* payoff expectation for a given value of his attribute vector c_i.

D ··· A cylinder set, defined by the condition $D = D_1 \times \cdots \times D_n$, where $D_1 \subseteqq C_1, \cdots, D_n \subseteqq C_n$.

$G(D)$ ··· For a given decomposable game G or G^*, $G(D)$ denotes the component game played in all cases where the vector c lies in cylinder D. D is called the defining cylinder of the component game $G(D)$.

Special Notation in Certain Sections

In section 3 (Part I):

a_{0i} denotes a vector consisting of those parameters of player i's payoff function U_i which (in player j's opinion) are *unknown* to all n players.

a_{ki} denotes a vector consisting of those parameters of the function U_i which (in j's opinion) are unknown to some of the players but are *known* to player k.

$a_0 = (a_{01}, \cdots, a_{0n})$ is a vector summarizing all information that (in j's opinion) none of the players have about the functions $U_1, \cdots, U_n$.

$a_k = (a_{k1}, \cdots, a_{kn})$ is a vector summarizing all information that (in j's opinion) player k has about the functions $U_1, \cdots, U_n$, except for the information that (in j's opinion) *all* n players have about these functions.

b_i is a vector consisting of all those parameters of player i's subjective probability distribution P_i which (in player j's opinion) are unknown to some or all of the players $k \neq i$.

In terms of these notations, player i's information vector (or attribute vector) c_i can be defined as

$c_i = (a_i, b_i)$.

V_i^* denotes player i's payoff function *before* vector a_0 has been integrated out. *After* elimination of vector a_0 the symbol V_i is used to denote player i's payoff function.

In sections 9–10 (Part II):

a^1 and a^2 denote the two possible values of player 1's attribute vector c_1.

b^1 and b^2 denote the two possible values of player 2's attribute vector c_2.

$r_{km} = R^*(c_1 = a^k \text{ and } c_2 = b^m)$ denotes the probability mass function corresponding to the basic probability distribution R^*.

$p_{km} = r_{km}/(r_{k1} + r_{k2})$ and $q_{km} = r_{km}/(r_{1m} + r_{2m})$ denote the corresponding *conditional* probability mass functions.

y^1 and y^2 denote player 1's two pure strategies.

z^1 and z^2 denote player 2's two pure strategies.

$y^{nt} = (y^n, y^t)$ denotes a normalized pure strategy for player 1, requiring the use of strategy y^n if $c_1 = a^1$, and requiring the use of strategy y^t if $c_1 = a^2$.

$z^{uv} = (z^u, z^v)$ denotes a normalized pure strategy for player 2, requiring the use of strategy z^u if $c_2 = b^1$, and requiring the use of strategy z^v if $c_2 = b^2$.

In section 11 (Part II):

a^1 and a^2 denote the two possible values that *either* player's attribute vector c_i can take.

$r_{km} = R^*(c_1 = a^k \text{ and } c_2 = a^m)$.

p_{km} and q_{km} have the same meaning as in sections 9–10.

y_i^* denotes player i's payoff demand.

y_i denotes player i's *gross* payoff.

x_i denotes player i's *net* payoff.

x_i^* denotes player i's net payoff in the case $(c_1 = a^1, c_2 = a^2)$.

x_i^{**} denotes player i's net payoff in the case $(c_1 = a^2, c_2 = a^1)$.

In section 13 (Part III):

$\alpha, \beta, \gamma, \delta$ denote specific values of vector c.

$\alpha_i, \beta_i, \gamma_i, \delta_i$ denote specific values of vector c_i.

$\alpha^i, \beta^i, \gamma^i, \delta^i$ denote specific values of vector c^i, etc.

$r_i(\gamma^i \mid \gamma_i) = R_i(c^i = \gamma^i \mid c_i = \gamma_i)$ denotes the probability mass function corresponding to player i's subjective probability distribution R_i (when R_i is a discrete distribution).

$r^*(\gamma) = R^*(c = \gamma)$ denotes the probability mass function corresponding to the basic probability distribution R^* (when R^* is a discrete distribution).

$\mathcal{R} = \{r^*\}$ denotes the set of all admissible probability mass functions r^*.

E denotes a similarity class, i.e., a set of nonnull points $c = \alpha, c = \beta, \cdots$ similar to one another (in the sense defined in Section 13).

In section 16 (Part III):

$R^{(i)}$ denotes the basic probability distribution R^* as assessed by player $i(i = 1, \cdots, n)$.

$R^{*\prime}$ denotes a given player's (player *j*'s) *revised* estimate of the basic probability distribution R^*.

$c'_i = (c_i, d_i)$ denotes player *j*'s *revised* definition of player *i*'s attribute vector c_i. (It is in general a larger vector than the vector c_i originally assumed by player *j*.)

R'_i denotes player *j*'s *revised* estimate of player *i*'s subjective probability distribution R_i.

1.

Following von Neumann and Morgenstern [7, p. 30], we distinguish between games with *complete information*, to be sometimes briefly called *C*-games in this paper, and games with *incomplete information*, to be called *I*-games. The latter differ from the former in the fact that some or all of the players lack full information about the "rules" of the game, or equivalently about its normal form (or about its extensive form). For example, they may lack full information about other players' or even their own payoff functions, about the physical facilities and strategies available to other players or even to themselves, about the amount of information the other players have about various aspects of the game situation, etc.

In our own view it has been a major analytical deficiency of existing game theory that it has been almost completely restricted to *C*-games, in spite of the fact that in many real-life economic, political, military, and other social situations the participants often lack full information about some important aspects of the "game" they are playing.[2]

It seems to me that the basic reason why the theory of games with incomplete information has made so little progress so far lies in the fact that these games give rise, or at least appear to give rise, to an infinite regress in reciprocal expectations on the part of the players, [3, pp. 30–32]. For example, let us consider any two-person game in which the players do not know each other's payoff functions. (To simplify our discussion I shall assume that each player knows his own payoff function. If we made the opposite assumption, then we would have to introduce even more complicated sequences of reciprocal expectations.)

In such a game player 1's strategy choice will depend on what he expects (or believes) to be player 2's payoff function U_2, as the nature of the latter will be an important determinant of player 2's behavior in the game. This expectation

[2] The distinction between games with *complete* and *incomplete* information (between *C*-games and *I*-games) must not be confused with that between games with *perfect* and *imperfect* information. By common terminological convention, the first distinction always refers to the amount of information the players have about the *rules* of the game, while the second refers to the amount of information they have about the other players' and their own previous *moves* (and about previous chance moves). Unlike games with *incomplete* information, those with *imperfect* information have been extensively discussed in the literature.

about U_2 may be called player 1's *first-order* expectation. But his strategy choice will also depend on what he expects to be player 2's first-order expectation about his own (player 1's) payoff function U_1. This may be called player 1's *second-order* expectation, as it is an expectation concerning a first-order expectation. Indeed, player 1's strategy choice will also depend on what he expects to be player 2's second-order expectation—that is, on what player 1 thinks that player 2 thinks that player 1 thinks about player 2's payoff function U_2. This we may call player 1's *third-order* expectation—and so on *ad infinitum*. Likewise, player 2's strategy choice will depend on an infinite sequence consisting of his first-order, second-order, third-order, etc., expectations concerning the payoff functions U_1 and U_2. We shall call any model of this kind a *sequential-expectations* model for games with incomplete information.

If we follow the Bayesian approach and represent the players' expectations or beliefs by subjective probablity distributions, then player 1's *first-order* expectation will have the nature of a subjective probablity distribution $P_1^1(U_2)$ over all alternative payoff functions U_2 that player 2 may possibly have. Likewise, player 2's first-order expectation will be a subjective probablity distribution $P_2^1(U_1)$ over all alternative payoff functions U_1 that player 1 may possibly have. On the other hand, player 1's *second-order* expectation will be a subjective probability distribution $P_1^2(P_2^1)$ over all alternative first-order subjective probability distributions P_2^1 that player 2 may possibly choose, etc. More generally, the kth-order expectation ($k > 1$) of either player i will be a subjective probability distribution $P_i^k(P_j^{k-1})$ over all alternative $(k - 1)$th-order subjective probability distributions P_j^{k-1} that the other player j may possibly entertain.[3]

In the case of n-person I-games the situation is, of course, even more complicated. Even if we take the simpler case in which the players know at least their own payoff functions, each player in general will have to form expectations about the payoff functions of the other $(n - 1)$ players, which means forming $(n - 1)$ different first-order expectations. He will also have to form expectations about the $(n - 1)$ first-order expectations entertained by each of the other $(n - 1)$ players, which means forming $(n - 1)^2$ second-order expectations, etc.

The purpose of this paper is to suggest an alternative approach to the analysis of games with incomplete information. This approach will be based on constructing, for any given I-game G, some C-game G^* (or possibly several different C-games G^*) game-theoretically *equivalent* to G. By this means we shall reduce the analysis of I-games to the analysis of certain C-games G^*; so that the problem of

[3] Probability distributions over some space of payoff functions or of probability distributions, and more generally probability distributions over function spaces, involve certain technical mathematical difficulties [5, pp. 355–357]. However, as Aumann has shown [1] and [2], these difficulties can be overcome. But even if we succeed in defining the relevant higher-order probability distributions in a mathematically admissible way, the fact remains that the resulting model—like *all* models based on the sequential-expectations approach—will be extremely complicated and cumbersome. The main purpose of this paper is to describe an alternative approach to the analysis of games with incomplete information, which completely avoids the difficulties associated with sequences of higher and higher-order reciprocal expectations.

such sequences of higher and higher-order reciprocal expectations will simply not arise.

As we have seen, if we use the Bayesian approach, then the sequential-expectations model for any given I-game G will have to be analyzed in terms of infinite sequences of higher and higher-order subjective probability distributions, i.e., subjective probability distributions over subjective probablity distributions. In contrast, under our own model, it will be possible to analyze any given I-game G in terms of *one* unique probability distribution R^* (as well as certain conditional probablity distributions derived from R^*).

For example, consider a two-person non-zero-sum game G representing price competition between two duopolist competitiors where neither player has precise information about the cost functions and the financial resources of the other player. This, of course, means that neither player i will know the true payoff function U_j of the other player j, because he will be unable to predict the profit (or the loss) that the other player will make with any given choice of strategies (i.e., price and output polices) s_1 and s_2 by the two players.

To make this example more realistic, we may also assume that each player has *some* information about the other player's cost functions and financial resources (which may be represented, e.g., by a subjective probability distribution over the relevant variables); but that each player i lacks exact information about *how much* the other player j actually knows about player i's cost structure and financial position.

Under these assumptions this game G will be obviously an I-game, and it is easy to visualize the complicated sequences of reciprocal expectations (or of subjective probablity distributions) we would have to postulate if we tried to analyze this game in terms of the sequential-expectations approach.

In contrast, the new approach we shall describe in this paper will enable us to reduce this I-game G to an equivalent C-game G^* involving four random events (i.e., chance moves) e_1, e_2, f_1, and f_2, assumed to occur *before* the two players choose their strategies s_1 and s_2. The random event $e_i (i = 1, 2)$ will determine player i's cost functions and the size of his financial resources; and so will completely determine his payoff function U_1 in the game. On the other hand, the random event f_i will determine the amount of information that player i will obtain about the cost functions and the financial resources of the other player $j (j = 1, 2 \text{ and } \neq i)$, and will thereby determine the actual amount of information[4] that player i will have about player j's payoff function U_j.

Both players will be assumed to know the joint probability distribution $R^*(e_1, e_2, f_1, f_2)$ of these four random events.[5] But, *e.g.*, player 1 will know the actual outcomes of these random events only in the case of e_1 and f_1, whereas

[4] In terms of the terminology we shall later introduce, the variables determined by the random events e_i and f_i will constitute the random vector c_i $(i = 1, 2)$, which will be called player i's information vector or attribute vector, and which will be assumed to determine player i's "type" in the game (cf. the third paragraph below).

[5] For justification of this assumption, see sections 4 and 5 below, as well as Part III of this paper.

player 2 will know the actual outcomes only in the case of e_2 and f_2. (In our model this last assumption will represent the facts that each player will know only his own cost functions and financial resources but will not know those of his opponent; and that he will, of course, know how much information he himself has about the opponent but will not know exactly how much information the opponent will have about him.)

As in this new game G^* the players are assumed to know the probability distribution $R^*(e_1, e_2, f_1, f_2)$, this game G^* will be a C-game. To be sure, player 1 will have no information about the outcomes of the chance moves e_2 and f_2, whereas player 2 will have no information about the outcomes of the chance moves e_1 and f_1. But these facts will not make G^* a game with "incomplete" information but will make it only a game with "imperfect" information (cf. Footnote 2 above). Thus, our approach will basically amount to replacing a game G involving *incomplete* information, by a new game G^* which involves *complete* but *imperfect* information, yet which is, as we shall argue, essentially equivalent to G from a game-theoretical point of view (see section 5 below).

As we shall see, this C-game G^* which we shall use in the analysis of a given I-game G will also admit of an alternative intuitive interpretation. Instead of assuming that certain important *attributes* of the players are determined by some hypothetical random events at the beginning of the game, we may rather assume that the players *themselves* are drawn at random from certain hypothetical populations containing a mixture of individuals of different "types", characterized by different attribute vectors (i.e., by different combinations of the relevant attributes). For instance, in our duopoly example we may assume that each player $i(i = 1, 2)$ is drawn from some hypothetical population Π_i containing individuals of different "types," each possible "type" of player i being characterized by a different attribute vector c_i, i.e., by a different combination of production costs, financial resources, and states of information. Each player i will know his own type or attribute vector c_i but will be, in general, ignorant of his opponent's. On the other hand, both players will again be assumed to know the joint probability distribution $R^*(c_i, c_2)$ governing the selection of players 1 and 2 of different possible types c_1 and c_2 from the two hypothetical populations Π_1 and Π_2.

It may be noted, however, that in analyzing a given I-game G, construction of an equivalent C-game G^* is only a partial answer to our analytical problem, because we are still left with the task of defining a suitable solution concept for this C-game G^* itself, which may be a matter of some difficulty. This is so because in many cases the C-game G^* we shall obtain in this way will be a C-game of unfamiliar form, for which no solution concept has been suggested yet in the game-theoretical literature.[6] Yet, since G^* will always be a game with complete information, its analysis and the problem of defining a suitable solution concept for it, will be at least amenable to the standard methods of modern game theory. We shall show in some examples how one actually can define appropriate solution concepts for such C-games G^*.

[6] More particularly, this game G^* will have the nature of a game with *delayed commitment* (see section 11 in Part II of this paper).

2.

Our analysis of I-games will be based on the assumption that, in dealing with incomplete information, every player i will use the Bayesian approach. That is, he will assign a *subjective* joint probability distribution P_i to all variables unknown to him—or at least to all unknown *independent* variables, i.e., to all variables not depending on the players' own strategy choices. Once this has been done he will try to maximize the mathematical expectation of his own payoff x_i in terms of this probability distribution P_i.[7] This assumption will be called the *Bayesian hypothesis*.

If incomplete information is interpreted as lack of full information by the players about the *normal form* of the game, then such incomplete information can arise in three main ways.

1. The players may not know the *physical outcome function* Y of the game, which specifies the physical outcome $y = Y(s_1, \cdots, s_n)$ produced by each strategy n-tuple $s = (s_1, \cdots, s_n)$ available to the players.

2. The players may not know their own or some other players' *utility functions* X_i, which specify the utility payoff $x_i = X_i(y)$ that a given player i derives from every possible physical outcome y.[8]

3. The players may not know their own or some other players' *strategy spaces* S_i, *i.e.*, the set of all strategies s_i (both pure and mixed) available to various players i.

All other cases of incomplete information can be reduced to these three basic cases—indeed sometimes this can be done in two or more different (but essentially equivalent) ways. For example, incomplete information may arise by some players' ignorance about the amount or the quality of physical resources (equipment, raw materials, etc.) available to some other players (or to themselves). This situation can be equally interpreted *either* as ignorance about the physical outcome function of the game (case 1), *or* as ignorance about the strategies available to various players (case 3). Which of the two interpretations we have to use will depend on how we choose to define the "strategies" of the players in question. For instance, suppose that in a military engagement our own side does not know the number of fire arms of a given quality available to the other side.

[7] A *subjective* probability distribution P_i entertained by a given player i is defined in terms of his own choice behavior, cf. [6]. In contrast, an *objective* probability distribution P^* is defined in terms of the long-run frequencies of the relevant events (presumably as established by an independent observer, say, the umpire of the game). It is often convenient to regard the subjective probabilities used by a given player i as being his personal *estimates* of the corresponding objective probabilities or frequencies unknown to him.

[8] If the physical outcome y is simply a vector of money payoffs $y_1, \cdots, y_n$ to the n players then we can usually assume that any player i's utility payoff $x_i = X_i(y_i)$ is a (strictly increasing) function of his money payoff y_i and that all players will know this. However, the other players j may not know the specific mathematical form of player i's utility function for money, X_i. In other words, even though they may know player i's *ordinal* utility function, they may not know his *cardinal* utility function. That is to say, they may not know how much *risk* he would be willing to take in order to increase his money payoff y_i by given amounts.

This can be interpreted as inability on our part to predict the *physical outcome* (*i.e.*, the amount of destruction) resulting from alternative strategies of the opponent, where any given "strategy" of his is defined as firing a given *percentage* of his fire arms (case 1). But it can also be interpreted as inability to decide whether certain strategies are *available* to the opponent at all, where now any given "strategy" of his is defined as firing a specified *number* of fire arms (case 3).

Incomplete information can also take the form that a given player i does not know whether another player j does or does not have information about the occurrence or non-occurrence of some specified event e. Such a situation will always come under case 3. This is so because in a situation of this kind, from a game-theoretical point of view, the crucial fact is player i's inability to decide whether player j is in a position to use any strategy s_j^0 involving *one* course of action in case event e does occur, and *another* course of action in case event e does not occur. That is, the situation will essentially amount to ignorance by player i about the availability of certain strategies s_j^0 to player j.

Going back to the three main cases listed above, cases 1 and 2 are both special cases of ignorance by the players about their own or some other players' *payoff functions* $U_i = X_i(Y)$ specifying the utility payoff $x_i = U_i(s_1, \cdots, s_n)$ a given player i obtains if the n players use alternative strategy n-tuples $s = (s_1, \cdots, s_n)$.

Indeed, case 3 can also be reduced to this general case. This is so because the assumption that a given strategy $s_i = s_i^0$ is not *available* to player i is equivalent, from a game-theoretical point of view, to the assumption that player i will never actually *use* strategy s_i^0 (even though it would be physically available to him) because by using s_i^0 he would always obtain some extremely low (*i.e.*, highly negative) payoffs $x_i = U_i(s_1, \cdots, s_i^0, \cdots, s_n)$, whatever strategies $s_1, \cdots, s_{i-1}, s_{i+1}, \cdots, s_n$ the other players $1, \cdots, i-1, i+1, \cdots, n$ may be using.

Accordingly, let $S_i^{(j)}$ ($j = 1$ or $j \neq 1$) denote the *largest* set of strategies s_i which in player j's opinion may be *conceivably* included in player i's strategy space S_i. Let $S_i^{(0)}$ denote player i's "true" strategy space. Then, for the purposes of our analysis, we shall define player i's strategy space S_i as

$$S_i = \bigcup_{k=0}^{n} S_i^{(k)}. \tag{2.1}$$

We lose no generality by assuming that this set S_i as defined by (2.1) is *known* to all players because any lack of information on the part of some player j about this set S_i can be represented within our model as lack of information about the numerical values that player i's payoff function $x_i = U_i(s_1, \cdots, s_i, \cdots, s_n)$ takes for some specific choices of s_i, and in particular whether these values are so low as completely to discourage player i from using these strategies s_i.[9]

Accordingly, we define an I-game G as a game where every player j *knows* the strategy spaces S_i of all players $i = 1, \cdots, j, \cdots, n$ but where, in general, he does *not* know the payoff functions U_i of these players $i = 1, \cdots, j, \cdots, n$.

[9] Likewise, instead of assuming that player j assigns subjective probabilities to events of the form $E = \{s_i^0 \notin S_i\}$, we can always assume that he assigns these probabilities to events of the form $E = \{U_i(s_1, \cdots, s_i, \cdots, s_n) < x_i^0$ whenever $s_i = s_i^0\}$, etc.

3.

In terms of this definition, let us consider a given I-game G from the point of view of a particular player j. He can write the payoff function U_i of each player i (including his own payoff function U_j for $i = j$) in a more explicit form as

$$(3.1)\quad x_i = U_i(s_1, \cdots, s_n) = V_i^*(s_1, \cdots, s_n; \quad a_{0i}, a_{1i}, \cdots, a_{ii}, \cdots, a_{ni}),$$

where V_i^*, unlike U_i, is a function whose mathematical form is (in player j's opinion) *known* to all n players; whereas a_{0i} is a vector consisting of those parameters of function U_i which (in j's opinion) are *unknown* to *all* players; and where each a_{ki} for $k = 1, \cdots, n$ is a vector consisting of those parameters of function U_i which (in j's opinion) are *unknown* to some of the players but are *known* to player k. If a given parameter α is known *both* to players k and m (without being known to all players), then this fact can be represented by introducing two variables α_{ki} and α_{mi} with $\alpha_{ki} = \alpha_{mi} = \alpha$, and then making α_{ki} a component of vector a_{ki} while making α_{mi} a component of vector a_{mi}.

For each vector a_{ki} ($k = 0, 1, \cdots, n$), we shall assume that its *range space* $A_{ki} = \{a_{ki}\}$, *i.e.*, the set of all possible values it can take, is the whole Euclidian space of the required number of dimensions. Then V_i^* will be a function from the Cartesian product $S_1 \times \cdots \times S_n \times A_{0i} \times \cdots \times A_{ni}$ to player i's utility line Ξ_i, which is itself a copy of the real line R.

Let us define a_k as the vector combining the components of all n vectors $a_{k1}, \cdots, a_{kn}$. Thus we write

$$(3.2)\qquad a_k = (a_{k1}, \cdots, a_{kn}),$$

for $k = 0, 1, \cdots, i, \cdots, n$. Clearly, vector a_0 summarizes the information that (in player j's opinion) none of the players has about the n functions $U_1, \cdots, U_n$, whereas vector $a_k (k = 1, \cdots, n)$ summarizes the information that (in j's opinion) player k has about these functions, except for the information that (in j's opinion) *all* n players share about them. For each vector a_k, its range space will be the set $A_k = \{a_k\} = A_{k1} \times \cdots \times A_{kn}$.

In equation (3.1) we are free to replace each vector $a_{ki} (k = 0, \cdots, n)$ by the larger vector $a_k = (a_{k1}, \cdots, a_{ki}, \cdots, a_{kn})$, even though this will mean that in each case the $(n - 1)$ sub-vectors $a_{k1}, \cdots, a_{k(i-1)}, a_{k(i+1)}, \cdots, a_{kn}$ will occur vacuously in the resulting new equation. Thus, we can write

$$(3.3)\qquad x_i = V_i^*(s_1, \cdots, s_n; \quad a_0, a_1, \cdots, a_i, \cdots, a_n).$$

For any given player i the n vectors $a_0, a_1, \cdots, a_{i-1}, a_{i+1}, \cdots, a_n$ in general will represent *unknown* variables; and the same will be true for the $(n - 1)$ vectors $b_i, \cdots, b_{i-1}, b_{i+1}, \cdots, b_n$ to be defined below. Therefore, under the Bayesian hypothesis, player i will assign a subjective joint probability distribution

$$(3.4)\quad P_i = P_i(a_0, a_1, \cdots, a_{i-1}, a_{i+1}, \cdots, a_n; \quad b_1, \cdots, b_{i-1}, b_{i+1}, \cdots, b_n)$$

to all these unknown vectors.

For convenience we introduce the shorter notations $a = (a_1, \cdots, a_n)$ and

$b = (b_1, \cdots, b_n)$. The vectors obtained from a and b by omitting the sub-vector a_i and b_i, respectively, will be denoted by a^i and b^i. The corresponding range spaces can be written as $A = A_1 \times \cdots \times A_n$; $B = B_1 \times \cdots \times B_n$; $A^i = A_1 \times \cdots \times A_{i-1} \times A_{i+1} \times \cdots \times A_n$; $B^i = B_1 \times \cdots \times B_{i-1} \times B_{i+1} \times \cdots \times B_n$.

Now we can write equations (3.3) and (3.4) as

$$x_i = V_i^*(s_1, \cdots, s_n; a_0, a) \tag{3.5}$$

$$P_i = P_i(a_0, a^i; b^i) \tag{3.6}$$

where P_i is a probability distribution over the vector space $A_0 \times A^i \times B^i$.

The other $(n - 1)$ players in general will not know the subjective probability distribution P_i used by player i. But player j (from whose point of view we are analyzing the game) will be able to write P_i for each player i (both $i = j$ and $i \neq j$) in the form

$$P_i(a_0, a^i; b^i) = R_i(a_0, a^i; b^i \mid b_i), \tag{3.7}$$

where R_i, unlike P_i, is a function whose mathematical form is (in player j's opinion) *known* to all n players; whereas b_i is a vector consisting of those parameters of function P_i which (in j's opinion) are *unknown* to some or all of the players $k \neq i$. Of course, player j will realize that player i himself will *know* vector b_i since b_i consists of parameters of player i's own subjective probability distribution P_i.

The vectors $b_1, \cdots, b_{i-1}, b_{i+1}, \cdots, b_n$ occurring in equation (3.4), which so far have been left undefined, are the parameter vectors of the subjective probability distributions $P_1, \cdots, P_{i-1}, P_{i+1}, \cdots, P_n$, unknown to player i. The vector b^i occurring in equations (3.6) and (3.7) is a combination of all these vectors $b_1, \cdots, b_{i-1}, b_{i+1}, \cdots, b_n$, and summarizes the information that (in player j's opinion) player i lacks about the other $(n - 1)$ players' subjective probability distributions $P_1, \cdots, P_{i-1}, P_{i+1}, \cdots, P_n$.

Clearly, function R_i is a function yielding, for each specific value of vector b_i, a probability distribution over the vector space $A^i \times B^i$.

We now propose to eliminate the vector a_0, unknown to all players, from equations (3.5) and (3.7). In the case of equation (3.5) this can be done by taking *expected values* with respect to a_0 in terms of player i's own subjective probability distribution $P_i(a_0, a^i; b^i) = R_i(a_0, a^i; b^i \mid b_i)$. We define

$$V_i(s_1, \cdots, s_n; a \mid b_i) = V_i(s_1, \cdots, s_n; a, b_i) = \int_{A_0} V_i^*(s_1 \cdots, s_n; a_0, a)\, d_{(a_0)} R_i(a_0, a^i; b^i \mid b_i). \tag{3.8}$$

Then we write

$$x_i = V_i(s_1, \cdots, s_n; a, b_i), \tag{3.9}$$

where x_i now denotes the *expected value* of player i's payoff in terms of his own subjective probability distribution.

In the case of equation (3.7) we can eliminate a_0 by taking the appropriate *marginal* probability distributions. We define

(3.10) $$P_i(a^i, b^i) = \int_{A_0} d_{(a_0)} P_i(a_0, a^i; b^i),$$

and

(3.11) $$R_i(a^i, b^i \mid b_i) = \int_{A_0} d_{(a_0)} R_i(a_0, a^i; b^i \mid b_i).$$

Then we write

(3.12) $$P_i(a^i, b^i) = R_i(a^i, b^i \mid b_i).$$

We now rewrite equation (3.9) as

(3.13) $$x_i = V_i(s_1, \cdots, s_n; a, b_i, b^i) = V_i(s_1, \cdots, s_n; a, b),$$

where vector b^i occurs only vacuously. Likewise we rewrite equation (3.12) as

(3.14) $$P_i(a^i, b^i) = R_i(a^i, b^i \mid a_i, b_i),$$

where on the right-hand side vector a_i occurs only vacuously.

Finally, we introduce the definitions $c_i = (a_i, b_i)$; $c = (a, b)$; and $c^i = (a^i, b^i)$. Moreover, we write $C_i = A_i \times B_i$; $C = A \times B$; and $C^i = A^i \times B^i$. Clearly, vector c_i represents the *total information* available to player i in the game (if we disregard the information available to all n players). Thus, we may call c_i player i's *information vector*.

From another point of view, we can regard vector c_i as representing certain physical, social, and psychological *attributes* of player i *himself*, in that it summarizes some crucial parameters of player i's own payoff function U_i as well as the main parameters of his beliefs about his social and physical environment. (The relevant parameters of player i's payoff function U_i again partly represent parameters of his subjective utility function X_i and partly represent parameters of his environment, *e.g.*, the amounts of various physical or human resources available to him, *etc.*) From this point of view, vector c_i may be called player i's *attribute vector*.

Thus, under this model, the players' incomplete information about the true nature of the game situation is represented by the assumption that in general the actual value of the attribute vector (or information vector) c_i of any given player i will be known only to player i himself, but will be *unknown* to the other $(n - 1)$ players. That is, as far as these other players are concerned, c_i could have any one of a number—possibly even of an infinite number—of alternative values (which together form the range space $C_i = \{c_i\}$ of vector c_i). We may also express this assumption by saying that in an I-game G, in general, the rules of the game as such allow any given player i to belong to any one of a number of possible "*types*", corresponding to the alternative values his attribute vector c_i could take (and so representing the alternative payoff functions U_i and the alternative subjective probability distributions P_i that player i

might have in the game). Each player is always assumed to know his own actual type but to be in general ignorant about the other players' actual types.

Equations (3.13) and (3.14) now can be written as

$$(3.15) \qquad x_i = V_i(s_1, \cdots, s_n; c) = V_i(s_1, \cdots, s_n; c_1, \cdots, c_n)$$

$$(3.16) \qquad P_i(c^i) = R_i(c^i | c_i)$$

or

$$(3.17) \quad P_i(c_1, \cdots, c_{i-1}, c_{i+1}, \cdots, c_n) = R_i(c_1, \cdots, c_{i-1}, c_{i+1}, \cdots, c_n | c_i).$$

We shall regard equations (3.15) and (3.17) [or (3.16)] as the standard forms of the equations defining an I-game G, considered from the point of view of some particular player j.

Formally we define the *standard form* of a given I-game G for some particular player j as an ordered set G such that

$$(3.18) \quad G = \{S_1, \cdots, S_n; C_1, \cdots, C_n; V_1, \cdots, V_n; R_1, \cdots, R_n\}$$

where for $i = 1, \cdots, n$ we write $S_i = \{s_i\}$; $C_i = \{c_i\}$; moreover, where V_i is a function from the set $S_1 \times \cdots \times S_n \times C_1 \times \cdots \times C_n$ to player i's utility line Ξ_i (which is itself a copy of the real line R); and where, for any specific value of the vector c_i, the function $R_i = R_i(c^i | c_i)$ is a probability distribution over the set $C^i = C_1 \times \cdots \times C_{i-1} \times C_{i+1} \times \cdots \times C_n$.

4.

Among C-games the natural analogue of this I-game G will be a C-game G^* with the same payoff functions V_i and the same strategy spaces S_i. However, in G^* the vectors c_i will have to be reinterpreted as being *random vectors* (chance moves) with an *objective* joint probability distribution

$$(4.1) \qquad R^* = R^*(c_1, \cdots, c_n) = R^*(c)$$

known to all n players.[10] (If some players did not know R^*, then G^* would not be a C-game.) To make G^* as similar to G as possible, we shall assume that each vector c_i will take its values from the same range space C_i in either game. Moreover, we shall assume that in game G^*, just as in game G, when player i chooses his strategy s_i, he will know only the value of his *own* random vector c_i but will not know the random vectors $c_1, \cdots, c_{i-1}, c_{i+1}, \cdots, c_n$ of the other $(n - 1)$ players. Accordingly we may again call c_i the *information vector* of player i.

Alternatively, we may again interpret this random vector c_i as representing certain physical, social, and psychological attributes of player i himself. (But, of course, now we have to assume that for all n players these attributes are determined by some sort of random process, governed by the probability distribution R^*.) Under this interpretation we may again call c_i the *attribute vector* of player i.

[10] Assuming that a joint probability distribution R^* of the required mathematical form exists (see section 5 below, as well as Part III of this paper).

We shall say that a given C-game G^* is in *standard form* if

1. the payoff functions V_i of G^* have the form indicated by equation (3.15);
2. the vectors $c_1, \cdots, c_n$ occurring in equation (3.15) are random vectors with a joint probability distribution R^* [equation (4.1)] known to all players;
3. each player i is assumed to know only his own vector c_i, and does not know the vectors $c_1, \cdots, c_{i-1}, c_{i+1}, \cdots, c_n$ of the other players when he chooses his strategy s_i.

Sometimes we shall again express these assumptions by saying that the rules of the game allow each player i to belong to any one of a number of alternative *types* (corresponding to alternative specific values that the random vector c_i can take); and that each player will always know his own actual type, but in general will not know those of the other players.

Formally we define a C-game G^* in standard form as an ordered set G^* such that

$$(4.2) \qquad G^* = \{S_1, \cdots, S_n; C_1, \cdots, C_n; V_1, \cdots, V_n; R^*\}.$$

Thus, the ordered set G^* differs from the ordered set G [defined by equation (3.18)] only in the fact that the n-tuple $R_1, \cdots, R_n$ occurring in G is replaced in G^* by the singleton R^*.

If we consider the normal form of a game as a special limiting case of a standard form (viz. as the case where the random vectors $c_1, \cdots, c_n$ are empty vectors without components), then, of course, every C-game has a standard form. But only a C-game G^* containing random variables (chance moves) will have a standard form non-trivially different from its normal form.

Indeed, if G^* contains more than one random variable, then it will have several different standard forms. This is so because we can always obtain new standard forms G^{**}—intermediate between the original standard form G^* and the normal form G^{***}—if we suppress *some* of the random variables occurring in G^*, without suppressing *all* of them (as we would do if we wanted to obtain the normal form G^{***} itself). This procedure can be called *partial* normalization as distinguished from the *full* normalization, which would yield the normal form G^{***}.[11]

5.

Suppose that G is an I-game (considered from player j's point of view) while G^* is a C-game, both games being given in standard form. To obtain complete similarity between the two games, it is not enough if the strategy spaces $S_1, \cdots, S_n$, the range spaces $C_1, \cdots, C_n$, and the payoff functions $V_1, \cdots, V_n$

[11] Partial normalization involves essentially the same operations as full normalization (see section 7 below). It involves taking the expected values of the payoff functions V_i with respect to the random variables to be suppressed, and redefining the players' strategies where necessary. However, in the case of partial normalization we also have to replace the probability distribution R^* of the original standard form G^*, by a *marginal* probability distribution not containing the random variables to be suppressed. (In the case of full normalization no such marginal distribution has to be computed because the normal form G^{***} will not contain random variables at all.)

of the two games are the same. It is necessary also that each player i in either game should always assign the same *numerical* proabability p to any given specific event E. Yet in game G player i will assess all probabilities in terms of his *subjective* probability distribution $R_i(c^i| c_i)$; whereas in game G^*—since vector c_i is known to him—he will assess all probabilities in terms of the objective *conditional* probability distribution $R^*(c^i| c_i)$ generated by the basic probability distribution $R^*(c)$ of the game G^*. Therefore, if the two games are to be equivalent, then numerically the distributions $R_i(c^i| c_i)$ and $R^*(c^i| c_i)$ must be identically equal.

This leads to the following definition. Let G be an I-game (as considered by player j), and let G^* be a C-game, both games being given in standard form. We shall say that G and G^* are *Bayes-equivalent* for player j if the following conditions are fulfilled:

1. The two games must have the same strategy spaces $S_1, \cdots, S_n$ and the same range spaces $C_1, \cdots, C_n$.
2. They must have the same payoff functions $V_1, \cdots, V_n$.
3. The subjective probability distribution R_i of each player i in G must satisfy the relationship

$$R_i(c^i| c_i) = R^*(c^i| c_i), \tag{5.1}$$

where $R^*(c) = R^*(c_i, c^i)$ is the basic probability distribution of game G^* and where

$$R^*(c^i | c_i) = R^*(c_i, c^i) \Big/ \int_{C^i} d_{(c^i)} R^*(c_i, c^i). \tag{5.2}$$

In view of equations (5.1) and (5.2) we can write

$$R^*(c) = R^*(c_i, c^i) = R_i(c^i | c_i) \cdot \int_{C^i} d_{(c^i)} R^*(c_i, c^i). \tag{5.3}$$

In contrast to equation (5.2), which ceases to have a clear mathematical meaning when the denominator on its right-hand side becomes zero, equation (5.3) always retains a clear mathematical meaning.

We propose the following postulate.

Postulate 1. Bayes-equivalence. Suppose that some I-game G and some C-game G^* are Bayes-equivalent for player j. Then the two games will be completely equivalent for player j from a game-theoretical standpoint; and, in particular, player j's strategy choice will be governed by the same decision rule (the same solution concept) in either game.

This postulate follows from the Bayesian hypothesis, which implies that every player will use his *subjective* probabilities exactly in the same way as he would use known *objective* probabilities numerically equal to the former. Game G (as assessed by player j) and game G^* agree in all defining characteristics, including the numerical probability distributions used by the players. The only difference is that in G the probabilities used by each player are subjective probabilities whereas in G^* these probabilities are objective (conditional) probabilities. But by the Bayesian hypothesis this difference is immaterial.

Of course, under the assumptions of the postulate, all we can say is that *for player j himself* the two games are completely equivalent for game-theoretical purposes. We cannot conclude on the basis of the information assumed that the two games are likewise equivalent also for some *other* players $k \neq j$. In order to reach this latter conclusion we would have to know that G and G^* would preserve their Bayes-equivalence even if G were analyzed in terms of the functions $V_1, \cdots, V_n$ and $R_1, \cdots, R_n$ postulated by these *other* players k, instead of being analyzed in terms of the functions $V_1, \cdots, V_n$ and $R_1, \cdots, R_n$ postulated by player j himself. But so long as we are interested only in the decision rules that player j himself will follow in game G, all we have to know are the functions $V_1, \cdots, V_n$ and $R_1, \cdots, R_n$ that player j will be using.

Postulate 1 naturally gives rise to the following questions. Given any I-game G, is it always possible to construct a C-game G^* Bayes-equivalent to G? And, in cases where this is possible, is this C-game G^* always unique? These questions are tantamount to asking whether for any arbitrarily chosen n-tuple of subjective probability distributions $R_1(c^1 | c_1), \cdots, R_n(c^n | c_n)$, there always *exists* a probability distribution $R^*(c_1, \cdots, c_n)$ satisfying the functional equation (5.3), and whether this distribution R^* is always *unique* in cases where it does exist. As these questions require an extended discussion, we shall answer them in Part III of this paper (see Theorem III and the subsequent heuristic discussion). We shall see that a given I-game G will have a C-game analogue G^* only if G itself satisfies certain consistency requirements. On the other hand, if such a C-game analogue G^* exists for G then it will be "essentially" unique (in the sense that, in cases where two different C-games G_1^*, and G_2^* are both Bayes-equivalent to a given I-game G, it will make no difference whether we use G_1^* or G_2^* for the analysis of G). In the rest of the present Part I of this paper, we shall restrict our analysis to I-games G for which a Bayes-equivalent C-game analogue G^* does exist.

As we shall make considerable use of Bayes-equivalence relationships between certain I-games G and certain C-games G^* given in standard form, it will be convenient to have a short designation for the latter. Therefore, we shall introduce the term *Bayesian games* as a shorter name for C-games G^* given in standard form. Depending on the nature of the I-game G we shall be dealing with in particular cases, we shall also speak of Bayesian two-person zero-sum games, Bayesian bargaining games, etc.

6.

In view of the important role that Bayesian games will play in our analysis, we shall now consider two alternative (but essentially equivalent) models for these games, which for some purposes will usefully supplement the model we have defined in Sections 4 and 5.

So far we have defined a Bayesian game G^* as a game where each player's payoff $x_i = V_i(s_1, \cdots, s_n; c_1, \cdots, c_n)$ depends, not only on the strategies $s_1, \cdots, s_n$ chosen by the n players, but also on some random vectors (information vectors or attribute vectors) $c_1, \cdots, c_n$. It has also been assumed that all players will know the joint probability distribution $R^*(c_1, \cdots, c_n)$ of these random vectors, but that in general the actual value of any given vector c_i will be known only to player i himself whose information vector (or attribute vector) it represents. This model will be called the *random-vector model* for Bayesian games.

An alternative model for Bayesian games can be described as follows. The actual individuals who will play the roles of players $1, \cdots, n$ in game G^* on any given occasion, will be selected *by lot* from certain populations $\Pi_1, \cdots, \Pi_n$ of potential players. Each population Π_i from which a given player i is to be selected will contain individuals with a variety of different attributes, so that *every* possible combination of attributes (i.e., every possible "type" of player i), corresponding to any specific value $c_i = c_i^0$ that the attribute vector c_i can take in the game, will be represented in this population Π_i. If in population Π_i a given individual's attribute vector c_i has the specific value $c_i = c_i^0$, then we shall say that he belongs to the *attribute class* c_i^0. Thus, each population Π_i will be partitioned into that many attribute classes as the number of different values that player i's attribute vector c_i can take in the game.

As to the random process selecting n players from the n populations $\Pi_1, \cdots, \Pi_n$, we shall assume that the probability of players $1, \cdots, n$ being selected from any specific n-tuple of attribute classes $c_1^0, \cdots, c_n^0$ will be governed[12] by the probability distribution $R^*(c_1, \cdots, c_n)$. We shall also retain the assumptions that this probability distribution R^* will be known to all n players, and that each player i will also know his own attribute class $c_i = c_i^0$ but, in general, will not know the other players' attribute classes $c_1 = c_1^0, \cdots, c_{i-1} = c_{i-1}^0$, $c_{i+1} = c_{i+1}^0, \cdots, c_n = c_n^0$. As in this model the lottery by which the players are selected occurs prior to any other move in the game, it will be called the *prior-lottery model* for Bayesian games.

Let G be a real-life game situation where the players have incomplete information, and let G^* be a Bayesian game Bayes-equivalent to G (as assessed by a given player j). Then this Bayesian game G^*, interpreted in terms of the prior-lottery model, can be regarded as a possible representation (of course a highly schematic representation) of the real-life random social process which has actually created this game situation G. More particularly, the prior-lottery model pictures this social process as it would be seen by an *outside observer* having information about some aspects of the situation but lacking information about some other aspects. He could not have enough information to predict the attribute vectors $c_1 = c_1^0, \cdots, c_n = c_n^0$ of the n individuals to be selected by this social process to play the roles of players $1, \cdots, n$ in game situation G. But he would have to have enough information to predict the joint probability distribution R^* of the attribute vectors $c_1, \cdots, c_n$ of these n individuals, and, of

12 Under our assumptions in general the selection of players $1, \cdots, n$ from the respective populations $\Pi_1, \cdots, \Pi_n$ will not be statistically independent events because the probability distribution $R^*(c_1, \cdots, c_n)$ in general will not permit of factorization into n independent probability distributions $R_1^*(c_1), \cdots, R_n^*(c_n)$. Therefore, strictly speaking, our model postulates *simultaneous* random selection of a whole player n-tuple from a population Π of all possible player n-tuples, where Π is the Cartesian product $\Pi = \Pi_1 \times \cdots \times \Pi_n$.

course, also to predict the mathematical form of the payoff functions $V_1, \cdots, V_n$. (But he could not have enough information to predict the payoff functions $U_1, \cdots, U_n$ because this would require knowledge of the attribute vectors of all n players.)

In other words, the hypothetical observer must have exactly all the information *common* to the n players, but must not have access to any additional information private to any one player (or to any sectional group of players—and, of course, he must not have access to any information inaccessible to all of the n players). We shall call such an observer a *properly informed* observer. Thus, the prior-lottery model for Bayesian games can be regarded as a schematic representation of the relevant real-life social process as seen by a properly informed outside observer.

As an example, let us again consider the price-competition game G with incomplete information, and the corresponding Bayesian game G^*, discussed in Section 1 above. Here each player's attribute vector c_i will consist of the variables defining his cost functions, his financial resources, and his facilities to collect information about the other player.[13] Thus, the prior-lottery model of G^* will be a model where each player is chosen at random from some population of possible players with different cost functions, different financial resources, and different information-gathering facilities. We have argued that such a model can be regarded as a schematic representation of the real-life social process which has actually produced the assumed competitive situation, and has actually determined the cost functions, financial resources, and information-gathering facilities, of the two players.

Dr. Selten has suggested[14] a third model for Bayesian games, which we shall call the *Selten model* or the *posterior-lottery model*. Its basic difference from the prior-lottery model consists in the assumption that the lottery selecting the active participants of the game will take place only *after* each potential player has chosen the strategy he would use in case he were in fact selected for active participation in the game.

More particularly, suppose that the attribute vector c_i of player i ($i = 1, \cdots, n$) can take k_i different values in the game. (We shall assume that all k_i's are finite but the model can be easily extended also to the infinite case.) Then, instead of having one randomly selected player i in the game, we shall assume that the role of player i will be played at the same time by k_i different players, each of them representing a different value of the attribute vector c_i. The set of all k_i individuals playing the role of player i in the game will be called the *role class* i. Different individuals in the same role class i will be distinguished by subscripts as players $i_1, i_2, \cdots$. Under these assumptions, obviously the total number of players in the game will not be n but rather will be the larger (usually much larger) number

$$K = \sum_{i=1}^{n} k_i . \tag{6.1}$$

[13] Cf. Footnote 4 above.

[14] In private communication (cf. Footnote 1 above).

It will be assumed that each player i_m from a given role class i will choose some strategy s_i from player i's strategy space S_i. Different members of the same role class i may (but need not) choose different strategies s_i from this strategy space S_i.

After all K players have chosen their strategies, one player i_m from each role class i will be randomly selected as an *active player*. Suppose that the attribute vectors of the n active players so selected will be $c_1 = c_1^0, \cdots, c_n = c_n^0$, and that these players, prior to their selection, have chosen the strategies $s_1 = s_1^0, \cdots, s_n = s_n^0$. Then each active player i_m, selected from role class i, will obtain a payoff

$$x_i = V_i(s_1^0, \cdots, s_n^0; c_1^0, \cdots, c_n^0). \tag{6.2}$$

All other $(K - n)$ players not selected as active players will obtain *zero* payoffs.

It will be assumed that, when the n active players are randomly selected from the n role classes, the probability of selecting individuals with any specific combination of attribute vectors $c_1 = c_1^0, \cdots, c_n = c_n^0$ will be governed by the probability distribution $R^*(c_1, \cdots, c_n)$.[15]

It is easy to see that in all three models we have discussed—in the random-vector model, the prior-lottery model, and the posterior-lottery model—the players' payoff functions, the information available to them, and the probability of any specific event in the game, are all essentially the same.[16] Consequently,

[15] In actual fact, we could just as well assume that each player would choose his strategy only *after* the lottery, and *after* being informed whether this lottery has selected him as an active player or not. (Of course if we made this assumption then players not selected as active players could simply forget about choosing a strategy at all.) From a game-theoretical point of view this assumption would make no real difference so long as each active player would have to choose his strategy without being told the names of the *other* players selected as active players, and in particular without being told the attribute classes to which these other active players would belong.

Thus the fundamental theoretical difference between our second and third models is not so much in the actual *timing* of the postulated lottery as such. It is not so much in the fact that in one case the lottery precedes, and in the other case it follows, the players' strategy choices. The fundamental difference rather lies in the fact that our second model (like our first) conceives of the game as an *n-person* game, in which only the n active players are formally "players of the game"; whereas our third model conceives of the game as a *K-person* game, in which both the active and the inactive players are formally regarded as "players". Yet, to make it easier to avoid confusion between the two models, it is convenient to assume also a difference in the actual timing of the assumed lottery.

[16] Technically speaking, the players' effective payoff functions under the posterior-lottery model are not quite identical with their payoff functions under the other two models, but this difference is immaterial for our purposes. Under the posterior-lottery model, let $r = r_i(c_i^0)$ be the probability (marginal probability) that a given player i_m with attribute vector $c_i = c_i^0$ will be selected as the active player from role class i. Then player i_m will have the probability r of obtaining a payoff corresponding to the payoff function V_i and will have the probability $(1 - r)$ of obtaining a zero payoff whereas under the other two models each player i will *always* obtain a payoff corresponding to the payoff function V_i. Consequently, under the posterior-lottery model player i_m's expected payoff will be only r times $(0 < r \leqq 1)$ the expected payoff he could anticipate under the other two models. However, under most game-theoretical solution concepts (and in particular under all solution concepts we would ourselves choose for analyzing game situations), the solution of the game

all three models can be considered to be essentially equivalent. But, of course, formally they represent quite different game-theoretical models, as the random-vector model corresponds to an n-person game G^* with complete information, whereas the posterior-lottery model corresponds to a *K-person* game G^{**} with complete information. In what follows, unless the contrary is indicated, by the term "Bayesian game" we shall always mean the n-person game G^* corresponding to the random-vector model, whereas the K-person game G^{**} corresponding to the posterior-lottery model will be called the *Selten game.*

In contrast to the other two models, the prior-lottery model formally does not qualify as a true "game" at all because it assumes that the n players are selected by a chance move representing the first move of the game, whereas under the formal game-theoretical definition of a game the identity of the players must always be known from the very beginning, *before* any chance move or personal move has occurred in the game.

Thus, we may characterize the situation as follows. The real-life social process underlying the I-game G we are considering is best represented by the prior-lottery model. But the latter does not correspond to a true "game" in a game-theoretical sense. The other two models are two alternative ways of converting the prior-lottery model into a true "game". In both cases this conversion entails a price in the form of introducing some unrealistic assumptions. In the case of the posterior-lottery model corresponding to the Selten game G^{**}, the price consists in introducing $(K - n)$ fictitious players in addition to the n real players participating in the game.[17]

In the case of the random-vector model corresponding to the Bayesian game G^*, there are no fictitious players, but we have to pay the price of making the unrealistic assumption that the attribute vector c_i of each player i is determined by a chance move *after* the beginning of the game—which seems to imply that player i will be in existence for some period of time, however short, during which *he will not know yet* the specific value $c_i = c_i^0$ his attribute vector c_i will take. So long as the Bayesian game G^* corresponding to the random-vector model is being considered in its standard form, this unrealistic assumption makes very little difference. But, as we shall see, when we convert G^* into its normal form this unrealistic assumption implied by our model does cause certain technical difficulties, because it seems to commit us to the assumption that each player can choose his normalized strategy (i.e., his strategy for the normal-form version of G^*) *before* he learns the value of his own attribute vector c_i. An important advantage of the Selten game G^{**} lies in the fact that it does not require this particular unrealistic assumption: we are free to assume that every player i_m

---

will remain invariant if the players' payoff functions are multiplied by positive constants r (even if different constants r are used for different players).

In any case, the posterior-lottery model can be made completely equivalent to the other two models if we assume that each active player i_m will obtain a payoff corresponding to the payoff function $V_i/r_i(c_i^0)$, instead of obtaining a payoff corresponding to the payoff function V_i as such [as prescribed by equation (6.2)].

[17] This will be true even if we change the timing of the assumed lottery in Selten's model (see Footnote 15 above).

will know his own attribute vector c_i from the very beginning of the game, and will always choose his own strategy in light of this information.[18]

Thus, as analytical tools used in the analysis of a given I-game G, both the Bayesian game G^* and the Selten game G^{**} have their own advantages and disadvantages.[19]

7.

Let G be an I-game given in standard form, and let G^* be a Bayesian game Bayes-equivalent to G. Then we define the normal form $\mathfrak{N}(G)$ of this I-game G as being the normal form $\mathfrak{N}(G^*)$ of the Bayesian game G^*.

To obtain this normal form we first have to replace the strategies s_i of each player i by *normalized strategies* s_i^*. A normalized strategy s_i^* can be regarded as a conditional statement specifying the strategy $s_i = s_i^*(c_i)$ that player i would use if his information vector (or attribute vector) c_i took any given specific value. Mathematically, a normalized strategy s_i^* is a function from the range space $C_i = \{c_i\}$ of vector c_i to player i's strategy space $S_i = \{s_i\}$. The set of all possible such functions s_i^* is called player i's normalized-strategy space $S_i^* = \{s_i^*\}$. In contrast to these normalized strategies s_i^*, the strategies s_i available to player i in the standard form of the game will be called his *ordinary strategies*.

If in a given game the information vector c_i of a certain player i can take only k different values (with k finite) so that we can write

$$c_i = c_i^1, \cdots, c_i^k, \tag{7.1}$$

then any normalized strategy s_i^* of this player can be defined simply as a k-tuple of ordinary strategies

$$s_i^* = (s_i^1, \cdots, s_i^k), \tag{7.2}$$

where $s_i^m = s_i^*(c_i^m)$, with $m = 1, \cdots, k$, denotes the strategy that player i would use in the standard form of the game if his information vector c_i took the specific value $c_i = c_i^m$. In this case player i's normalized strategy space $S_i^* = \{s_i^*\}$ will be the set of all such k-tuples s_i^*, that is, it will be the k-times repeated Cartesian product of player i's ordinary strategy space S_i by itself. Thus we can write $S_i^* = S_i^1 \times \cdots \times S_i^k$ with $S_i^1 = \cdots = S_i^k = S_i$.

Under either of these definitions, the normalized strategies s_i^* will not have the nature of *mixed* strategies but rather that of *behavioral* strategies. Never-

[18] Moreover, as Selten has pointed out, his model also has the advantage that it can be extended to the case where the subjective probability distributions $R_1, \cdots, R_n$ of a given I-game G fail to satisfy the required consistency conditions, so that no probability distribution R^* satisfying equation (5.3) will exist, and therefore no Bayesian game G^* Bayes-equivalent to G can be constructed at all. In other words, for any I-game G we can always define an equivalent Selten game G^{**}, even in cases where we cannot define an equivalent Bayesian game G^*. (See Section 15, Part III.)

[19] We have given intuitive reasons why a Bayesian game G^* and the corresponding Selten game G^{**} are essentially equivalent. For a more detailed and more rigorous game-theoretical proof the reader is referred to a forthcoming paper by Reinhard Selten.

theless, these definitions are admissible because any game G^* in standard form is a game of perfect recall, and so it will make no difference whether the players are assumed to use behavioral strategies or mixed strategies [4].

Equation (3.15) can now be written as

$$x_i = V_i(s_1^*(c_1), \cdots, s_n^*(c_n); \quad c_1, \cdots, c_n) = V_i(s_1^*, \cdots, s_n^*; c). \tag{7.3}$$

In order to obtain the normal form $\mathfrak{N}(G) = \mathfrak{N}(G^*)$, all we have to do now is to take expected values in equation (7.3) with respect to the whole random vector c, in terms of the basic probability distribution $R^*(c)$ of the game. We define

$$\mathcal{E}(x_i) = W_i(s_1^*, \cdots, s_n^*) = \int_C V_i(s_1^*, \cdots, s_n^*; c)\, d_c R^*(c). \tag{7.4}$$

Since each player will treat his expected payoff as his effective payoff from the game, we can replace $\mathcal{E}(x_i)$ simply by x_i and write

$$x_i = W_i(s_1^*, \cdots, s_n^*). \tag{7.5}$$

We can now define the normal form of games G and G^* as the ordered set

$$\mathfrak{N}(G) = \mathfrak{N}(G^*) = \{S_1^*, \cdots, S_n^*; W_1, \cdots, W_n\}. \tag{7.6}$$

Compared with equations (3.18) and (4.2) defining the standard forms of these two games, in equation (7.6) the ordinary strategy spaces S_i have been replaced by the normalized strategy spaces S_i^*, and the ordinary payoff functions V_i have been replaced by the normalized payoff functions W_i. On the other hand, the range spaces C_i as well as the probability distributions R_i or R^* have been omitted because the normal form $\mathfrak{N}(G) = \mathfrak{N}(G^*)$ of games G and G^* does not any more involve the random vectors $c_1, \cdots, c_n$.

This normal form, however, has the disadvantage that it is defined in terms of the players' *unconditional* payoff expectations $\mathcal{E}(x_i) = W_i(s_1^*, \cdots, s_n^*)$, though in actual fact each player's strategy choice will be governed by his *conditional* payoff expectation $\mathcal{E}(x_i | c_i)$, because he will always know his own information vector c_i at the time of making his strategy choice. This conditional expectation can be defined as

$$\mathcal{E}(x_i | c_i) = Z_i(s_1^*, \cdots, s_n^* | c_i) = \int_{C^i} V_i(s_1^*, \cdots, s_n^*; c_i, c^i) d_{(c^i)} R^*(c^i | c_i). \tag{7.7}$$

To be sure, it can be shown (see Theorem I of Section 8, Part II) that if any given player i maximizes his *unconditional* payoff expectation W_i, then he will also be maximizing his *conditional* payoff expectation $Z_i(\cdot | c_i)$ for each specific value of c_i, with the possible exception of a small set of c_i values which can occur only with probability zero. In this respect our analysis bears out von Neumann and Morgenstern's Normalization Principle [7, pp. 79–84], according to which the players can safely restrict their attention to the normal form of the game when they are making their strategy choices.

However, owing to the special nature of Bayesian games, the Normalization

Principle has only restricted validity for them, and their normal form $\mathfrak{N}(G^*)$ must be used with special care, because solution concepts based on uncritical use of the normal form may give counterintuitive results (see Section 11 of Part II of this paper). In view of this fact, we shall introduce the concept of a semi-normal form. The *semi-normal form* $\mathcal{S}(G) = \mathcal{S}(G^*)$ of games G and G^* will be defined as a game where the players' strategies are the normalized strategies s_i^* described above, but where their payoff functions are the *conditional* payoff-expectation functions $Z_i(\cdot \mid c_i)$ defined by equation (7.7). Formally we define the semi-normal form of the games G and G^* as the ordered set

$$(7.8) \quad \mathcal{S}(G) = \mathcal{S}(G^*) = \{S_1^*, \cdots, S_n^*; C_1, \cdots, C_n; Z_1, \cdots, Z_n; R^*\}.$$

As the semi-normal form, unlike the normal form, does involve the random vectors $c_1, \cdots, c_n$, now the range spaces $C_1, \cdots, C_n$, and the probability distribution R^*, which have been omitted from equation (7.6), reappear in equation (7.8).

Instead of von Neumann and Morgenstern's Normalization Principle, we shall use only the weaker Semi-normalization Principle (Postulate 2 below), which is implied by the Normalization Principle but which does not itself imply the latter:

Postulate 2. Sufficiency of the Semi-normal Form. The solution of any Bayesian game G^*, and of the Bayes-equivalent I-game G, can be defined in terms of the semi-normal form $\mathcal{S}(G^*) = \mathcal{S}(G)$, without going back to the standard form of G^* or of G.

References

1. Robert J. Aumann, "On Choosing a Function at Random," in Fred B. Wright (editor), *Symposium on Ergodic Theory*, New Orleans: Academic Press, 1963, pp. 1–20.
2. ——, "Mixed and Behavior Strategies in Infinite Extensive Games," in M. Dresher, L. S. Shapley, and A. W. Tucker (editors), *Advances in Game Theory*, Princeton: Princeton University Press, 1964, pp. 627–650.
3. John C. Harsanyi, "Bargaining in Ignorance of the Opponent's Utility Function," *Journal of Conflict Resolution*, 6 (1962), pp. 29–38.
4. H. W. Kuhn, "Extensive Games and the Problem of Information," in H. W. Kuhn and A. W. Tucker (editors), *Contributions to the Theory of Games*, Vol. II, Princeton: Princeton University Press, 1953, pp. 193–216.
5. J. C. C. McKinsey, *Introduction to the Theory of Games*, New York: McGraw-Hill, 1952.
6. Leonard J. Savage, *The Foundations of Statistics*, New York: John Wiley and Sons, 1954.
7. John von Neumann and Oskar Morgenstern, *Theory of Games and Economic Behavior*, Princeton: Princeton University Press, 1953.

CHAPTER 7

GAMES WITH INCOMPLETE INFORMATION PLAYED BY "BAYESIAN" PLAYERS

PART II. BAYESIAN EQUILIBRIUM POINTS*†[1]

Part I of this paper has described a new theory for the analysis of games with incomplete information. It has been shown that, if the various players' subjective probability distributions satisfy a certain mutual-consistency requirement, then any given game with incomplete information will be equivalent to a certain game with complete information, called the "Bayes-equivalent" of the original game, or briefly a "Bayesian game."

Part II of the paper will now show that any Nash equilibrium point of this Bayesian game yields a "Bayesian equilibrium point" for the original game and conversely. This result will then be illustrated by numerical examples, representing two-person zero-sum games with incomplete information. We shall also show how our theory enables us to analyze the problem of exploiting the opponent's erroneous beliefs.

However, apart from its indubitable usefulness in locating Bayesian equilibrium points, we shall show it on a numerical example (the Bayes-equivalent of a two-person cooperative game) that the normal form of a Bayesian game is in many cases a highly unsatisfactory representation of the game situation and has to be replaced by other representations (e.g., by the semi-normal form). We shall argue that this rather unexpected result is due to the fact that Bayesian games must be interpreted as games with "delayed commitment" whereas the normal-form representation always envisages a game with "immediate commitment."

8.

Let G be an I-game (as assessed by player j), and let G^* be a Bayesian game Bayes-equivalent to G and having the function $R^*(c_1, \cdots, c_n)$ as its basic probability distribution. Let s_i^* be a normalized strategy of player i in game G. Suppose that, for some specific value $c_i = c_i^0$ of player i's attribute vector c_i, the (ordinary) strategy $s_i = s_i^*(c_i)$ selected by this normalized strategy s_i^* maximizes player i's conditional payoff expectation

$$(8.1) \qquad \mathcal{E}(x_i \mid c_i^0) = Z_i(s_1^*, \cdots, s_i^*, \cdots, s_n^* \mid c_i^0)$$

if the normalized strategies $s_1^*, \cdots, s_{i-1}^*, s_{i+1}^*, \cdots, s_n^*$ of the other $(n-1)$ players are kept constant. Then we shall say that this normalized strategy s_i^* of player i is a *best reply at the point* $c_i = c_i^0$ to the other players' normalized strategies $s_1^*, \cdots, s_{i-1}^*, s_{i+1}^*, \cdots, s_n^*$.

* Received June 1965 and revised May 1966.

† Part I of "Games with Incomplete Information Played by 'Bayesian' Players" appeared in the preceding issue of *Management Science*. Part III will appear in the next issue. A "Table of Contents" and "Glossary" appeared with Part I.

[1] The numbering of sections and theorems will be consecutive in Parts I, II, and III of this paper. The author wishes again to express his indebtedness to the persons and institutions listed in Footnote 1 of Part I.

Now suppose that s_i^* actually possesses this best-reply property at *all* possible values of the attribute vector c_i, with the possible exception of a small set C_i^* of c_i values, having a total probability mass zero. That is, we are assuming that the event $E = \{c_i \in C_i^*\}$ is being assigned zero probability by the marginal probability distribution

$$(8.2) \qquad R^*(c_i) = \int_{C^i} d_{(c^i)} R^*(c_i, c^i)$$

derived from the basic probability distribution $R^*(c_1, \cdots, c_i, \cdots, c_n) = R^*(c_i, c^i)$. Then we shall say that s_i^* is *almost uniformly* a best reply to $s_1^*, \cdots, s_{i-1}^*, s_{i+1}^*, \cdots, s_n^*$.

Finally suppose that in a given normalized strategy n-tuple $s^* = (s_1^*, \cdots, s_n^*)$ *every* component s_i^* $(i = 1, \cdots, n)$ is almost uniformly a best reply to the other $(n - 1)$ components $s_1^*, \cdots, s_{i-1}^*, s_{i+1}^*, \cdots, s_n^*$. Then we shall say that s^* is a *Bayesian equilibrium point in game G.*

We can now state the following theorem.

Theorem I. Let G be an I-game, and let G^* be a Bayesian game Bayes-equivalent to G (as assessed by player j). In order that any given n-tuple of normalized strategies $s^* = (s_1^*, \cdots, s_n^*)$ be a Bayesian equilibrium point in game G, it is both *sufficient* and *necessary* that in the normal form $\mathfrak{N}(G^*)$ of game G^* this n-tuple s^* be an equilibrium point in Nash's sense [3, 5].

Proof. In view of equations (5.3), (7.4) and (7.7) we can write

$$(8.3) \qquad W_i(s_1^*, \cdots, s_i^*, \cdots, s_n^*) = \int_{C_i} Z_i(s_1^*, \cdots, s_i^*, \cdots, s_n^* \mid c_i)\, d_{(c_i)} R^*(c_i)$$

where $R^*(c_i)$ is again the marginal probability distribution defined by equation (8.2).

To prove the sufficiency part of the theorem, suppose that s^* is not a Bayesian equilibrium point in G. This means that at least one of the components of s^*, say, s_i^*, is not an almost uniformly best reply to the other components of s^*. Consequently, there must be some set C_i^* of possible c_i values at which the function $Z_i(\cdot \mid c_i)$ could be increased if we replaced the ordinary strategies $s_i = s_i^*(c_i)$ selected by the normalized strategy s_i^*, with some alternative ordinary strategies $s'_i = s_i^{**}(c_i)$. That is, for all $c_i \in C_i^*$ we must have

$$(8.4) \qquad Z_i[s_i^{**}(c_i), (s^*)^i] > Z_i[s_i^*(c_i), (s^*)^i],$$

where $(s^*)^i$ denotes the $(n - 1)$-tuple

$$(8.5) \qquad (s^*)^i = (s_1^*, \cdots, s_{i-1}^*, s_{i+1}^*, \cdots, s_n^*).$$

Moreover, the probability distribution $R^*(c_i)$ must associate a non-zero total probability mass with this set C_i^*.

Now let s_i^{***} be a normalized strategy of player i, selecting the strategies

$s_i^{**}(c_i)$ for all points c_i in set C_i^*, but coinciding with s_i^* everywhere else. That is,

(8.6) $$s_i^{***}(c_i) = s_i^{**}(c_i) \qquad \text{for all } c_i \in C_i^*$$
$$s_i^{***}(c_i) = s_i^{*}(c_i) \qquad \text{for all } c_i \notin C_i^*.$$

Then, in view of equations (8.3) and (8.4), we must have

(8.7) $$W_i[s_i^{***}, (s^*)^i] > W_i[s_i^*, (s^*)^i].$$

Consequently, in game $\mathfrak{N}(G^*)$ the n-tuple s^* is not an equilibrium point in Nash's sense. Thus, if s^* is not a Bayesian equilibrium point in G, then it cannot be a Nashian equilibrium point in $\mathfrak{N}(G^*)$, contrary to our hypothesis.

To prove the necessity part of the theorem, suppose that s^* is not an equilibrium point in Nash's sense in game $\mathfrak{N}(G^*)$. This means that at least one component of s^*, say, s_i^*, can be replaced by some alternative normalized strategy s_i^{***} in such a way as to increase the numerical value of the function W_i. That is, there must exist some normalized strategy s_i^{***} satisfying (8.4). But in view of equation (8.3) this implies that, for some set C_i^* of possible c_i values, we must have

(8.8) $$Z_i[s_i^{***}(c_i), (s^*)^i] > Z_i[s_i^*(c_i), (s^*)^i].$$

Moreover, this set C_i^* must have a non-zero total probability mass associated with it under the probability distribution $R^*(c_i)$. Consequently, s^* will not be a Bayesian equilibrium point in game G. Thus, if s^* is not a Nashian equilibrium point in $\mathfrak{N}(G^*)$ then it cannot be a Bayesian equilibrium point in game G, contrary to our hypothesis. This completes the proof.

In view of Nash's [3, 5] equilibrium-point theorem and its various generalizations, any I-game satisfying some rather mild regularity conditions will have at least one Bayesian equilibrium point. In particular:

Theorem II. Let G be a standard-form I-game (as assessed by player j), for which a Bayes-equivalent Bayesian game G^* exists. Suppose that G is a finite game, in which each player i has only a finite number of (ordinary) pure strategies s_i. Then G will have at least one Bayesian equilibrium point.

Proof. Even if each player i has only a finite number of pure strategies s_i in the standard form of game G, he will have an infinity of pure normalized strategies s_i^* in the normal form $\mathfrak{N}(G)$ of the game if his information vector c_i can take an infinity of different values. But it is easy to see that the normalized behavioral strategies of $\mathfrak{N}(G)$ will even then always satisfy the continuity and contractibility requirements of Debreu's generalized equilibrium-point theorem [1]. Hence, $\mathfrak{N}(G)$ will always contain an equilibrium point s^* in Nash's sense, and so by Theorem I s^* will be a Bayesian equilibrium point for game G itself as originally given in standard form.

9.

Now we propose to discuss two numerical examples, partly to illustrate the concept of Bayesian equilibrium points as defined in the last section, and partly to illustrate the nature of the solution concept our approach yields in the simplest class of I-games, viz., two-person zero-sum games with incomplete information.

Suppose that in a given two-person zero-sum game G player 1 can belong to either of two attribute classes $c_1 = a^1$ and $c_1 = a^2$, where his belonging to class a^1 may be intuitively interpreted that he is in a "weak" position (e.g., in terms of military equipment, or man power, or the economic resources available to him, etc.) whereas his belonging to class a^2 may be interpreted that he is in a "strong" position. Likewise, player 2 can belong to either of two attribute classes $c_2 = b_1$ and $c_2 = b^2$, where class b^1 again may be taken to indicate that he is in a "weak" position whereas class b^2 indicates that he is in a "strong" position. Thus, we have altogether four possible cases in that the two players may belong to classes (a^1, b^1) or (a^1, b^2) or (a^2, b^1) or (a^2, b^2).

In any one of these cases, each player has two pure strategies, called $s_1 = y^1$ and $s_1 = y^2$ in the case of player 1, and called $s_2 = z^1$ and $s_2 = z^2$ in the case of player 2. (In terms of an intuitive interpretation, y^1 and z^1 may represent "more aggressive" strategies while y^2 and z^2 may represent "less aggressive" ones.) These strategies will yield the following payoffs to player 1 in the four possible cases (player 2's payoff is always the negative of player 1's).

TABLE 1

In case (a^2, b^1): Saddle point at (y^1, z^1).

	z^1	z^2
y^1	2	5
y^2	−1	20

TABLE 2

In case (a^1, b^2): Saddle point at (y^2, z^1).

	z^1	z^2
y^1	−24	−36
y^2	0	24

TABLE 3

In case (a^2, b^1): Saddle point at (y^1, z^2).

	z^1	z^2
y^1	28	15
y^2	40	4

TABLE 4

In case (a^2, b^2): Saddle point at (y^1, z^1).

	z^1	z^2
y^1	12	20
y^2	2	13

(For the sake of simplicity, in all four cases we have chosen payoff matrices with saddle points in pure strategies.)

Since both players' attribute vectors c_1 and c_2 can take only a finite number (viz. two) alternative values, all probability distributions of the game can be represented by the corresponding probability-mass functions. In particular, the basic probability distribution $R^*(c_1, c_2)$ can be represented by the matrix of joint probabilities r_{km} = Prob ($c_1 = a^k$ and $c_2 = b^m$) with $k, m = 1, 2$. We shall assume that this matrix $\{r_{km}\}$ will be as follows.

TABLE 5

	$c_2 = b^1$	$c_2 = b^2$
$c_1 = a^1$	$r_{11} = 0.40$	$r_{12} = 0.10$
$c_1 = a^2$	$r_{21} = 0.20$	$r_{22} = 0.30$

This matrix yields the following conditional probabilities, corresponding to the two *rows*:

TABLE 6

	$c_2 = b^1$	$c_2 = b^2$
Prob ($c_2 = b^m \mid c_1 = a^1$)	$p_{11} = 0.8$	$p_{12} = 0.2$
Prob ($c_2 = b^m \mid c_1 = a^2$)	$p_{21} = 0.4$	$p_{22} = 0.6$

Here $p_{km} = r_{km}/(r_{k1} + r_{k2})$. The first row of this matrix states the conditional probabilities (subjective probabilities) that player 1 will assign to the two alternative hypotheses about player 2's attribute class (viz. to $c_2 = b^1$ and to $c_2 = b^2$) when his own attribute class is $c_1 = a^1$. The second row states the probabilities he will assign to these two hypotheses when his own attribute class is $c_1 = a^2$.

In contrast, the conditional probabilities corresponding to the two *columns* of matrix $\{r_{km}\}$ are:

TABLE 7

	Prob ($c_1 = a^k \mid c_2 = b^1$)	Prob ($c_1 = a^k \mid c_2 = b^2$)
$c_1 = a^1$	$q_{11} = 0.67$	$q_{12} = 0.25$
$c_1 = a^2$	$q_{21} = 0.33$	$q_{22} = 0.75$

Here $q_{km} = r_{km}/(r_{1m} + r_{2m})$. The first column of this matrix states the conditional probabilities (subjective probabilities) that player 2 will assign to the two alternative hypotheses about player 1's attribute class (viz. to $c_1 = a^1$ and to $c_1 = a^2$) when his own attribute class is $c_2 = b^1$. The second column states the probabilities he will use when his own attribute class is $c_2 = b^2$.

All three probability matrices stated in Tables 5 to 7 show that in our example the two players assume a positive correlation between their attribute classes, in the sense that it will tend to *increase* the probability a given player will assign to the hypothesis that his opponent is in a strong (or weak) position if he himself is in a strong (or weak) position. (Such a situation may arise if both players' power positions are determined largely by the same environmental influences; or if each player is making continual efforts to keep his own military, industrial, or other kind of strength more or less on a par with the other player's, etc. Of course in other situations there may be a negative correlation, or none at all.)

In this game, player 1's normalized pure strategies will be of the form $s_1^* = y^{nt} = (y^n, y^t)$, where $s_1 = y^n$ $(n = 1, 2)$ is the ordinary pure strategy player 1 would use if $c_1 = a^1$, whereas y^t $(t = 1, 2)$ is the ordinary pure strategy he would use if $c_1 = a^2$. Likewise, player 2's normalized pure strategies will be of the form $s_2^* = z^{uv} = (z^u, z^v)$, where $s_2 = z^u$ $(u = 1, 2)$ is the ordinary pure strategy player 2 would use if $c_2 = b^1$, whereas $s_2 = z^v$ is the ordinary pure strategy he would use if $c_2 = b^2$.

It is easy to verify that the payoff matrix W for the normal form $\mathfrak{N}(G)$ will be as follows.

TABLE 8

	z^{11}	z^{12}	z^{21}	z^{22}
y^{11}	7.6	8.8	6.2	7.4
y^{12}	7.0	9.1	1.0	3.1
y^{21}	8.8	13.6	14.6	19.4
y^{22}	8.2	13.9	9.4	15.1

The entries of this matrix are player 1's *total* (i.e., unconditional) payoff expectations corresponding to alternative pairs of normalized strategies (y^{nt}, z^{uv}). For instance, the entry for the strategy pair (y^{12}, z^{11}) is:

$$0.4 \times 2 + 0.1 \times (-24) + 0.2 \times 40 + 0.3 \times 2 = 7.0, \quad \text{etc.}$$

The only saddle point of this matrix W in pure strategies is at the point (y^{21}, z^{11}). (There is none in mixed strategies.) This means that in this game player 1's optimal strategy will be to use strategy y^2 if his attribute vector takes the value $c_1 = a^1$, and to use strategy y^1 if his attribute vector takes the value $c_1 = a^2$. On the other hand, player 2's optimal strategy will always be z^1, irrespective of the value of his attribute vector c_2.

It is easy to see that, in general, these optimal strategies are quite different from the optimal strategies associated with the four matrices stated in Tables 1 to 4. For instance, if the payoff matrix for case (a^1, b^1) is considered in isolation (see Table 1) then player 1's optimal strategy is y^1, though if we consider the game as a whole then his optimal normalized strategy y^{21} actually requires him to use strategy y^2 whenever his attribute vector has the value $c_1 = a^1$. Likewise, if the payoff matrix for case (a^2, b^1) is considered in isolation (see Table 3), then player 2's optimal strategy is z^2, though his optimal normalized strategy z^{11} for the game as a whole always requires him to use strategy z^1. These facts, of course, are not really surprising: they show only that the players will have to use different strategies if they do not know each other's attribute vectors, and different strategies if they do know them. For example, in case (a^1, b^1) player 1's optimal strategy would be y^1 only if he did know that his opponent's attribute vector is $c_2 = b^1$. But if all his information about player 2's attribute vector c_2 is what he can infer from the fact that his own attribute vector is $c_1 = a^1$, on the basis of the probability matrices of Tables 5 to 7, then his optimal strategy will be y^2 as prescribed by his optimal normalized strategy y^{21}, etc.

The strategy pair (y^{21}, z^{11}), being a saddle point, will also be an equilibrium point in Nash's sense for the normal form $\mathfrak{N}(G)$ of the game. Thus, so long as player 2 is using strategy z^{11}, player 1 will be maximizing his *total* payoff expectation W_1 by choosing strategy y^{21}. Conversely, so long as player 1 is using strategy y^{21}, player 2 will be maximizing his total payoff expectation W_2 by choosing strategy z^{11}.

What is more important, by Theorem I of the previous section, the strategy pair (y^{21}, z^{11}) will also be a *Bayesian equilibrium point* in the standard form of game G (indeed it will be the only one since it is the only saddle point of the normal-form game). Hence, so long as player 2 uses strategy z^{11}, player 1 will maximize, not only his *total* payoff expectation W_1, but also his *conditional* payoff expectation $Z_1(\cdot \mid c_1)$, *given* the value of his own attribute vector c_1. To verify this we first list player 1's conditional payoff expectations when his attribute vector takes the value $c_1 = a^1$:

TABLE 9

	z^{11}	z^{12}	z^{21}	z^{22}
y^1	−3.2	−5.6	−0.8	−3.2
y^2	−0.8	4.0	16.0	20.8

We shall call this matrix $Z_1(\cdot \mid a^1)$. We do find that the highest entry of column z^{11} is in row y^2, corresponding to the *first* component of the normalized strategy $y^{21} = (y^2, y^1)$.

Next we list player 1's conditional payoff expectations when his attribute vector takes the value $c_1 = a^2$:

TABLE 10

	z^{11}	z^{12}	z^{21}	z^{22}
y^1	18.4	23.2	13.2	18.0
y^2	17.2	23.8	2.8	9.4

We shall call this matrix $Z_1(\cdot \mid a^2)$. Now we find that the highest entry of column z^{11} is in row y^1, corresponding to the *second* component of the normalized strategy $y^{21} = (y^2, y^1)$, as desired. Thus in both cases the ordinary strategy representing the appropriate component of the normalized strategy y^{21} maximizes player 1's conditional payoff expectation against the normalized strategy z^{11} of player 2, in accordance with Theorem I.

We leave it to the reader to verify by a similar computation that conversely the ordinary strategy z^1 [which happens to be both the first and the second component of the normalized strategy $z^{11} = (z^1, z^1)$] does maximize player 2's conditional payoff expectation against the normalized strategy y^{21} of player 1, both in case player 2's attribute vector takes the value $c_2 = b^1$ and in case it takes the value $c_2 = b^2$, as required by Theorem I.

In this particular example, each player's optimal strategy not only maximizes his security level in terms of his *total* payoff expectations W_i (which is true by definition), but also maximizes his security level in terms of his *conditional* payoff expectations $Z_i(\cdot \mid c_i)$. For instance, in Table 9, strategy y^2 (which is the ordinary strategy prescribed for him by his optimal normalized strategy y^{21} in the case $c_1 = a^1$) is not only player 1's *best reply* to the normalized strategy z^{11} of his opponent, but is also his *maximin* strategy. Likewise, in Table 10, strategy y^1 is not only player 1's best reply to z^{11} but is also his maximin strategy.

However, we now propose to show by a counter example that this relationship has no general validity for two-person zero-sum games with incomplete information. Consider a game G^0 where each player's attribute vector can again take two possible values, viz. $c_1 = a^1, a^2$ and $c_2 = b^1, b^2$. Suppose the payoff matrices in the resulting four possible cases are as follows:

TABLE 11

In case (a^1, b^1):

	z^1	z^2
y^1	8	−8
y^2	0	−4

Saddle point at (y^2, z^2).

TABLE 12

In case (a^1, b^2):

	z^1	z^2
y^1	−4	8
y^2	0	12

Saddle point at (y^2, z^1).

TABLE 13

In case (a^2, b^1):

	z^1	z^2
y^1	−8	12
y^2	−12	16

Saddle point at (y^1, z^1).

TABLE 14

In case (a^2, b^2):

	z^1	z^2
y^1	4	−4
y^2	0	−8

Saddle point at (y^1, z^2).

Suppose that basic probability matrix of this game G^0 is as follows:

TABLE 15

	$c_2 = b^1$	$c_2 = b^2$
$c_1 = a^1$	$r_{11} = 0.25$	$r_{12} = 0.25$
$c_1 = a^2$	$r_{21} = 0.25$	$r_{22} = 0.25$

The payoff matrix W for the normal form $\mathfrak{N}(G^0)$ of G^0, listing player 1's *total* payoff expectations, will now be:

TABLE 16

	z^{11}	z^{12}	z^{21}	z^{22}
y^{11}	0	1	1	2
y^{12}	−2	−1	1	2
y^{21}	−1	0	3	4
y^{22}	−3	−2	3	4

The only saddle point of this matrix is at (y^{11}, z^{11}). Hence player 1's optimal normalized strategy is $y^{11} = (y^1, y^1)$ whereas player 2's is $z^{11} = (z^1, z^1)$. Player 1's *conditional* payoff expectations $Z_1(\cdot \mid c_1)$ in the case $c_1 = a^1$ will be as follows:

TABLE 17

	z^{11}	z^{12}	z^{21}	z^{22}
y^1	2	8	−6	0
y^2	0	6	−2	4

In accordance with Theorem I, as the table shows, player 1's best reply to the opponent's optimal normalized strategy z^{11} is strategy y^1, the ordinary strategy corresponding to his own optimal normalized strategy y^{11}. But player 1's security level would be maximized by y^2, and not by y^1.

This result has the following implication. Let G be a two-person zero-sum game with incomplete information. Then we may use the von Neumann-Morgenstern solution of the normal form $\mathfrak{N}(G)$ of this game as the solution for the game G itself as originally defined in standard form. But, in general, the only real justification for this procedure will lie in the fact that every pair of optimal normalized strategies s_1^* and s_2^* will represent a Bayesian equilibrium point in G. However, we cannot justify this approach by the maximin and minimax properties of s_1^* and s_2^* because in general these strategies will not have such properties in terms of the players' *conditional* payoff expectations $Z_i(\cdot \mid c_i)$, which are the quantities the players will be really interested in and which they will treat as their true anticipated payoffs from the game.

10.

We now propose to illustrate on a third numerical example how our model enables us to deal with the problem of *exploiting the opponent's erroneous beliefs*. Let G again be a two-person zero-sum game where each player can belong to two different attribute classes, viz. to class $c_1 = a^1$ or to class $c_1 = a^2$, and to class $c_2 = b^1$ or to class $c_2 = b^2$, respectively. We shall assume that in the four possible combinations of attribute classes the payoffs of player 1 will be again the same as stated in Tables 1 to 4 of the previous section. However, the basic probability matrix $\{r_{km}\}$ of the game will now be assumed to be:

TABLE 18

	$c_2 = b^1$	$c_2 = b^2$
$c_1 = a^1$	$r_{11} = 0.01$	$r_{12} = 0.00$
$c_1 = a^2$	$r_{21} = 0.09$	$r_{22} = 0.90$

Accordingly, player 1 will now use the following conditional probabilities (subjective probabilities):

TABLE 19

	$c_2 = b^1$	$c_2 = b^2$
Prob $(c_2 = b^m \mid c_1 = a^1)$	$p_{11} = 1.00$	$p_{12} = 0.00$
Prob $(c_2 = b^m \mid c_1 = a^2)$	$p_{21} = 0.09$	$p_{22} = 0.91$

On the other hand, player 2 will now use the following conditional probabilities (subjective probabilities):

TABLE 20

	$\text{Prob}(c_1 = a^k \mid c_2 = b^1)$	$\text{Prob}(c_1 = a^k \mid c_2 = b^2)$
$c_1 = a^1$	$q_{11} = 0.10$	$q_{12} = 0.00$
$c_1 = a^2$	$q_{21} = 0.90$	$q_{22} = 1.00$

The payoff matrix W for the normal form of the game, stating player 1's *total* payoff expectations, will now be:

TABLE 21

	z^{11}	z^{12}	z^{21}	z^{22}
y^{11}	13.34	20.54	11.20	19.40
y^{12}	5.42	15.32	2.21	12.11
y^{21}	13.31	20.51	12.35	19.55
y^{22}	5.39	15.29	2.36	12.26

The only saddle point of this game is at (y^{21}, z^{21}). Thus player 1's optimal strategy is $y^{21} = (y^2, y^1)$ whereas player 2's optimal strategy is $z^{21} = (z^2, z^1)$.

Now suppose that player 1 actually belongs to attribute class $c_1 = a^1$. Then our result concerning the players' optimal strategies can be interpreted as follows. As $c_1 = a^1$, by Table 19 player 1 will be able to infer that player 2 must belong to attribute class $c_2 = b^1$. Hence, by Table 20 player 1 will be able to conclude that player 2 is assigning a near-unity probability (viz. $q_{21} = .90$) to the *mistaken hypothesis* that player 1 belongs to class $c_1 = a^2$, which would mean that case (a^2, b^1) would apply (since player 2's own attribute class is $c_2 = b^1$). Therefore, player 1 will expect player 2 to choose strategy z^2, which is the strategy the latter would use by Table 3 if he thought that case (a^2, b^1) represented the actual situation.

If case (a^2, b^1) did represent the actual situation, then by Table 3 player 1's best reply to strategy z^2 would be strategy y^1. But under our assumptions player 1 will know that the actual situation is (a^1, b^1), rather than (a^2, b^1). Hence, by Table 1 his actual best reply to z^2 will be y^2, and not y^1. Thus we can say that by choosing strategy y^2 player 1 will be able to *exploit* player 2's *mistaken belief* that player 1 probably belongs to class $c_1 = a^2$.

As Table 1 shows, if player 2 knew the true situation and acted accordingly then player 1 could obtain no more than the payoff $x_1 = 2$. If player 2 did make the mistaken assumption that player 1's attribute vector was $c_1 = a^2$, but if player 1 did not exploit this mistake (i.e., if player 1 used the same strategy y^1 he would use against a better informed opponent) then his payoff would be $x_1 = 5$. Yet if he does exploit player 2's mistake and counters the latter's stragety z^2 by his strategy y^2, then he will obtain the payoff $x_1 = 20$.

11.

The purpose of our last numerical example will be to show that the normal form $\mathfrak{N}(G) = \mathfrak{N}(G^*)$ is often an unsatisfactory representation of the corresponding

I-game G (as originally defined in standard form), and is an unsatisfactory representation even for the corresponding Bayesian game G^*. More particularly, we shall argue that the solutions we would naturally associate with the normal-form game $\mathfrak{N}(G) = \mathfrak{N}(G^*)$ may often give highly counter-intuitive results for G or G^*.

Let G be a two-person bargaining game in the sense of Nash [4] in which the two players have to divide \$100. If they cannot agree on their shares, then both of them will obtain zero payoffs. It will be assumed that both players have linear utility functions for money, so that their dollar payoffs can be taken to be the same as their utility payoffs.

Following Nash, we shall assume that the game will be played as follows. Each player i ($i = 1, 2$) has to state his payoff demand y_i^* without knowing what the other player's payoff demand y_j^* will be. If $y_1^* + y_2^* \leqq 100$, then each player will obtain a payoff $y_i = y_i^*$ equal to his own demand. If $y_1^* + y_2^* > 100$, then both players will obtain zero payoffs $y_1 = y_2 = 0$.

To introduce incomplete information in the game, we shall assume that either player i may have to pay half of his gross payoff y_i to a certain secret organization or he may not; and that he himself will always know in advance whether this is the case or not, but that neither player will know whether the other has to make this payment or not. We shall say that player i's attribute vector takes the value $c_i = a^1$ if he has to make this payment, and takes the value $c_i = a^2$ if this is not the case.

Thus, we can define the net payoffs x_i of both players i as:

$$(11.1) \qquad x_i = \tfrac{1}{2}y_i \quad \text{if} \quad c_i = a^1$$

$$x_i = y_i \quad \text{if} \quad c_i = a^2 \qquad i = 1, 2.$$

Let X be the *payoff space* of the game, that is, the set of all possible net-payoff vectors $x = (x_1, x_2)$ that the players can achieve in the game.

If the players' attribute vectors are (a^1, a^1), then $X = X^{11}$ will be the triangular area satisfying the inequalities

$$(11.2) \qquad x_i \geqq 0 \qquad i = 1, 2$$

and

$$(11.3) \qquad x_1 + x_2 \leqq 50.$$

If these vectors are (a^1, a^2), then $X = X^{12}$ will be the triangle satisfying (11.2) as well as

$$(11.4) \qquad 2x_1 + x_2 \leqq 100.$$

If these vectors are (a^2, a^1), then $X = X^{21}$ will be the triangle satisfying (11.2) as well as

$$(11.5) \qquad x_1 + 2x_2 \leqq 100.$$

Finally, if these vectors are (a^2, a^2), then $X = X^{22}$ will be the triangle satisfy-

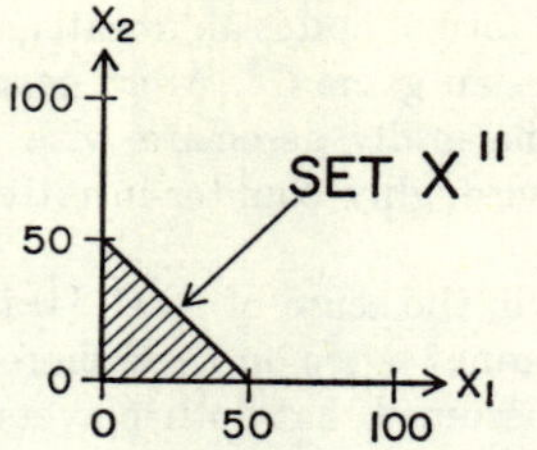

FIGURE 1

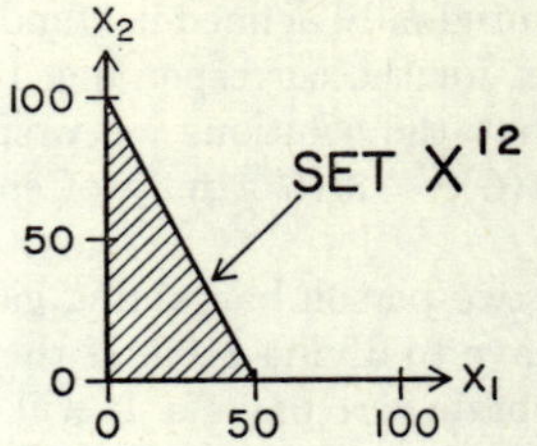

FIGURE 2

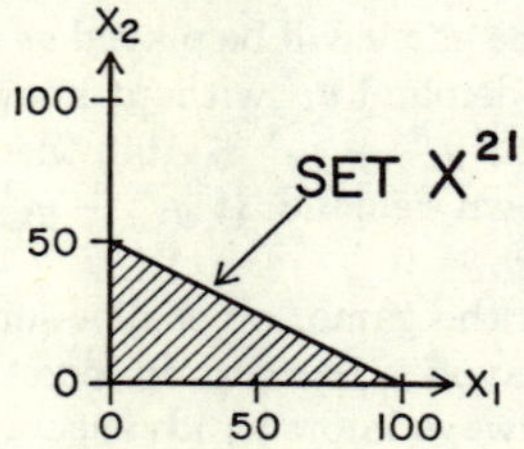

FIGURE 3

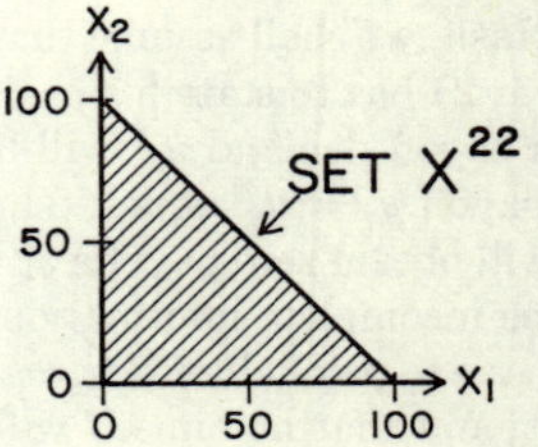

FIGURE 4

ing (11.2) as well as

(11.6) $$x_1 + x_2 \leqq 100$$

(see Figures 1 to 4).

Now suppose that the basic probability matrix $\{r_{km}\}$ of the game is as follows:

TABLE 22

	$c_2 = a^1$	$c_2 = a^2$
$c_1 = a^1$	$r_{11} = \frac{1}{2}\epsilon$	$r_{12} = \frac{1}{2} - \frac{1}{2}\epsilon$
$c_1 = a^2$	$r_{21} = \frac{1}{2} - \frac{1}{2}\epsilon$	$r_{22} = \frac{1}{2}\epsilon$

Here $r_{km} = \text{Prob}\ (c_1 = a^k \text{ and } c_2 = a^m)$ while ϵ is a very small positive number which for practical purposes can be taken to be zero.

Accordingly, both players i will use the following conditional probabilities (subjective probabilities):

TABLE 23

	$c_j = a^1$	$c_j = a^2$
$\text{Prob}(c_j = a^m \mid c_i = a^1)$	$p_{11} = q_{11} = \epsilon$	$p_{12} = q_{21} = 1 - \epsilon$
$\text{Prob}(c_j = a^m \mid c_i = a^2)$	$p_{21} = q_{12} = 1 - \epsilon$	$p_{22} = q_{22} = \epsilon$

That is, each player will assume a virtually complete negative correlation between his own attribute vector and the other player's: he will expect the other player to be free from the side-payment obligation if he himself is subject to it, and will expect the other player to be subject to this obligation if he himself is free from it.

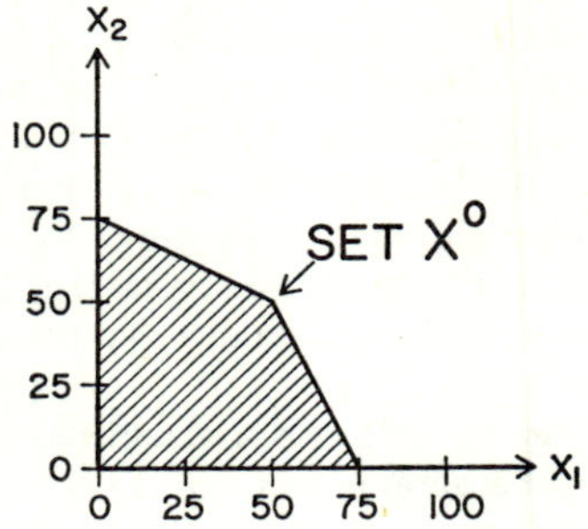

FIGURE 5

In constructing the normal form of this game, we shall take $\epsilon = 0$. Hence, the players will have $\frac{1}{2}$ probability of being able to select a payoff vector $x = x^*$ from the set X^{12}, and have $\frac{1}{2}$ probability of being able to select a payoff vector $x = x^{**}$ from the set X^{21}. Thus, their payoff vector from the normal-form game as a whole will be

$$x = \tfrac{1}{2}x^* + \tfrac{1}{2}x^{**} \quad \text{with} \quad x^* \in X^{12}, \qquad x^{**} \in X^{21}. \tag{11.7}$$

The set $X^0 = \{x\}$ of all such vectors x is the area satisfying condition (11.2) as well as the two further inequalities:

$$2x_1 + x_2 \leqq 75 \tag{11.8}$$

$$x_1 + 2x_2 \leqq 75. \tag{11.9}$$

(See Figure 5.)
Thus the normal form $\mathfrak{N}(G)$ of this game will be a two-person bargaining game in which the players can choose any payoff vector $x = (x_1, x_2)$ from the set X^0, and obtain the payoffs $x_1 = 0$ and $x_2 = 0$ if they cannot agree on which particular vector x to choose.

Any solution concept satisfying the usual Efficiency and Symmetry Postulates will yield the solution $x_1 = x_2 = 50$. This, however, means that if case (a^1, a^2) arises, then the players will obtain the payoffs $x_1^* = 0$, $x_2^* = 100$, whereas if case (a^2, a^1) arises, then they will obtain the payoffs $x_1^{**} = 100$ and $x_2^{**} = 0$. This is so because, in view of condition (11.7), the only way we can obtain the vector $x = (50, 50)$ is by choosing $x^* = (0, 100)$ and $x^{**} = (100, 0)$.

If we interpret the normal form of the game in the usual way, and assume that the players will choose their strategies and will agree on the outcome *before* they know whether case (a^1, a^2) or case (a^2, a^1) will obtain, then as a solution of the normal form of the game this result makes very good sense. It means that in accordance with the Efficiency Postulate the players will try to *minimize* the expected value of the amount they would have to pay out to outsiders, and therefore will agree to assign a zero payoff to that player who would be under an obligation to hand over half of his payoff to the secret organization in question.

However, if we stick to the original definition of the game, under which the players can choose their strategies and can conclude agreements only *after* each player has observed the value of his own attribute vector c_i, then this result is highly counter-intuitive. This is so because in this game there is near-perfect negative correlation between the two players' attribute vectors; so that each player will not only know the value of his own attribute vector but will also be able to infer with virtual certainty what the value of the other player's attribute vector is. Hence, even though formally the game is a game with *incomplete* information, it will always essentially reduce to a two-person bargaining game with *complete* information.

More particularly, if case (a^1, a^2) arises, then this game G will be essentially equivalent to a Nash-type bargaining game G^{12} where the payoff space is the set $X = X^{12}$, and where this fact is known to both players. Likewise, if case (a^2, a^1) arises, then game G will be essentially equivalent to a Nash-type bargaining game G^{21} where the payoff space is the set $X = X^{21}$, and where again this fact is known to both players. Yet, it would be quite counter-intuitive to suggest that in game G^{12} player 1 would agree to a solution giving him a zero payoff $x_1^* = 0$, or that in game G^{21} player 2 would agree to a solution giving him a zero payoff $x_2^{**} = 0$.

In terms of the corresponding Bayesian game G^*, we can state our argument as follows. If the players can reach an agreement *before* a chance move decides whether case (a^1, a^2) or case (a^2, a^1) will occur, then each player may be quite willing to agree to an arrangement under which he would obtain \$100 in one of the two possible cases, and would obtain \$0 in the other case, both cases having the same probability $\frac{1}{2}$. But if the players can conclude an agreement only *after* the outcome of this chance event has come to be known to both of them, then neither player will voluntarily accept a payoff of \$0 in order that the other player can obtain all the \$100—even if such an arrangement would minimize the two players' total payment to the secret organization.

For definiteness let us assume that in actual fact in games G^{12} and G^{21} the players will use the Nash solution [4]. Then in G^{12} their payoffs will be $x_1^* = 25$ and $x_2^* = 50$, whereas in G^{21} their payoffs will be $x_1^{**} = 50$ and $x_2^{**} = 25$. Hence, by equation (11.7) the players' expected payoffs from game G^* as a whole will be $x_1 = \frac{1}{2}x_1^* + \frac{1}{2}x_1^{**} = 37.5$ and $x_2 = \frac{1}{2}x_2^* + \frac{1}{2}x_2^{**} = 37.5$.[2]

Of course, for game G^* as a whole, the vector $x = (37.5, 37.5)$ is an inefficient payoff vector because it is strongly dominated, e.g., by the payoff vector $x = (50, 50)$ previously considered. Yet in the case of game G^* the use of an inefficient solution is justified by the fact that G^* is not a cooperative game in the full sense of this term, precisely because the players are not allowed to conclude agreements at the very beginning of the game, before any chance move or personal move has occurred—but rather can conclude agreements only *after* the chance move deciding between cases (a^1, a^2) and (a^2, a^1) has been completed. Thus, by using the inefficient payoff vector $x = (37.5, 37.5)$ as the solution for game G^*,

[2] In terms of the terminology to be introduced in Part III, the bargaining game G under discussion can be *decomposed* into G^{12} and G^{21} as component games. Hence, according to Postulate 3 of Part III, whatever solution concept we may wish to use, the solution of game G must be equivalent to using the solution of game G^{12} or of game G^{21} depending on which of these two games represents the game really played by the two players.

we do not violate the principle that each fully cooperative game should have an efficient solution.[3, 4]

We can state our conclusion also as follows. Most games discussed in the game-theoretical literature are *games with immediate commitment*, in the sense that each player is free to commit himself to some particular strategy *before* any chance move or personal move has taken place. (In the special case where the game is a cooperative game, the players are also free to enter into fully binding and enforceable agreements at that point.)

In contrast, the Bayesian games G^* we are using in the analysis of games with incomplete information, must be interpreted as *games with delayed commitment*,[5] where the players can commit themselves to specific strategies (and can enter into binding agreements if the game they are playing is a cooperative game), only *after* certain moves have occurred in the game, viz., the chance moves determining the players' attribute vectors.[6]

By the very definition of the normal form of a game, von Neumann and Morgenstern's Normalization Principle [6] can apply only to games with immediate commitment. Hence, in general, we cannot expect that the normal form $\mathfrak{N}(G^*) = \mathfrak{N}(G)$ of a Bayesian game G^* will be a fully satisfactory representation of this game G^*—or of the incomplete-information game G Bayes-equivalent to G^*. This is the reason why in the analysis of these games we have found it necessary to replace von Neumann and Morgenstern's Normalization Principle by our weaker Semi-normalization Principle (Postulate 2 in section 7 of Part I of this paper).

REFERENCES

1. GERARD DEBREU, "A Social Equilibrium Existence Theorem," *Proceedings of the National Academy of Sciences*, 38 (1952), 886–893.
2. JOHN C. HARSANYI AND REINHARD SELTEN, "A Generalized Nash Solution for Two-Person Bargaining Games with Incomplete Information," 1967 (unpublished).
3. JOHN F. NASH, "Equilibrium Points in n-Person Games," *Proceedings of the National Academy of Sciences*, 36 (1950), 48–49.
4. ——, "The Bargaining Problem," *Econometrica*, 18 (1950), 155–162.
5. ——, "Non-Cooperative Games," *Annals of Mathematics*, 54 (1951), 286–295.
6. JOHN VON NEUMANN AND OSKAR MORGENSTERN, *Theory of Games and Economic Behavior*, Princeton: Princeton University Press, 1953.

3 To be sure, game G itself if a fully cooperative game, yet it is a cooperative game with incomplete information. But our point has been that, as we replace this I-game G by a Bayes-equivalent C-game G^*, the fully cooperative nature of the game is lost since the players' attribute vectors c_i come to be reinterpreted as chance moves *preceding* their choices of strategies.

4 Our present discussion has been restricted to a very special case of two-person bargaining games with incomplete information, viz., to the case where almost perfect correlation exists between the two players' attribute vectors c_1 and c_2. The problem of defining a suitable solution concept in the general case will be discussed in [2].

5 Quite apart from this connection with Bayesian games, the general class of games with delayed commitment seems to have many other applications, and would deserve further study.

6 Note that the corresponding Selten games G^{**} do not give rise to this problem: they can always be interpreted as games with immediate commitment and therefore can be adequately represented by their normal forms $\mathfrak{N}(G^{**})$.

CHAPTER 8

GAMES WITH INCOMPLETE INFORMATION PLAYED BY "BAYESIAN" PLAYERS

PART III. THE BASIC PROBABILITY DISTRIBUTION OF THE GAME*†[1]

Parts I and II of this paper have described a new theory for the analysis of games with incomplete information. Two cases have been distinguished: consistent games in which there exists some basic probability distribution from which the players' subjective probability distributions can be derived as conditional probability distributions; and inconsistent games in which no such basic probability distribution exists. Part III will now show that in consistent games, where a basic probability distribution exists, it is essentially unique.

It will also be argued that, in the absence of special reasons to the contrary, one should try to analyze any given game situation with incomplete information in terms of a consistent-game model. However, it will be shown that our theory can be extended also to inconsistent games, in case the situation does require the use of an inconsistent-game model.

12.

We now propose to turn back to the questions raised in Section 5 of Part I of this paper. Given any arbitrarily chosen I-game G, is it always possible to construct some Bayesian game G^* Bayes-equivalent to G? And in cases where this is possible, is this Bayesian game G^* always unique? As we have seen in Section 5, these questions are equivalent to asking whether, for any arbitrarily chosen n-tuple of subjective probability distributions $R_1(c^1 \mid c_1), \cdots, R_n(c^n \mid c_n)$, there always *exists* some probability distribution $R^*(c_1, \cdots, c_n)$ satisfying the functional equation(5.3), and whether this distribution R^* is always *unique* in cases in which it does exist. The answers to these two questions are given by Theorem III below. But before stating the Theorem we shall have to introduce a few definitions.

In Section 3 of Part I we formally defined the range space $C_i = \{c_i\}$ of each attribute vector c_i $(i = 1, \cdots, n)$ as being the *whole* Euclidean space E^r of the required number of dimensions. For convenience we shall now relax this definition, and shall allow the range space C_i to be restricted to any proper *subset* C_i^* of the space E^r. (Obviously it makes little difference whether we say that in a given game G the vector c_i ranges only over some subset C_i^* of the Euclidean space E^r, or whether we say that formally the range of this vector c_i is the whole

* Received June 1965 and revised May 1966.

† Parts I and II of "Games with Incomplete Information Played by 'Bayesian' Players" have appeared in the preceding two issues of *Management Science: Theory*. A "Table of Contents" and a "Glossary" appeared with Part I.

[1] The author wishes again to express his indebtedness to the persons and institutions listed in Footnote 1 of Part I of this paper.

space E^r but that outside this set C_i^* the vector c_i has zero probability density everywhere.)

A given subset D of the range space $C = \{c\}$ for the composite vector $c = (c_1, \cdots, c_n)$ is called a *cylinder* if it is a Cartesian product of the form $D = D_1 \times \cdots \times D_n$, where $D_1, \cdots, D_n$ are subsets of the range spaces $C_1 = \{c_1\}, \cdots, C_n = \{c_n\}$ of the vectors $c_1, \cdots, c_n$. These sets $D_1, \cdots, D_n$ are called the *factor sets* of cylinder D. Two cylinders D and D' are said to be *separated* if all their corresponding factor sets D_i and D'_i $(i = 1, \cdots, n)$ are disjoint.

Suppose that in a given I-game G (as assessed by player j) each of the range spaces C_i $(i = 1, \cdots, n)$ can be partitioned into two or more non-empty subsets $D_i^1, D_i^2, \cdots$, such that each player i can always infer with probability one that every other player's information vector c_k has taken a value $c_k = \gamma_k$ from the set D_k^m $(m = 1, 2, \cdots)$ whenever his own information vector c_i has taken a value $c_i = \gamma_i$ from set D_i^m. In symbols this means that

$$(12.1) \qquad R_i(c_k \in D_k^m \mid c_i) = 1 \quad \text{for all} \quad c_i \in D_i^m,$$

and for all pairs of players i and k. In this case we shall say that game G can be *decomposed* into the component games $G^1, G^2, \cdots$, in which G^m $(m = 1, 2, \cdots)$ is defined as the game resulting from G if the range space C_i of each information vector c_i $(i = 1, \cdots, n)$ is restricted to the set $C_i^m = D_i^m$. Accordingly, the range space C of the composite vector $c = (c_1, \cdots, c_n)$ in each component game G^m is restricted to the cylinder $C^m = D^m = D_1^m \times \cdots \times D_n^m$. We shall call D^m the *defining cylinder* of game G^m and shall write $G^m = G(D^m)$.

If G can be decomposed into two or more component games $G^1, G^2, \cdots$ then it is said to be *decomposable*, while in the opposite case it is said to be *indecomposable*.

This terminology is motivated by the fact that whenever a given decomposable game G is being played, the players in actual fact will always play one of its component games $G^1, G^2, \cdots$. Moreover, they will always know, *before* making any move in the game, and before choosing their strategies s_i, *which* of these component games has to be played on that particular occasion, since each player can always observe whether his own information vector c_i lies in set D_i^1, or D_i^2, etc.

We shall use the same terminology also in the case of C-games G^* given in standard form, except that condition (12.1) will then have to be replaced by its analogue:

$$(12.2) \qquad R^*(c_k \in D_k^m \mid c_i) = 1 \quad \text{for all} \quad c_i \in D_i^m.$$

The concept of decomposition suggests the following postulate.

Postulate 3. Decomposition of games. Suppose that a given I-game G (as assessed by player j)—or a given C-game G^*—can be decomposed into two or more component games $G^1 = G(D^1), G^2 = G(D^2), \cdots$. Let $\sigma^1, \sigma^2, \cdots$ be the solutions (in terms of any particular solution concept) of $G^1, G^2, \cdots$ if they are regarded as independent games. Then the solution σ of the composite game G—or G^*—

itself will be equivalent to using solution σ^1 whenever the players' information vectors c_i take values from the sets D_i^1, to using solution σ^2 whenever these vectors c_i take values from the sets D_i^2, etc.

In other words, playing game G—or G^*—in practice always means playing one of its component games G^m. Hence in each case the players will use strategies appropriate for that particular component game G^m, and their strategy choices will be unaffected by the fact that, had their information vectors c_i taken different values, they might now have to play a different component game $G^{m'} \neq G^m$. (To repeat, the justification of this postulate is wholly dependent on the fact that, under our definition of games G and G^*, at the time of their strategy choices the players will always be assumed to know which particular component game G^m they are actually playing. If the players could choose their strategies *before* they were informed about the values of their own information vectors c_i—and so before they knew which component game G^m they would be playing—then Postulate 3 would be inappropriate.)

Let $R_1 = R_1(c^1 \mid c_1), \cdots, R_n = R_n(c^n \mid c_n)$ be the subjective probability distributions that player j uses in analyzing a given I-game G. We shall say that these distributions $R_1, \cdots, R_n$ are *mutually consistent* if there exists some probability distribution $R^* = R^*(c)$ [not necessarily unique] satisfying equation (5.3) with respect to $R_1, \cdots, R_n$. In this case, we shall say that game G (as assessed by player j) is itself *consistent*. Moreover, each probability distribution R^* satisfying (5.3) will be called *admissible*.

On the basis of these definitions we can now state the following theorem.

Theorem III. Let G be an I-game (as assessed by player j) in which the player's subjective probability distributions are discrete or absolutely continuous or mixed (but have no singular components). Then three cases are possible:

1. Game G may be inconsistent, in which case no admissible probability distribution R^* will exist.

2. Game G may be consistent and indecomposable, in which case exactly one admissible probability distribution R^* will exist.

3. Game G may be consistent and decomposable. In this case the following statements hold.

(a) Game G can be decomposed into a finite or infinite number of component games $G^1 = G(D^1)$, $G^2 = G(D^2), \cdots$, themselves indecomposable.

(b) There will be an infinite set of different admissible probability distributions R^*.

(c) But for each indecomposable component game $G^m = G(D^m)$, every admissible distribution R^* will generate the same uniquely determined conditional probability distribution

$$R^m = R^m(c) = R^*(c \mid c \in D^m), \tag{12.3}$$

whenever this conditional probability distribution R^m generated by this admissible distribution R^* is well defined.

(d) Every admissible probability distribution R^* will be a convex combination of the conditional probability distributions $R^1, R^2, \cdots$, and different admissible distributions R^* will differ only in the probability weights $R^*(D^m) = R^*(c \in D^m)$ assigned to each component R^m. Conversely, every convex combination of the distributions $R^1, R^2, \cdots$ will be an admissible distribution R^*.

13.

We shall prove the theorem only for games where the distributions $R_1, \cdots, R_n$ are discrete (so that all admissible distributions R^*, if there are any, will also be discrete).[2] For games with absolutely continuous probability distributions the proof is essentially the same, except that the probability mass functions used in the proof have to be replaced by probability density functions. Once the theorem has been proven for both discrete and continuous distributions, the proof for mixed distributions is quite straightforward. Before proving the theorem itself, we shall prove a few lemmas.

Let G be an I-game (as assessed by player j) where all n players i have discrete subjective probability distributions $R_i = R_i(c^i \mid c_i)$. For each probability distribution R_i we can define the corresponding probability mass function r_i as

$$r_i = r_i(\gamma^i \mid \gamma_i) = R_i(c^i = \gamma^i \mid c_i = \gamma_i) \qquad i = 1, \cdots, n, \tag{13.1}$$

where $\gamma^i = (\gamma_1, \cdots, \gamma_{i-1}, \gamma_{i+1}, \cdots, \gamma_n)$ and γ_i stand for specific values of the vectors c^i and c_i, respectively. Likewise, for each admissible probability distribution R^* (if there is one) we can define the corresponding probability mass function r^* as

$$r^* = r^*(\gamma) = R^*(c = \gamma) \tag{13.2}$$

where $\gamma = (\gamma_1, \cdots, \gamma_n)$ stands for specific values of vector c. If R^* is admissible then the corresponding probability mass function r^* will also be called admissible. The set of all admissible functions r^* will be called $\mathcal{R}$.

Equation (5.3) now can be written in the form

$$r^*(\gamma) = r_i(\gamma^i \mid \gamma_i) \cdot r^*(\gamma_i) \qquad \text{for all } r^* \in \mathcal{R} \tag{13.3}$$

and for all values $c = \gamma$, $c^i = \gamma^i$, and $c_i = \gamma_i$. Here $r^*(\gamma_i)$ denotes the marginal probability

$$r^*(\gamma_i) = \sum_{(\gamma^i \varepsilon C^i)} r^*(\gamma_i, \gamma^i). \tag{13.4}$$

We shall call any point $c = \gamma = (\gamma_i, \gamma^i)$ of the range space C a *null point* if one or more of the expressions $r_1(\gamma^1 \mid \gamma_1), \cdots, r_n(\gamma^n \mid \gamma_n)$ are zero. We shall call a point $c = \gamma$ a *nonnull point* if all these expressions are positive.

In a consistent game G a given value $c_i = \gamma_i$ of vector c_i will be called an *impossible value* if every admissible function r^* assigns to this value $c_i = \gamma_i$ the marginal probability $r^*(\gamma_i) = 0$. We shall adopt the convention that in any consistent game G each range space $C_i = \{c_i\}$, $i = 1, \cdots, n$, will be defined with

[2] Section 13 contains only the proof of Theorem III and may be omitted without loss of continuity.

the exclusion of all impossible values $c_i = \gamma_i$ of vector c_i, from C_i. Therefore, we shall assume that

$$\text{(13.5)} \qquad \gamma_i \in C_i \quad \text{implies} \quad r^*(\gamma_i) > 0 \quad \text{for } some \quad r^* \in \mathcal{R}.$$

Consequently at a null point $c = \gamma$ *all* n expressions $r_1(\gamma^1 \mid \gamma_1), \cdots, r_n(\gamma^n \mid \gamma_n)$ will be zero because, in view of (13.3) and (13.5), $r_i(\gamma^i \mid \gamma_i) = 0$ implies $r_k(\gamma^k \mid \gamma_k) = 0$ for $k = 1, \cdots, n$.

Lemma 1. Suppose that in a consistent game G at a given point $c = \gamma$ all admissible functions r^* identically take the value

$$\text{(13.6)} \qquad r^*(\gamma) = 0.$$

This will be the case if and only if $c = \gamma$ is a null point.

The lemma again follows from (13.3) and (13.5).

Let $c = \alpha = (\alpha_1, \cdots, \alpha_n)$, $c = \beta = (\beta_1, \cdots, \beta_n)$, $c = \gamma = (\gamma_1, \cdots, \gamma_n)$, and $c = \delta = (\delta_1, \cdots, \delta_n)$ be four nonnull points, such that $\alpha_i = \beta_i$ and $\gamma_i = \delta_i$, whereas $\alpha_k = \gamma_k$ and $\beta_k = \delta_k$. Let r^* be an admissible function taking a positive value at the point $c = \alpha$. Then, by equation (13.3), the quantities $r^*(\alpha_i) = r^*(\beta_i)$ and $r^*(\alpha_k) = r^*(\gamma_k)$ will also be positive, and, therefore, the same will be true of the quantities $r^*(\beta)$ and $r^*(\gamma)$. This in turn implies that the quantities $r^*(\beta_k) = r^*(\delta_k)$ and $r^*(\gamma_i) = r^*(\delta_i)$, as well as the quantity $r^*(\delta)$, will also be positive. Consequently in view of equation (13.3) we have:

$$\text{(13.7)} \qquad \frac{r^*(\alpha)}{r^*(\delta)} = \frac{r^*(\alpha)}{r^*(\beta)} \cdot \frac{r^*(\beta)}{r^*(\delta)} = \frac{r_i(\alpha^i \mid \alpha_i)}{r_i(\beta^i \mid \beta_i)} \cdot \frac{r_k(\beta^k \mid \beta_k)}{r_k(\delta^k \mid \delta_k)},$$

but also

$$\text{(13.8)} \qquad \frac{r^*(\alpha)}{r^*(\delta)} = \frac{r^*(\gamma)}{r^*(\delta)} \cdot \frac{r^*(\alpha)}{r^*(\gamma)} = \frac{r_i(\gamma^i \mid \gamma_i)}{r_i(\delta^i \mid \delta_i)} \cdot \frac{r_k(\alpha^k \mid \alpha_k)}{r_k(\gamma^k \mid \gamma_k)}.$$

If the probability mass functions r_i and r_k—or, equivalently, the probability distributions R_i and R_k—are chosen arbitrarily, then equations (13.7) and (13.8) may become inconsistent, so that no admissible probability mass function r^* taking a positive value $r^*(\gamma) > 0$ at the nonnull point $c = \gamma$ will exist. But then, by Lemma 1, the game will be inconsistent. Thus, we can state:

Lemma 2. If the subjective probability distributions $R_1, \cdots, R_n$ of a given game G are chosen arbitrarily, then they may turn out to be mutually inconsistent, so that no admissible probability distribution R^* will exist.

Now suppose again that G is a consistent game with discrete probability distributions. Any pair of nonnull points $c = \gamma$ and $c = \delta$ will be called *similar* if all admissible probability mass functions r^* yield the same unique ratio $r^*(\gamma)/r^*(\delta)$ whenever this ratio is well defined. (That is, this must be true for all admissible functions r^* not taking zero values $r^*(\gamma) = r^*(\delta) = 0$ at *both* of these two points.) Clearly the relation of similarity partitions the set C^* of all nonnull points into equivalence classes, which we shall call *similarity classes.* All points $c = \gamma$ belonging to the same similarity class E will be similar, and all points $c = \gamma$ and $c = \delta$ belonging to two different similarity classes E and E' will be dissimilar.

Lemma 3. Let $c = \gamma$ and $c = \delta$ be two nonnull points of a consistent game G, agreeing in their ith component $c_i = \gamma_i = \delta_i$. Then γ and δ will belong to the same similarity class E.

Proof. By Lemma 1, we can find some admissible function $r^* = r^0$ such that

$$r^0(\gamma) > 0. \tag{13.9}$$

In view of (13.3) this implies that

$$r^0(\gamma_i) = r^0(\delta_i) > 0. \tag{13.10}$$

As both γ and δ are nonnull points we also have

$$r_i(\gamma^i \mid \gamma_i) > 0 \quad \text{and} \quad r_i(\delta^i \mid \delta_i) > 0. \tag{13.11}$$

Now let $r^* = r^{00}$ be any other admissible function for which the ratio $r^{00}(\delta)/r^{00}(\gamma)$ is well defined. In view of (13.3) and (13.10), we can write

$$r^0(\delta)/r^0(\gamma) = r_i(\delta^i \mid \delta_i)/r_i(\gamma^i \mid \gamma_i) = r^{00}(\delta)/r^{00}(\gamma). \tag{13.12}$$

By (13.9) and (13.11), the first two ratios in the equation will also be well defined. As this equation will hold for *any* admissible function $r^* = r^{00}$ for which the last ratio is well defined, we can infer that the points δ and γ will belong to the same similarity class E, as desired.

For a given set $E \subseteq C$, let D_i be the set of all vectors $c_i = \gamma_i$ occurring as the ith component of some vector(s) $c = \gamma = (\gamma_1, \cdots, \gamma_i, \cdots, \gamma_n)$ in set E. Then D_i is called the *projection* of set E on vector space C_i. Moreover, if $D_1, \cdots, D_n$ are the projections of set E on the vector spaces $C_1, \cdots, C_n$ then the cylinder $D = D_1 \times \cdots \times D_n$ is called the cylinder *spanned* by set E. Obviously, $E \subseteq D$.

Lemma 4. Let $D^1, D^2, \cdots$ be the cylinders spanned by the similarity classes $E^1, E^2, \cdots$ of a given consistent game G. Then these cylinders $D^1, D^2, \cdots$ will be mutually separated.

Proof. All we have to prove is that for any given value of i $(i = 1, \cdots, n)$ the projections D_i^m and $D_i^{m'}$ of two different similarity classes E^m and $E^{m'}$ are always disjoint. Now suppose this would not be the case. Then D_i^m and $D_i^{m'}$ would have some common point $c_i = \gamma_i$. This in turn would imply that γ_i would be the ith component of some point $c = \gamma = (\gamma_i, \gamma^i)$ in set E^m and also of some point $c = \delta = (\delta_i, \delta^i) = (\gamma_i, \delta^i)$ in set $E^{m'}$. But then, by Lemma 3, γ and δ would belong to the same similarity class, and could not belong to two different similarity classes E^m and $E^{m'}$, which is inconsistent with the assumptions made. Hence, D^m and $D^{m'}$ cannot have any common point $c_i = \gamma_i$.

Lemma 5. If the nonnull points of a given consistent game G can be partitioned into two or more similarity classes $E^1, E^2, \cdots$, then game G itself can be decomposed into corresponding component games $G^1 = G(D^1)$, $G^2 = G(D^2), \cdots$, where the defining cylinder D^m $(m = 1, 2, \cdots)$ of component game $G^m = G(D^m)$ is the cylinder spanned by the similarity class E^m.

Proof. Let $c_i = \alpha_i$ and $c_k = \beta_k$ be specific values of vectors c_i and c_k, such that $\alpha_i \in D_i^m$ but $\beta_k \notin D_k^m$. Let $c = \gamma = (\gamma_1, \cdots, \alpha_i, \cdots, \beta_k, \cdots, \gamma_n)$ be any point having $\gamma_i = \alpha_i$ as its ith component and having $\gamma_k = \beta_k$ as its kth component.

Then, by Lemma 4, γ will be a null point so that

$$(13.13) \qquad r_i(\gamma^i \mid \alpha_i) = r_i(\beta_k, \gamma^{ik} \mid \alpha_i) = 0,$$

where γ^{ik} is the vector which remains if the ith and k^{th} components are both omitted from $\gamma = (\gamma_1, \cdots, \gamma_n)$. Therefore,

$$(13.14) \qquad r_i(\beta_k \mid \alpha_i) = R_i(c_k = \beta_k \mid c_i = \alpha_i) = 0$$

whenever $\alpha_i \in D_i^m$ but $\beta_k \notin D_k^m$. Consequently,

$$(13.15) \qquad R_i(c_k \notin D_k^m \mid c_i) = 0 \qquad \text{if } c_i \in D_i^m$$

and so

$$(13.16) \qquad R_i(c_k \in D_k^m \mid c_i) = 1 \quad \text{if} \quad c_i \in D_i^m.$$

Hence, for each cylinder D^m ($m = 1, 2, \cdots$), every pair of factor sets D_i^m and D_k^m ($i, k = 1, \cdots, n$) will satisfy condition (12.1), and so game G can be decomposed into the component games $G^m = G(D^m)$, $m = 1, 2, \cdots$.

Lemma 6. A given consistent game G is decomposable if and only if it has more than one admissible probability distribution R^*.

Proof of the "if" part of the lemma. If all nonnull points of game G belong to the same similarity class E, then, for all pairs of nonnull points γ and δ, the ratios $r^*(\gamma)/r^*(\delta)$ are uniquely determined. Hence, the game can have only one admissible probability mass function r^* and only one admissible probability distribution R^*.

Thus, if game G does have more than one admissible probability distribution R^*, then it must contain two or more similarity classes E^1, E^2, $\cdots$; and, therefore, by Lemma 5 it will be decomposable.

Proof of the "only if" part of the lemma. We have to show that G will have more than one admissible probability distribution R^* (or more than one admissible probability mass function r^*) if it is a decomposable game.

Suppose to the contrary that a given game G can be decomposed into the component games $G^1 = G(D^1)$, $G^2 = G(D^2)$, $\cdots$ but has only one admissible probability mass function $r^* = r^0$. Then, by Lemma 1, this function r^0 must take a positive value $r^0(\gamma) > 0$ at each nonnull point $c = \gamma$ of G. Let r^m ($m = 1, 2, \cdots$) be the conditional probability mass function

$$(13.17) \qquad \begin{aligned} r^m(\gamma) &= r^0(\gamma \mid \gamma \in D^m) \\ &= r^0(\gamma) / \textstyle\sum_{(c \in D^m)} r^0(c) \qquad \text{for all } \gamma \in D^m, \\ r^m(\gamma) &= 0 \qquad \text{for all } \gamma \notin D^m. \end{aligned}$$

Since $r^0(c) > 0$ at all nonnull points c, this function r^m will be everywhere well defined. In view of (13.3) and (13.17) we can write

$$(13.18) \qquad r^m(\gamma) = r_i(\gamma^i \mid \gamma_i) \cdot r^m(\gamma_i),$$

which means that the function $r^* = r^m$ ($m = 1, 2, \cdots$) satisfies condition (13.3). Hence, the functions r^1, r^2, $\cdots$ are just as much admissible probability mass functions as r^0 itself is, which contradicts the assumption that $r^* = r^0$ is the only admissible probability mass function of the game. This completes the proof of Lemma 6.

Lemma 7. A consistent game G is indecomposable if and only if all its nonnull points belong to the same similarity class E.

Proof of the "if" part. If all nonnull points are similar than the ratios $r^*(\gamma)/r^*(\delta)$ are uniquely determined for all nonnull points γ and δ. Hence, the game can have only one admissible probability mass function r^*, and so, by Lemma 6, it is indecomposable.

Proof of the "only if" part. If game G is indecomposable, then, by Lemma 6, it will have only one admissible function r^*. Hence, the ratios $r^*(\gamma)/r^*(\delta)$ will be uniquely determined for all pairs of nonnull points γ and δ, and so all of these points will be similar.

Lemma 8. Let G be a decomposable consistent game, and let $G^1 = G(D^1)$, $G^2 = G(D^2), \cdots$ be its component games as defined by Lemma 5. Then

1. Each component game G^m will be itself an indecomposable game.
2. If $r^* = r^0$ and $r^* = r^{00}$ are two admissible probability mass functions for G then, for each component game G^m $(m = 1, 2, \cdots)$, both $r^* = r^0$ and $r^* = r^{00}$ will generate the same conditional probability mass function

$$r^m(\gamma) = r^0(\gamma \mid \gamma \in D^m) = r^{00}(\gamma \mid \gamma \in D^m) \tag{13.19}$$

whenever these conditional probability mass functions generated by r^0 and r^{00} are well defined—that is, whenever the two functions r^0 and r^{00} assign a positive probability mass to the event $E = \{\gamma \in D^m\}$.

3. Every admissible probability mass function r^* for the game will be a convex combination of these functions $r^1, r^2, \cdots$ corresponding to the component games $G^1, G^2, \cdots$; and any convex combination of these functions $r^1, r^2, \cdots$ will be an admissible probability mass function r^*.

Proof. Parts 1 and 2 follow from Lemma 7 in view of the fact that all nonnull points γ lying in any given cylinder D^m belong to the same similarity class E^m.

Part 3 follows from equations (13.17) and (13.3).

Lemmas 2, 5, 6, 7, and 8 together imply Theorem I as restricted to games with discrete probability distributions. We have already indicated (in the first paragraph of this section) how the proof can be extended to the general case.

14.

By Theorem III, if a given player j participating in some I-game G does not take special care to base his analysis of the game on mutually consistent subjective probability distributions $R_1, \cdots, R_j, \cdots, R_n$, then, in general, no admissible basic probability distribution R^* will exist; and, therefore, there will be no Bayesian game G^* Bayes-equivalent to G. In view of this fact we now propose the following postulate.

Postulate 4. Mutual consistency. Unless he has special reasons to believe that the subjective probability distributions $R_1, \cdots, R_j, \cdots, R_n$ used by the n players are, in fact, mutually inconsistent, every player j of an I-game G will always analyze the game in terms of a set of some mutually consistent probability distributions $R_1, \cdots, R_j, \cdots, R_n$.

Postulate 4 can be supported by the following considerations:

(i) Player j can greatly simplify his analysis of the game situation by using n mutually consistent distributions $R_1, \cdots, R_n$, because this will enable him to reduce the problem of analyzing the I-game G to the much easier problem of analyzing an equivalent C-game G^*. Therefore, player j will be well advised to assume n mutually consistent distributions $R_1, \cdots, R_n$, unless he feels that the information he has about the game situation is incompatible with this assumption.

(ii) Admittedly, if player j does not restrict his choice to mutually consistent distributions R_i, then he will have a freer hand in postulating suitable differences between the different players' subjective probability distributions R_i, and so will have a freer hand in providing mathematical representation for any differences he may wish to assume between different players i as to their beliefs and expectations about the game situation. However, even if he does restrict his choice to mutually consistent R_i's, this need not prevent him from finding suitable representation for such inter-player differences in his mathematical model of the game; for such differences can often be given fully adequate representation by assuming corresponding differences between the relevant players' attribute vectors c_i, and by choosing a basic probability distribution $R^* = R^*(c_1, \cdots, c_n)$ which is a sufficiently asymmetric function of the different players' attribute vectors c_i; so that the conditional probability distributions $R_i = R^*(c^i \mid c_i)$ for different players i will have appropriately different mathematical forms. Mutually inconsistent R_i's will have to be assumed only in cases in which player j comes to the definite conclusion that the inter-player differences in question could not be represented by mutually consistent R_i's.

(iii) According to the prior-lottery model (see Section 6 of Part I of this paper), every I-game G can be considered to be a result of a random social process which has selected the n actual players of the game from some hypothetical populations $\Pi_1, \cdots, \Pi_n$ of possible players, where the probability that this random process will select n players with any given combination of specific attribute vectors $c_1 = c_1^0, \cdots, c_n = c_n^0$ is governed by some (objective) probability distribution $R^* = R^*(c_1, \cdots, c_n)$. However, in general, the true distribution R^* governing this random process will be unknown to the players.

Let $R_i^* = R^*(c^i \mid c_i)$, for $i = 1, \cdots, n$, be the conditional probability distributions generated by this unknown distribution R^*. Clearly the n distributions $R_1^*, \cdots, R_n^*$ will always be mutually consistent, since by definition they are conditional distributions generated by the same basic distribution R^*. The n players' subjective probability distributions $R_1, \cdots, R_n$ can be regarded as their *estimates* of these mutually consistent, but unknown, conditional probability distributions $R_1^*, \cdots, R_n^*$.

Accordingly, it is natural to argue (at least in cases where player j has no special reasons for using inconsistent distributions R_i) that, instead of trying to estimate each player's subjective probability distribution R_i separately, player j

should rather try to estimate directly the objective probability distribution R^* governing the random social process in question; and then he should choose the n distributions R_i he will use in the analysis of the game situation, by setting $R_i = R^*(c^i \mid c_i)$, where R^* now refers to his *estimate* of the true distribution R^*. Obviously, if player j uses this estimation procedure, he will always come up with n mutually consistent distributions $R_1, \cdots, R_n$.

We have seen in Section 6 (Part I) that the prior-lottery model pictures the game situation in the way in which it would be seen by an intelligent and "properly informed" outside observer.[3] Consequently, when player j is trying to estimate the probability distribution R^*, he will have to ask himself the question, "What probability distribution R^* would be chosen by a randomly selected intelligent and properly informed outside observer, as his estimate of the true probability distribution governing the random social process underlying this game situation G?" (If player j feels that he cannot give a unique answer to this question because different intelligent and properly informed observers are likely to choose different probability distributions R^*, then he can construct his own probability distribution R^* by some averaging process over the various distributions R^* he thinks different observers might choose.)

To put it differently, in trying to estimate the probability distribution R^*, player j should try to use only the information *common* to all n players. Of course, in his analysis at some point or another he should make use of *all* information he has about the game situation, including any special information which may be available to him but may not be available to the other players. But under our model all such special information must be represented by incorporating it into player j's own information vector c_j. In contrast, the basic probability distribution R^* itself is meant to represent only information common to all n players. (The same is true for the n subjective probability distributions $R_1, \cdots, R_n$ and the n payoff functions $V_1, \cdots, V_n$: by definition they are meant to contain only such information that, in player j's opinion, is common property of the n players —see Section 3 in Part I.)

To be sure, if player j follows the suggested estimation procedure this will guarantee only that his own subjective probability distribution R_j and the subjective probability distributions $R_1, \cdots, R_{j-1}, R_{j+1}, \cdots, R_n$ he will *ascribe* to the other $(n - 1)$ players will satisfy equation (5.3). But, of course, it will not guarantee that his own subjective probability distribution R_j and the subjective probability distributions $R_1, \cdots, R_{j-1}, R_{j+1}, \cdots, R_n$ the other $(n - 1)$ players will in *actual fact* use in the game will satisfy equation (5.3). In other words, the suggested estimation procedure will guarantee *internal consistency* among the n probability distributions $R_1, \cdots, R_n$ that player j himself will use in analyzing the game, but will not and cannot guarantee *external consistency* among the probability distributions used by the different players. Indeed, so long as each player has to choose his subjective probabilities (probability estimates)

[3] We have defined a "properly informed" observer as a person possessing all information *shared* by all n players but having no extra information unavailable to the players, or available to some particular player(s) yet unavailable to the other player(s).

independently of the other players, no conceivable estimation procedure can ensure consistency among the different players' subjective probabilities.

More particularly, there is no way of ensuring that the different players will choose the *same* subjective probability distribution as their estimate of the objective probability distribution R^* underlying the game situation. For, by the very nature of subjective probabilities, even if two individuals have exactly the same information and are at exactly the same high level of intelligence, they may very well assign different subjective probabilities to the very same events.

At the same time, while the suggested estimation procedure does not in any way guarantee external consistency among the different players' probability estimates, we feel it does go as far as an estimation procedure can go in promoting such external consistency. By asking each player to take the point of view of an outside observer in estimating the objective probability distribution R^*, it asks him to choose an estimate as close as possible to the estimates other intelligent people might be expected to choose, and to make his estimate as independent as possible of his own personal prejudices and idiosyncrasies.

15.

So far our analysis has been restricted to the case in which player j has no special information suggesting mutual inconsistencies among the probability distributions $R_1, \cdots, R_j, \cdots, R_n$ of the players. Now what happens if player j does have such information? For instance, what happens if in an international conflict situation country j obtains quite convincing reports from its diplomatic representatives, intelligence agencies, newspaper correspondents, etc., to the effect that in analyzing the international situation country i is using a very different political and social philosophy, and a very different model of the world, from country j's own; and that as a result the subjective probability distribution R_i used by country i is clearly inconsistent with the subjective probability distribution R_j used by country j?

In a situation such as this, one obvious possibility is to take this information at face value and to conclude that the players' subjective probability distributions *are* in fact inconsistent, which means that every player j simply has to resign himself to the necessity of analyzing the game situation in terms of n mutually inconsistent subjective probability distributions $R_1, \cdots, R_j, \cdots, R_n$.

As Reinhard Selten has pointed out,[4] his posterior-lottery model (see Section 6, Part I of this paper) can be extended to this case of mutually inconsistent subjective probability distributions. All we have to do is to assume that when all K players have chosen their strategies there will be a *separate* lottery $L(i_m)$ for every player i_m, instead of there being merely *one* grand lottery L^* for all K players. For each player i_m in role class i his lottery $L(i_m)$ will choose $(n - 1)$ players [one player from each of the $(n - 1)$ role classes $1, \cdots, i - 1, i + 1, \cdots, n$] as his "partners" in the game. If player i_m himself has the attribute vector $c_i = c_i^0$, then the probability of his $(n - 1)$ partners' having any specific

[4] In private communication (see Footnote 1 in Part I).

combination of attribute vectors $c_1 = c_1^0; \cdots; c_{i-1} = c_{i-1}^0; c_{i+1} = c_{i+1}^0; \cdots;$ $c_n = c_n^0$ will be governed by the subjective probability distribution $R_i = R_i(c^i \mid c_i = c_i^0)$ which player i himself entertains. Player i_m will receive the payoff:

$$(15.1) \qquad x_i = V_i(s_1^0, \cdots, s_i^0, \cdots, s_n^0; c_1^0, \cdots, c_i^0, \cdots, c_n^0)$$

where $s_1^0, \cdots, s_n^0$ are the strategies chosen by player i_m and by his partners, whereas $c_1^0, \cdots, c_n^0$ are the attribute vectors of these same n players.

Note that, under this model, "partnership" is not necessarily a symmetric relationship. If lottery $L(i_m)$ of player i_m chooses a given player k_r as a partner for player i_m, then it does not follow that lottery $L(k_r)$ of player k_r will likewise choose player i_m as a partner[5] for player k_r. As partnership in general will not be a symmetric relationship, and as the lotteries associated with different players will be quite independent, this model does not presuppose any mutual consistency in the sense of Postulate 4 among the n probability distributions $R_1 = R_1(c^1 \mid c_1), \cdots, R_n = R_n(c^n \mid c_n)$ governing the lotteries conducted for the players in the n role classes $1, \cdots, n$.

Thus, a Selten game G^{**} will exist even for an inconsistent I-game G, for which no Bayesian game G^* exists.

16.

It is, however, always questionable whether any given information suggesting inconsistencies among the different players' subjective probability distributions $R_1, \cdots, R_n$ should really be taken at face value.

First of all, information about other players' assessment of probabilities (and indeed about their internal beliefs and attitudes in general) tends to be very unreliable, because the players will often have a vested interest in misleading the other players about their real ways of thinking. But even in the empirical facts themselves about the other players' probability judgments are quite correct, they will be usually open to alternative interpretations.

More particularly, any inconsistency among the various players' subjective probability distributions R_i is always a result of discrepancies among the basic probability distributions $R^* = R^{(i)}$ used by different players i (see Theorem IV below). On the other hand, these discrepancies among the probability distributions $R^{(i)}$ will themselves often admit of explanation in terms of the *differences in the information* available to different players i—in which case, as we shall see, the

[5] This fact does not give rise to any logical difficulty. It only means that, owing to the outcome of lottery $L(i_m)$, player i_m's payoff may come to depend on the strategy s_k^0 and the attribute vector c_k^0 of some player k_r—even though, owing to the outcome of lottery $L(k_r)$, player k_r's own payoff will not be dependent on the strategy s_i^0 and the attribute vector c_i^0 of player i_m, but rather will be dependent on the strategy s_i^{00} and the attribute vecto- c_i^{00} of another player $i_t \neq i_m$ in role class i. (Of course, when the players choose their strater gies they will not know who will be whose partner and whether any such relation of partnership will be reciprocal or not.)

game can be *reinterpreted* as a game involving mutually *consistent* subjective probability distributions $R_1, \cdots, R_n$ on the part of the n players.

We now propose to state the following simple theorem.

Theorem IV. Let

$$R^*(c_1, \cdots, c_i, \cdots, c_n) = R^{(i)}(c_1, \cdots, c_i, \cdots, c_n) \tag{16.1}$$

be player i's basic probability distribution ($i = 1, \cdots, n$), defined as his estimate of the probability distribution governing the random social process which gave rise to the current game situation. Suppose that

$$R^{(1)} = \cdots = R^{(n)} = R^0. \tag{16.2}$$

Then the n players' subjective probability distributions $R_1, \cdots, R_n$ will be mutually consistent.

Proof. Suppose player i thinks that the social process in question is governed by the probability distribution $R^* = R^{(i)}$, and knows that his own attribute vector has the value $c_i = c_i^0$. Then he must use the conditional probability distribution $R^*(c^i \mid c_i) = R^{(i)}(c^i \mid c_i)$, generated by this distribution $R^* = R^{(i)}$, for evaluating the probability that the other $(n - 1)$ players will have any specific combination of attribute vectors $c_1 = c_1^0; \cdots; c_{i-1} = c_{i-1}^0; c_{i+1} = c_{i+1}^0; \cdots; c_n = c_n^0$. Hence, player i's subjective probability distribution R_i will be defined by this conditional distribution[6]

$$R_i = R_i(c^i \mid c_i) = R^*(c^i \mid c_i) = R^{(i)}(c^i \mid c_i). \tag{16.3}$$

This will be true for all players $i = 1, \cdots, n$. Consequently, in view of equation (16.2) we can write

$$R_i = R^0(c^i \mid c_i) \qquad i = 1, \cdots, n. \tag{16.4}$$

Hence, all n functions $R_1, \cdots, R_n$ will have the nature of conditional probability distributions generated by the same basic probability distribution R^0, which means that they will be mutually consistent.

By Theorem IV, if the functions $R_1, \cdots, R_n$ are in fact mutually inconsistent, then the basic probability distributions $R^{(1)}, \cdots, R^{(n)}$ used by the n players cannot be all identical; and it will be the discrepancies existing among these basic probability distributions $R^{(1)}, \cdots, R^{(n)}$ that account for the inconsistency among the subjective probability distributions $R_1, \cdots, R_n$. Now, our contention is that the discrepancies existing among the different players' basic probability distributions $R^{(i)}$ will often admit of explanation in terms of differences in the information that different players have about the nature of the social process underlying the current game situation.

For instance, in our previous example, if the subjective probability distributions R_i and R_j used by countries i and j are mutually inconsistent, this will indi-

[6] In the case $k \neq i$, the relationship $R_i = R^{(k)}(c^i \mid c_i)$ will hold in general only if $R^{(k)} = R^{(i)}$ as required by equation (16.2). However, in the case $k = i$, the relationship will always hold, by the very definitions of the subjective probability distributions R_i and of the basic probability distributions $R^* = R^{(i)}$.

cate that they assess probabilities in terms of two quite different basic probability distributions $R^{(i)}$ and $R^{(j)}$. More particularly, *e.g.*, country i may assign much higher probabilities than country j does to the success of violent social revolutions in various parts of the world. Now it will often be a very natural hypothesis that these differences in the two countries' assessment of probabilities are due to differences in the information they have acquired about the world as a result of their divergent historical experiences—more particularly as a result of their divergent experience about the prospects of violent social revolutions on the one hand, and about the chances of achieving social reforms by peaceful non-revolutionary methods on the other hand.

Of course, if player j makes the assumption that the discrepancies among the probability distributions $R^{(i)}$ used by different players i are due to differences in the information available to them, this will mean under our terminology that these distributions $R^{(i)}$ are not really the *basic* probability distributions of these players, because in this case, obviously, these distributions $R^{(i)}$ cannot be assumed any longer to represent exclusively the information *common* to all n players. Instead, each distribution $R^{(i)}$ will now have to be interpreted as a *conditional* probability distribution, conditioned by some special information available to player i himself but not available to all n players.

On the other hand, if player j is willing to assume that *all* discrepancies among the distributions $R^{(i)}$ used by different players i are due to these differences in the information available to them, then he can consider all these distributions $R^{(i)}$ to be conditional distributions derived from *one* and the *same* unconditional distribution $R^{*\prime}$, which means that this distribution $R^{*\prime}$ will be the true basic distribution of the game.

In symbols, let d_i be a vector summarizing all special information about the relevant variables which makes any given player i adopt the probability distribution $R^{(i)} = R^{(i)}(c_1, \cdots, c_i, \cdots, c_n)$—instead of adopting some probability distribution $R^{(k)} = R^{(k)}(c_1, \cdots, c_i, \cdots, c_n)$ used by another player k. Then player i's true information vector (or attribute vector) will not be the vector c_i but rather will be the larger vector $c'_i = (c_i, d_i)$ summarizing the information contained in *both* vectors c_i and d_i.

Consequently, the true basic probability distribution $R^{*\prime}$ of the game will be the joint probability distribution of these larger vectors c'_i, and so will have the form

$$(16.5) \qquad R^{*\prime} = R^{*\prime}(c'_1, \cdots, c'_n) = R^{*\prime}(c_1, d_1; \cdots; c_n, d_n).$$

In contrast, each distribution $R^* = R^{(i)}$ will have the nature of a conditional probability distribution (or more exactly of a conditional marginal probability distribution) of the form

$$(16.6) \quad R^{(i)} = R^{(i)}(c_1, \cdots, c_i, \cdots, c_n) = R^{*\prime}(c_1, \cdots, c_i, \cdots, c_n \mid d_i).$$

In other words, player j will find that, as soon as he redefines the players' attribute vectors as the larger vectors $c'_i = (c_i, d_i)$, then the information he has about the other players' subjective probabilities will become compatible with the hypothesis that all n players are actually using the *same* basic probability distribution $R^{*\prime} = R^{*\prime}(c'_1, \cdots, c'_j, \cdots, c'_n)$. Consequently, by Theorem IV, this information will be compatible also with the hypothesis that the n players are using *mutually consistent* subjective probability distributions $R'_1, \cdots, R'_j, \cdots, R'_n$ of the fcrm

$$(16.7) \qquad R'_i = R'_i((c^i)' \mid c'_i) = R^{*\prime}((c^i)' \mid c'_i) \qquad i = 1, \cdots, j, \cdots, n,$$

where

$$(c^i)' = (c'_1, \cdots, c'_{i-1}, c'_{i+1}, \cdots, c'_n).$$

To sum up, if the information player j has about the other players' subjective probabilities seems to suggest that the n players are using mutually *inconsistent* subjective probability distributions $R_1, \cdots, R_j, \cdots, R_n$, then more careful reanalysis of the game situation may very well lead him to the conclusion that, in actual fact, the players' subjective probability distributions can be better represented by another set of probability distributions $R'_1, \cdots, R'_j, \cdots, R'_n$, mutually *consistent* with one another.

Note that, in reanalyzing the game situation in this way, player j must use not only the information he has obtained independently about the game, but also any information he can infer from what he knows about the other players' assessment of probabilities. For instance, suppose that player j himself would be inclined to assign the probability $P_j(e)$ to a given event e, but then obtains the information that another player i assigns a very different probability $P_i(e) \neq P_j(e)$ to this event. Then player j cannot rest the matter with the conclusion that he and player i have apparently been assessing the probability of this event on the basis of two rather different sets of information. Instead, he also has to reach a decision on whether his own assessment of this probability is likely to be based on more correct and more complete information than player i's assessment is, or whether the opposite is the case; and, in particular, he has to decide whether player i's quite different assessment of this probability is or is not a good enough reason for him himself also to change his own assessment of this probability from $P_j(e)$ to some number closer to the probability $P_i(e)$ which player i is assigning to event e.

Of course, this method of reconciling the different players' assessments of probabilities in terms of the same basic probability distribution $R^{*\prime}$ will work only if player j feels that the empirical facts which he knows about the other players' probability assessments are favorable to, or at least are compatible with, the hypothesis that the discrepancies among the various players' probability judgments can be reasonably explained in terms of differences in their information. In our own view, in most cases the empirical facts will be at least compatible with this hypothesis. But if in any given I-game G player j reaches the opposite conclusion, then he will have to fall back upon Selten's model (see Section 15), which permits mutually inconsistent subjective probability distributions on the part of the players.

17.

According to Theorem III (Section 12), even if player j uses mutually consistent subjective probability distributions $R_1, \cdots, R_n$ in his analysis of a given I-game G, he may still have the problem of there being an infinite set of equally admissible probability distributions R^* satisfying equation (5.3) [Part I of this paper]; which means that there will be an infinite number of Bayesian games G^* Bayes-equivalent to this I-game G.

However, by Postulate 3, this fact will not give rise to any difficulties because, if player j uses the solution σ of *any* of these games G^* as the solution of game G, he will always obtain the *same* solution σ. This is so because, by Theorem III, even if there are many Bayesian games G^* Bayes-equivalent to G, there will always be only *one* Bayesian game $(G^m)^*$ Bayes-equivalent to any given indecomposable *component game* G^m of G (since G^m will have only one unique admissible probability distribution $R^*(G^m) = R^m$. Hence, no ambiguity will result if he uses the solution σ^m of this Bayesian game $(G^m)^*$ as the solution of this component game G^m. On the other hand, by Postulate 3, the solutions σ^m of these component games G^m will completely determine the solution σ of the whole game G.

Yet, even though the possible multiplicity of Bayesian games G^* Bayes-equivalent to a given I-game G causes no real problem, for some purposes it will be convenient if we can make the assumption that there is always only *one* Bayesian game G^* Bayes-equivalent to each I-game G. We can achieve this by using the fact that, under the procedure proposed in Section 14, each player j will start his analysis of a given I-game G by first choosing a basic probability distribution $R^* = R^*(c_1, \cdots, c_n)$, and then will define the n probability distributions $R_1 = R_1(c^1 \mid c_1), \cdots, R_n = R_n(c^n \mid c_n)$ as the conditional distributions $R_1 = R^*(c^1 \mid c_1); \cdots; R_n = R^*(c^n \mid c_n)$ derived from this distribution R^*. This make it natural to define the I-game G itself (as assessed by player j) in terms of this one distribution R^* chosen by player j at the beginning of his analysis, rather than in terms of the n distributions $R_1, \cdots, R_n$ derived from this distribution R^*.

Formally, this will mean that not only a C-game G^* in standard form (i.e., a Bayesian game G^*), but also any consistent I-game G in standard form, will now be defined as the ordered set

$$(17.1) \qquad G = G^* = \{S_1, \cdots, S_n; C_1, \cdots, C_n; V_1, \cdots, V_n; R^*\},$$

instead of being defined as the ordered set

$$(17.2) \quad G = \{S_1, \cdots, S_n; C_1, \cdots, C_n; V_1, \cdots, V_n; R_1, \cdots, R_n\}$$

in accordance with equation (3.18), which was our original suggestion. (Of course, in the case of an inconsistent I-game G we have to go on using equation (17.2) as our formal definition of the game.)

This now completes our general discussion of incomplete-information games G and of their Bayesian-game analogues G^*, which represent games with complete but imperfect information involving certain chance moves. We have discussed the actual solution concepts our approach selects for these games only in the case of a few simple illustrative examples (see Sections 9–11, Part II). More systematic discussion of these solution concepts for a wider range of games with incomplete information must be left for other forthcoming publications.

CHAPTER 9

USES OF BAYESIAN PROBABILITY MODELS IN GAME THEORY[1]

1 *Objective and subjective probabilities and rational behaviour.* Operationally, we can distinguish between *objective* and *subjective* probabilities as follows. Some probabilities are assigned the same numerical values by virtually all expert observers: for example, virtually all experts will agree that a fair coin will show either side with the same probability 1/2; or that a fair die will show any one of its six sides with the same probability 1/6; *etc.* Other probabilities will typically be assessed differently by different people: for example, different experts on horse racing will usually assign different numerical probabilities to a particular horse coming in first in a given race. Probabilities of the former type may be called objective probabilities, while those of the latter type may be called subjective probabilities.

Objective probabilities can be regarded as objective *structural properties* of certain physical (or biological or psychological or social) systems; and they manifest themselves in the actual statistical behaviour of these systems, *i.e.* in the long-run frequencies of various alternative outcomes (*e.g.* in the fact that the long-run frequency of HEADS in tossing a fair coin tends to be very close to 50 per cent). In contrast, subjective probabilities express *psychological attitudes* (expectations) held by various observers and decision makers, rather than structural properties of the objects observed and acted on.

[1] The author wants to thank the National Science Foundation for supporting this research through Grant SOC 77–06394 to the Center for Research in Management Science, University of California, Berkeley. He also wants to express his gratitude to the Center for Interdisciplinary Research (Zentrum für interdisziplinäre Forschung) of the University of Bielefeld, West Germany, for providing ideal working conditions for him during the academic year 1978–9.

The objective probabilities characterizing the behaviour of many physical systems can be ascertained only by detailed statistical observations. But for other physical systems these probabilities can often be predicted on the basis of theoretical considerations, more particularly on the basis of possible *symmetry* properties displayed by some of these systems (as in the case of a fair coin, or a fair die, *etc.*).

The distinction between objective and subjective probabilities gives rise to the following threefold classification of *individual* decision problems (*i.e.* of problems of *individual* rational behaviour). We say that a given decision maker acts under *certainty* when he can predict the actual outcome (*i.e.* the actual consequences) of any action he can take. We say that he acts under *risk* when he can at least predict the objective probabilities associated with different possible outcomes. Finally, we will say that he acts under *uncertainty* when some or all of these probabilities are unknown to him (or when they do not even exist as well-defined objective probabilities).

Since the subject of this paper will be the uses of subjective probabilities in game theory, it may be helpful to say something about the relationship between such *individual* decision problems and *game-theoretical* decision problems. (I will make additional comments on this relationship in Section **3** below.) Decision problems in general can be divided into those of *individual* decision, and those arising in a *social* setting. The latter in turn can be divided into *moral* decision problems and *game-theoretical* ones. A *moral* decision problem is a problem of how the different individual members of society can best promote the *common good*, *i.e.* can best promote the common interests of society as a whole. In contrast, a *game-theoretical* decision problem is a problem of how two or more rational individuals, with possibly very dissimilar interests, can best promote their *own* interests *against*, or possibly in cooperation *with*, one another. (Note that game situations involving conflicts of interest can arise not only among *egoists*, each of them trying to promote his personal self-interest, but rather also among *altruists*, trying to promote some altruistic objectives – as long as the latter are trying to promote different and more or less incompatible altruistic objectives.)

Of course, all three categories of individual, moral, and game-theoretical, decision problems come under the general heading of problems of *practical* rationality, *i.e.* of problems of rational choice among alternative *actions*; and thus contrast with problems of *theoretical* rationality, *i.e.* with problems of rational choice among alternative theoretical *beliefs*. (The problems of theoretical rationality in turn comprise both problems of *deductive*, and of *inductive*, rationality.) Thus, our classification of various problems of rationality can be summarized like this:

(A) *Theoretical* rationality.
- (A1) *Deductive* rationality.
- (A2) *Inductive* rationality.

(B) *Practical* rationality (rational behaviour).
- (B1) *Individual* rational behaviour.
  - (B1a) Under certainty
  - (B1b) Under risk
  - (B1c) Under uncertainty.
- (B2) *Moral* rationality.
- (B3) *Game-theoretical* rationality.

Leaving the problems of theoretical rationality on one side,[2] the rationality criteria (decision rules) appropriate for the various specific subcases of practical rationality can be described as follows. Already classical economic theory has established the fact that rational behaviour under *certainty* (case B1a) can be characterized as *utility maximization* (see, *e.g.* Debreu 1959: 55–9). Modern Bayesian decision theory has extended this result by showing that rational behaviour under *risk* and under *uncertainty* (cases B1b and B1c) is equivalent to *expected-utility maximization*. More particularly, in the case of risk, where all relevant probabilities are known to the decision maker, he must maximize an expected-utility expression that uses these known *objective* probabilities as probability weights; whereas in the case of uncertainty he must use his own *subjective* probabilities in lieu of the unknown objective probabilities. (For a discussion of the rationality axioms actually needed to establish the expected-utility maximization theorem, see Harsanyi 1978.)

In the case of *moral* rationality (case B2), I have tried to show that the appropriate rationality criterion is maximizing the *average utility*

[2] Modern logic provides a very specific and very convincing characterization of deductive rationality. In contrast, in my opinion, we still lack a satisfactory theory of inductive rationality – though it *might* be possible to develop such a theory by suitable extension of Bayesian decision theory.

level of all individuals in society (Harsanyi 1953; 1955; 1976, Chapters I to V; 1977a, Chapter 4; for a brief summary of my theory, see Harsanyi 1977b). Finally, in the case of *game-theoretical* rationality, the required rationality criteria are presumably provided by the various game-theoretical solution concepts. (For a critique of the solution concepts of classical game theory, and for suggestion of an alternative solution theory, see Harsanyi 1977e.)

Even though the five classes of practical decision problems (the problems of rational behaviour under certainty, risk, and uncertainty, as well as the problems of morality, and of game-theoretical rationality) all represent quite distinct analytical problems, in actual fact there has been a good deal of interaction among individual decision theory, moral theory, and game theory. For example, von Neumann and Morgenstern (1947) already found it useful to provide an axiomatic analysis of rational behaviour under *risk* in order to clarify certain game-theoretical problems. My own work on moral theory (see my papers quoted above) has used the same theory of rational behaviour under risk as a foundation for a utilitarian theory of morality. In this paper I now propose to discuss how the Bayesian theory of rational behaviour under *uncertainty* has been used to solve certain game-theoretical problems.

2 *Bayesian probability models for games with incomplete information.* Classical game theory uses probabilities for two main purposes. One is to characterize *random moves*. The other is to characterize various classes of randomized strategies, such as *mixed strategies* (both individually randomized and jointly randomized mixed strategies) and *behaviour strategies*. In both cases, the numerical probabilities employed are interpreted as *objective* probabilities, and are assumed to be generated by suitable random devices with fully known statistical properties.

The first major use of *subjective* probabilities in game theory has been, I believe, in analysing games with *incomplete information* (Harsanyi 1967–8). A notable feature of this approach has been that, whereas the games to be studied are originally described in terms of certain *subjective* probability distributions (which express the players' expectations about some game parameters unknown to them), in the actual course of the analysis these subjective probability distributions are replaced by *objective* probability distributions of the same mathematical forms.

To describe the nature of the problem, I have to discuss two related, but actually quite different, taxonomic distinctions: that between games with *perfect* and with *imperfect* information, and that between games with *complete* and with *incomplete* information. A game is said to involve *perfect* or *imperfect* information according as the players do or do not have full information at every stage of the game about all *past moves*, including all personal moves made by the various players, and all random moves decided by chance. On the other hand, a game is said to involve *complete* or *incomplete* information according as all players do or do not have full information about the very *nature* of the game as specified by the extensive form (*i.e.* the game tree), or at least as specified by the normal form (*i.e.* the payoff matrix).

Reflection will show that almost all real-life social situations are games with incomplete information. In most cases the players will have less than full information about each other's *payoff functions*. (First of all, they may be unable to predict the physical outcome to be produced by various possible strategy combinations. Secondly, they may not know other players' preference rankings among different possible physical outcomes. Finally, even if they do know these preference rankings (and, therefore, know the other players' ordinal utility functions), they may not know the other players' attitudes to risk taking (*i.e.* may not know the other players' cardinal utility functions).) Likewise, the players may not know the specific *strategies* available to other players. Finally, they may not know how much *information* the various players have about the other players' payoff functions and strategy sets.

Yet, in spite of the great importance that games with incomplete information have in empirical applications – as well as from a theoretical point of view – classical game theory was totally unable to analyse such games. (But it had no difficulty in dealing with games involving either *perfect* or *imperfect* information, as long as these were games with *complete* information.)

The problem, however, can easily be overcome with the help of Bayesian decision theory.

(i) Let G be an n-person game with incomplete information in normal form. We can always model the players' incomplete information by assuming it to be incomplete information about the *other* players. More specifically, we can use the following model. The role of each player i in the game can be played by any one of a number of individuals with different personal characteristics, different power positions, different resource endowments, *etc.* These different individuals will be called the possible *types* (or *subplayers*) of player i. They will be numbered as type $1, 2, \ldots, k_i, \ldots, K_i$, of player i. Incomplete information can now be modelled by the assumption that, whereas each player i will know his *own* actual type (*i.e.* he *will* know the number k_i), he will not know the *other* players' actual types (he will *not* know the numbers $k_1, \ldots, k_{i-1}, k_{i+1}, \ldots, k_n$).

In a game with complete information, the payoff function of any given player i can be written as $U_i(s_1, \ldots, s_i, \ldots, s_n)$, indicating that his payoff will depend on the strategies $s_1, \ldots, s_i, \ldots, s_n$ of players $1, \ldots, i, \ldots, n$. But, in our incomplete-information game, each player i's payoff function will have to be written as $U_i(s_1, \ldots, s_i, \ldots, s_n; k_1, \ldots, k_i, \ldots, k_n)$, to indicate that in general his payoff may depend, not only on the various players' strategies, but also on their actual types.

(ii) Assuming that all players follow the rationality postulates of Bayesian decision theory, each player i will express his expectations about the actual values of the unknown variables $k_1, \ldots, k_{i-1}, k_{i+1}, \ldots, k_n$ in the form of a *subjective* joint probability distribution $P_i = P_i(k_1, \ldots, k_{i-1}, k_{i+1}, \ldots, k_n)$. For more convenient notation, I will write $\overline{k_i} = (k_1, \ldots, k_{i-1}, k_{i+1}, \ldots, k_n)$, so that P_i can be written as $P_i = P_i(\overline{k_i})$.

Of course, the other players in general *will not know* the subjective probability distribution P_i entertained by player i. Within our model this fact can be expressed by the assumption that the different types of player i would in general entertain *different* subjective probability distributions P_i. (As in our model the other players will not know what player i's type actually is, this assumption ensures that they will not know the distribution P_i either.) To indicate this dependence of P_i on player i's actual type, we will denote the subjective probability distribution used by any given type k_i of player i as $P_i(\overline{k_i}|k_i)$. We will also use the notation

(2.1) $\Pi_i = \{P_i(\overline{k_i}|1), P_i(\overline{k_i}|2), \ldots, P_i(\overline{k_i}|K_i)\}$

for $i = 1, \ldots, n$.

Thus, Π_i is the set of *all* subjective probability distributions that would be used by the various types $1, 2, \ldots, K_i$ of a given player i.

Full characterization of the game will now obviously require, not only specification of game G itself, but also specification of the sets

$\Pi_1, \ldots, \Pi_n$. Accordingly, we will write $G^\circ = \{G; \Pi_1, \ldots, \Pi_n\}$. Thus, G° is simply game G, as enriched by information about the subjective probability distributions that would be used by the various possible types of every player.

(iii) The next step of our analysis will involve construction of a new game G^*, similar to game G°, except that all *subjective* probability distributions $P_i(\overline{k_i}|k_i)$ entertained by the various types k_i of any given player i will be replaced by *objective* probability distributions $R_i(\overline{k_i}|k_i)$ of the same mathematical forms.

To be more specific, in many social situations it will be natural to choose the subjective probability distributions $P_i(\overline{k_i}|k_i)$ in such a way that all of them will be obtainable as *conditional* probability distributions

(2.2) $$P_i(\overline{k_i}|k_i) = P_i(k_1, \ldots, k_{i-1}, k_{i+1}, \ldots, k_n|k_i) = R(k_1, \ldots, k_{i-1}, k_{i+1}, \ldots, k_n|k_i),$$

derived from the *same* basic joint probability distribution

(2.3) $$R = R(k_1, \ldots, k_{i-1}, k_i, k_{i+1}, \ldots, k_n).$$

If this is the case then the desired game G^* can be obtained by using the following simple model.

At the beginning of the game, nature conducts a "lottery" in order to choose the types $k_1, \ldots, k_n$ who will play the roles of players $1, \ldots, n$ in the game. This lottery will be governed by the joint probability distribution R (which, therefore, will have the nature of an *objective* probability distribution). This distribution R will be known to all players (and to all types of every player). Each player i will also know his own type but will be ignorant of the *other* players' actual types. Yet, since he will know the probability distribution R and will know his own type k_i, his information about the player types likely to represent the other players will correspond to the *conditional* probability distribution $R(k_1, \ldots, k_{i-1}, k_{i+1}, k_n|k_i)$. In view of (2.2) this *objective* conditional probability distribution will have the same mathematical form as the corresponding *subjective* probability distribution $P_i(k_1, \ldots, k_{i-1}, k_{i+1}, \ldots, k_n|k_i)$ occurring in game G° did.

Of course, in other social situations we may have to postulate subjective probability distributions P_i which *cannot* be derived from any given basic distribution R as conditional distributions. In such cases, a more complicated construction is required in order to obtain a suitable game G^*. (In particular, instead of having one "lottery" at the beginning of the game, we may need several "lotteries" at the end.) But it can be shown that it is *always* possible to construct a game G^* in which the subjective probability distributions $P_i(\overline{k_i}|k_i)$ are replaced by objective probability distributions of the same mathematical forms (see Harsanyi 1967–8: 486–502).

(iv) I now propose to argue that, once a game G^* with the desired properties has been constructed, this game will be game-theoretically *equivalent* to game $G°$. This follows from the general principle that Bayesian decision makers (including the players of a game if they follow the Bayesian rationality postulates) will always act on any given set of numerical probabilities in *exactly the same way*, regardless of whether these probabilities are *objective* probabilities based on the best information available to them, or are their own carefully chosen *subjective* probabilities (when the information available does not include a knowledge of the relevant objective probabilities).

Yet, this means that the problem of analysing a game with *incomplete* information, G (or $G°$), has been reduced to the problem of analysing a game with *complete* (even though *imperfect*) information, G^*. (G^* is a game with complete, yet with imperfect, information because the players' original ignorance about some basic parameters of the game has been replaced by an ignorance about the outcomes of certain random moves or "lotteries".)

These modelling techniques, involving an introduction of suitably chosen random moves, have opened up a completely new dimension of flexibility for game theory in the study of game situations. We now can produce games with any desired distribution of knowledge and of ignorance among the various players, and can conveniently study how alternative informational assumptions will change the nature of any given game. We can also study how any given player can *infer* some pieces of information originally denied to him, by observing the moves of some other players who already possess this information, and whose moves may in fact express this information. We can also investigate how each player can optimally *convey* information to some other players, or can optimally *withhold* information from them, in accordance with his own strategic interests in the game (see Harsanyi 1977d).

3 *The tracing procedure: a Bayesian approach to defining solutions for noncooperative games.* Another application of Bayesian ideas in game theory has been in the *tracing procedure*, which is an attempt to extend the Bayesian approach from one-person decision problems to the n-person decision problem of finding a solution for an n-person noncooperative game (Harsanyi 1975).

Most game theorists agree that the solution of any noncooperative game must be an *equilibrium point.*[3] This is so because any reasonable solution must be a strategy combination $s = (s_1, \ldots, s_n)$ with the property that no player i will have a clear incentive to *deviate* from his solution strategy s_i merely because he expects the other players to *follow* their own solution strategies $s_1, \ldots, s_{i-1}, s_{i+1}, \ldots, s_n$. But if any given player i's solution strategy s_i were *not* a best reply to the other players' solution strategies then he *would* have such an incentive. Consequently, *every* player's solution strategy *must* be in fact a best reply to the other players'. But this means that the solution must be an equilibrium point.

Nash (1951) has shown that every finite game (every game with a finite number of players, and with a finite number of pure strategies for each player) has at least one equilibrium point. But, unfortunately, equilibrium points are very far from being unique: almost every interesting game has more than one equilibrium point; indeed, it will often have a very large number, and may even have infinitely many. The purpose of the tracing procedure is to choose *one* specific equilibrium point of any noncooperative game as the solution of this game.

Let me now introduce some notations. Suppose that G is an n-person noncooperative game in which player i ($i = 1, \ldots, n$) has K_i different pure strategies, to be denoted as $a_{i1}, a_{i2}, \ldots, a_{iK_i}$. A mixed strategy s_i of player i will be a probability vector of the form $s_i = (s_{i1}, s_{i2}, \ldots, s_{iK_i})$, where $s_{ik} (k = 1,2, \ldots, K_i)$ is the probability that this mixed strategy s_i assigns to the pure strategy a_{ik}. A strategy combination of the n players will be written as $s = (s_1, \ldots, s_n)$. A strategy combination of the $(n-1)$ players *other* than player i will be written as $\bar{s}_i = (s_1, \ldots, s_{i-1}, s_{i+1}, \ldots, s_n)$. We will also write $s = (s_i, \bar{s}_i)$. Player i's payoff function will be written as $U_i(s_i, \bar{s}_i)$.

The tracing procedure is based on the following model. Any player $j \neq i$ will have to form expectations about the strategy that

[3] Let me remind the reader of the relevant definitions. A given strategy s_i of player i is a *best reply* to the strategies $s_1, \ldots, s_{i-1}, s_{i+1}, \ldots, s_n$ used by the other $(n-1)$ players if s_i maximizes player i's payoff $U_i(s_1, \ldots, s_{i-1}, s_i, s_{i+1}, \ldots, s_n)$ when all other players' strategies are kept constant. A given strategy combination $s = (s_1, \ldots, s_n)$ is called an *equilibrium point* if the strategy s_i of *every* player i is a best reply to the strategies $s_1, \ldots, s_{i-1}, s_{i+1}, \ldots, s_n$ of the other players.

player i will use in the game. In accordance with Bayesian theory, it is assumed that these expectations will take the form of a *subjective* probability distribution $p_i = (p_{i1}, p_{i2}, \ldots, p_{iK_i})$ over player i's pure strategies, where p_{ik} ($k = 1, 2, \ldots, K_i$) is the subjective probability that player j assigns to the hypothesis that player i will use his pure strategy a_{ik}.

It is assumed that *any* player $j \neq i$ will derive his subjective probability distribution p_i over i's pure strategies from the basic parameters of the game (*e.g.* from the payoff functions) by the *same* computation procedure, so that different players j will arrive at the *same* distribution p_i. This distribution p_i will be called the *prior probability distribution* over player i's pure strategies.

Mathematically, such a prior distribution p_i, being a probability distribution over i's pure strategies, will have the nature of a *mixed strategy* by player i. But of course it will have a very different intuitive interpretation from an ordinary mixed strategy. Whereas an ordinary mixed strategy s_i always involves the assumption that player i *himself* will randomize among his alternative pure strategies, a prior probability distribution p_i does not involve any such assumption. Rather, it merely expresses an *uncertainty* in the *other* players' minds about the actual strategy that player i will use in the game.

I will write $p = (p_1, \ldots, p_n)$, $\overline{p_i} = (p_1, \ldots, p_{i-1}, p_{i+1}, \ldots, p_n)$, and $p = (p_i, \overline{p_i})$. Thus, $\overline{p_i}$ is a probability vector summarizing player i's expectations about the behaviour of the *other* $(n-1)$ players' behaviour in the game, whereas p summarizes *all n* players' expectations about one another's behaviour.

Suppose we know the vector p. How can we then predict the outcome of the game? The simplest answer would be this. Since the expectations of each player i about his fellow players' behaviour are expressed by subvector $\overline{p_i}$ of vector p, he will use that particular strategy s_i^0 that is his *best reply* to $\overline{p_i}$, *i.e.* he will use that particular strategy $s_i = s_i^0$ that maximizes his expected payoff, defined by the expression

$$U_i(s_i, \overline{p_i}). \tag{3.1}$$

Accordingly, one may try to argue that the outcome (solution) of game G should be defined as the strategy combination $s^0 = (s_1^0, \ldots, s_i^0, \ldots, s_n^0)$, consisting of these best-reply strategies $s_1^0, \ldots, s_i^0, \ldots, s_n^0$. I will describe this view as the *naive Bayesian approach*.

Regrettably, this simple-minded naive Bayesian approach is

unacceptable because the strategy combination s^0 in general *will not be an equilibrium point* of game G.

Yet, even though this naive Bayesian approach would be an unsatisfactory solution theory, it can be used as a starting point for a more satisfactory theory. In fact, the tracing procedure always starts with this best-reply strategy combination s^0, but then systematically and continuously transforms this strategy combination until in the end it *converges* to an equilibrium point $s^* = (s_1^*, \ldots, s_n^*)$ of the game, which then will be chosen as the solution. Indeed, the tracing procedure is meant to model the process of *convergent expectations* by which rational players, starting from a state of expectational disequilibrium, can make their expectations converge upon a specific equilibrium point of the game.

The tracing procedure is based on a one-parameter family of auxiliary games G^t, with $0 \leq t \leq 1$. In any given game G^t, the payoff function U_i^t of each player i will be defined as

(3.2) $$U_i^t(s_i,\bar{s}_i) = (1-t)U_i(s_i,\bar{p}_i) + tU_i(s_i,\bar{s}_i).$$

Clearly, for $t = 1$, game $G^t = G^1$ will have the same payoff functions

(3.3) $$U_i^1(s_i,\bar{s}_i) = U_i(s_i,\bar{s}_i)$$

as the original game G has. Therefore, we can write $G^1 = G$. On the other hand, for $t = 0$, game $G^t = G^0$ will have payoff functions of form (3.1) since

(3.4) $$U_i^0(s_i,\bar{s}_i) = U_i(s_i,\bar{p}_i)$$

Equation (3.4) shows that game G^0 is a game of very special form which makes each player's payoff depend only on his *own* strategy s_i, and makes it independent of the other players' strategy combination $\bar{s}_i$. Consequently, G^0 can be factored into n independent maximization problems, one for each player, since each player's task in the game will be simply to maximize the expression $U_i(s_i,\bar{p}_i)$, regardless of what the other players are doing.

For each game G^t, let E^t be the set of all equilibrium points of G^t. By Nash's theorem (Nash 1951), each of these sets E^t will be nonempty. Let P be the graph of the correspondence $t \rightarrow E^t$, $0 \leq t \leq 1$. For any fixed t, this graph will have as many points as the number of equilibrium points in E^t. Accordingly, in nondegenerate cases, in the region $t = 0$, this graph P will have only *one* point x^0, because each player i will have a *unique* (pure) strategy $s_i = s_i^0$ maximizing the payoff function $U_i^0 = U_i(s_i,\bar{p}_i)$, so that the equilibrium-point set E^0 will contain only the *one* point $s^0 = (s_1^0, \ldots, s_n^0)$. Moreover, one can show that, in nondegenerate cases, graph P will contain exactly *one* continuous path L, leading from this one point x^0 in region $t = 0$ to some point x^1 in region $t = 1$. This point x^1 will always correspond to a specific equilibrium point $s^* = (s_1^*, \ldots, s_n^*)$ of the original game $G = G^1$, and this equilibrium point s^* will be chosen as the *solution* of game G.

The mathematical procedure selecting s^* as the solution is called the *tracing procedure* because it is based on tracing (*i.e.* following) this continuous path L from its starting point x^0 (corresponding to the strategy combination s^0) to its end point x^1 (corresponding to the solution s^*).

This version of the tracing procedure, called the *linear* tracing procedure, will always yield a unique solution s^* in all nondegenerate cases. But in degenerate cases it may yield several (or even infinitely many) solutions because

(i) Graph P may contain more than one point x^0 in region $t = 0$ (in which case it will always contain infinitely many points x^0).

(ii) Moreover, even if x^0 is unique, there may be several (or even infinitely many) continuous paths leading from point x^0 to points in region $t = 1$.

It can be shown, however, that a modified form of the procedure, called the *logarithmic* tracing procedure, will *always* yield a unique solution, even in highly degenerate cases.

The solution defined by the tracing procedure always has the form $s^* = T(G,p)$. In other words, the solution s^* defined by the tracing procedure will not only depend on game G for which the solution is defined, but will also depend, and often quite sensitively depend, on the vector $p = (p_1, \ldots, p_n)$ of prior probability distributions chosen as a starting point. This of course means that the tracing procedure in itself does not provide a complete solution theory for noncooperative games. It yields a complete solution theory only if it is supplemented by a theory of how to choose suitable *prior probability distributions* p_i for any given game G. But, in conjunction with such a theory, it does represent a very attractive approach to defining solutions for noncooperative games. (For a more detailed discussion of the tracing procedure, and for the proofs of the relevant mathematical theorems, see Harsanyi 1975. For a discussion of the prior distributions see Harsanyi 1977c. The last-mentioned paper also contains a general discussion of the Harsanyi–Selten solution theory, based on the tracing procedure, and on suitable procedures for constructing prior probability distributions. See also Harsanyi 1977e.)

4 *Conclusion*. I have tried to show on two examples that, even though individual decision theory and game theory are dealing with very different classes of decision problems, a use of Bayesian decision-theoretical tools in game theory can produce valuable results.

University of California, Berkeley
University of Bielefeld

REFERENCES

Debreu, Gerard. 1959. *Theory of Value*. New York.

Harsanyi, John C. 1953. Cardinal utility in welfare economics and in the theory of risk-taking. *Journal of Political Economy* **61**, 434–5. Reprinted in Harsanyi 1976: 3–5.

Harsanyi, John C. 1955. Cardinal welfare, individualistic ethics, and interpersonal comparisons of utility. *Journal of Political Economy* **63**, 309–21. Reprinted in Harsanyi 1976: 6–23.

Harsanyi, John C. 1967–8. Games with incomplete information played by 'Bayesian' players. Parts I–III. *Management Science* **14**, 159–82, 320–34, and 486–502.

Harsanyi, John C. 1975. The tracing procedure. *International Journal of Game Theory* **4**, 61–94.

Harsanyi, John C. 1976. *Essays on Ethics, Social Behavior, and Scientific Explanation*. Dordrecht.

Harsanyi, John C. 1977a. *Rational Behavior and Bargaining Equilibrium in Games and in Social Situations*. Cambridge.

Harsanyi, John C. 1977b. Morality and the theory of rational behavior. *Social Research* **44**, 623–56.

Harsanyi, John C. 1977c. A solution concept for *n*-person noncooperative games. *International Journal of Game Theory* **5**, 211–25.

Harsanyi, John C. 1977d. Time and the flow of information in noncooperative games. In *Quantitative Wirtschaftforschung*, ed. H. Albach *et al.*, pp. 255–67. Tübingen.

Harsanyi, John C. 1977e. A solution theory for noncooperative games and its implications for cooperative games. *Working Paper CP–401 Center for Research in Management Science*. Berkeley.

Harsanyi, John C. 1978. Bayesian decision theory and utilitarian ethics. *American Economic Review Papers and Proceedings* **68**, 223–8.

Nash, John F. 1951. Noncooperative games. *Annals of Mathematics* **54**, 286–95.

von Neumann, John & Morgenstern, Oskar. 1947. *Theory of Games and Economic Behavior*, 2nd ed., Princeton.

SECTION D

NONCOOPERATIVE MODELS FOR COOPERATIVE GAMES

CHAPTER 10

AN EQUILIBRIUM-POINT INTERPRETATION OF STABLE SETS AND A PROPOSED ALTERNATIVE DEFINITION*

The paper argues that the von Neumann-Morgenstern definition of stable sets is unsatisfactory because it neglects the destabilizing effect of *indirect* dominance relations. This argument is supported both by heuristic considerations and by construction of a bargaining game $B(G)$, formalizing the bargaining process by which the players agree on their payoffs from an n-person cooperative game G. (G itself is assumed to be given in characteristic-function form allowing side payments.) The strategies ρ_i the players use in this bargaining game will determine which payoff vectors x will be *stationary*, i.e., will have the property that, should such a payoff vector x be proposed to the players, all further bargaining will stop and x will be accepted as the outcome of the game. It will be suggested that a stable set should be defined as the set V of all stationary payoff vectors x, on the assumption that the players' bargaining strategies ρ_i will form a canonical equilibrium point ρ in the bargaining game $B(G)$.

In a certain class of games, to be called absolutely stable games, indirect dominance relations turn out to be irrelevant, and in these games the suggested definition for stable sets is equivalent to the von Neumann-Morgenstern definition. But in general this is not the case. However, if we assume that the players will adopt bargaining strategies deliberately discouraging any use of indirect dominance relations, then, in *every* game, the stable sets our model yields will always be von Neumann-Morgenstern stable sets.

At the end of the paper, we briefly discuss a possible modification in the suggested definition for stable sets, based on a modified bargaining game $B^0(G)$, in which the players are permitted to accept cuts in their provisional payoffs if they think that this move will increase their final payoffs.

I. HEURISTIC DISCUSSION

1. Introduction

The von Neumann-Morgenstern stable sets (solutions) of an n-person cooperative game G are based on certain implicit assumptions about the nature of the bargaining process among the n players. We shall try to make these assumptions explicit, and shall argue that they neglect the destabilizing effect of *indirect* dominance relations. We shall support this argument by constructing a bargaining game $B(G)$ to formalize the bargaining process through which the players are supposed to agree on their payoffs x_i from game G.

* Processed by Professor Melvin F. Shakun, Departmental Editor for Game Theory and Gaming and Associate Editor William Lucas; received December 28, 1971, revised January 2, 1973. This paper has been with the author 6 months for revision.

The author wishes to express his thanks to the National Science Foundation for supporting this research by grants GS-2091 and GS-3222 through the Center for Research in Management Science, University of California, Berkeley. Special thanks are due also to Mr. Robert J. Weber, of Cornell University, who was one of the referees for this paper. He made many valuable suggestions, which include suggesting Part (b) of Theorem 1 in its present form, and suggesting the proof of Lemma 9. For helpful comments received, the author also wants to thank Professor Robert J. Aumann, of the Hebrew University, Jerusalem, Israel; Professor Reinhard Selten, of the University of Bielefeld, West Germany; Dr. Lloyd S. Shapley, of RAND Corporation, Santa Monica; and Professor Pravin P. Varaiya, of the University of California, Berkeley.

The strategies ρ_i the players use in this bargaining game $B(G)$ will determine which particular payoff vectors x will be "stationary," that is, which will have the property that, should such a payoff vector x be proposed to the players, all further bargaining will stop, and this payoff vector x will become the outcome of the game. The same strategies ρ_i will also select one specific stationary payoff vector x as the actual outcome. We shall argue that a *stable set* should be defined as the set V of all stationary payoff vectors x, on the assumption that the players will use bargaining strategies ρ_i corresponding to an equilibrium point ρ with suitable properties (to be called a canonical equilibrium point) in the bargaining game $B(G)$.

Thus, under this model, the strategies ρ_i the players use in this bargaining game $B(G)$ play two roles. One is to select one specific imputation x as the outcome of the game. The other is to define a stable set V, typically containing many other imputations $y \neq x$, besides the one imputation x actually achieved by these strategies. In this sense, the players' bargaining strategies select, not only one specific imputation, but rather a whole "standard of behavior" (von Neumann and Morgenstern, [8, 1947, pp. 31–45]).

We shall find that in a certain class of games, to be called absolutely stable games, indirect dominance relations are irrelevant; and in these games our definition will yield the same stable sets V as the von Neumann-Morgenstern definition does. But, in other games, in general, this will not be the case.

However, as we shall see, if we assume that the players will adopt bargaining strategies deliberately discouraging any use of indirect dominance relations, then, in *every* game, the stable sets our model yields will be stable sets corresponding to the von Neumann-Morgenstern definition.

Nash [5, 1951, p. 295] has suggested that we can obtain a clearer understanding of the alternative solution concepts proposed for cooperative games, and can better identify and evaluate the assumptions they make about the players' bargaining behavior, if we reconstruct them as equilibrium points in suitably defined bargaining games—treating the latter formally as noncooperative games. This paper is meant to be a step towards implementing Nash's program.

For convenience, we shall restrict our discussion to "classical" cooperative games *with* side payments, though our main results can be extended also to games *without* side payments (Aumann and Peleg, [2, 1960]). Unlike von Neumann and Morgenstern, we shall assume that each cooperative game G is given in *characteristic-function form*. That is, we shall assume that the characteristic function v is directly specified by the very definition of game G, instead of being derived from the normal form of G.

2. Definitions

The set of all n players will be called N. The characteristic function v is a function assigning a real number $v(S)$ to each subset S of N. We shall assume that v has the usual properties:

(1) $$v(\emptyset) = 0,$$

where $\emptyset$ is the empty set, and

(2) $$v(S \cup T) \geqq v(S) + v(T)$$

if S and T are disjoint subsets of N. Property (2) is called *superadditivity*.

We shall also assume that v is given in *zero-reduced* form so that

(3) $$v(S) = 0 \quad \text{if} \quad |S| = 1,$$

where $|S|$ is the number of players in S. (This number will sometimes be denoted also as s.) Finally, we shall assume that v is *essential* so that

(4) $$v(N) > \sum_{i\in N} v(\{i\}) = 0.$$

A given coalition $S \subseteq N$ is called *effective* (Maschler, [4, 1963, p. 233]) if, for all possible nonempty subcoalitions R and T with $R \cup T = S$ and $R \cap T = \emptyset$, we have

(5) $$v(S) > v(R) + v(T).$$

Every one-person coalition $S = \{i\}$ is always effective because it vacuously satisfies (5). Our essentiality requirement (4) is equivalent to requiring that the game should contain at least one effective coalition S containing *more* than one player.

By (2) and (3), the highest value $v(S)$ any coalition S can have is

(6) $$v(S) = v(N) - \sum_{i\notin S} v(\{i\}) = v(N),$$

whereas the lowest value it can have is

(7) $$v(S) = \sum_{i\in S} v(\{i\}) = 0.$$

S is called a *winning coalition* if it satisfies (6), and is called a *losing coalition* if it satisfies (7). A given game G is called a *simple game* (von Neumann and Morgenstern, [8, 1947, Chap. X]) if every coalition in G is either winning or losing.

A game G in characteristic-function form is defined as follows. The task of the players is to form the all-player coalition N, or else to partition themselves into two or more disjoint coalitions $S_1, S_2, \cdots$. The s members of any coalition S can achieve any *coalitional payoff vector* $x^S = (x_i, \cdots, x_j)$, where $S = \{i, \cdots, j\}$, subject to

(8) $$\sum_{i\in S} x_i \leqq v(S)$$

and to

(9) $$x_i \geqq v(\{i\}) = 0 \quad \text{for all} \quad i \in S.$$

The resulting payoff vector $x = (x_1, \cdots, x_n)$ for the n players will always satisfy

(10) $$\sum_{i\in N} x_i \leqq v(N)$$

and

(11) $$x_i \geqq v(\{i\}) = 0 \quad \text{for all} \quad i \in N.$$

A payoff vector x satisfying (10) and (11) is called a *semi-imputation*. In what follows, whenever we refer to a payoff vector for the n players we shall always mean a semi-imputation.

If a semi-imputation x also satisfies the efficiency condition

(12) $$\sum_{i\in N} x_i = v(N),$$

then it is called an *imputation*. A given imputation x is called an *interior* or an *exterior* imputation, according as it is an interior or an exterior point of the set X of all imputations. Equivalently, an imputation x is called *interior* if it assigns positive payoffs $x_i > 0$ to all n players; and it is called *exterior* if it assigns zero payoff(s) $x_i = 0$ to some player(s).

We say that a given payoff vector $y = (y_1, \cdots, y_n)$ *dominates* some payoff vector $x = (x_1, \cdots, x_n)$ through a particular coalition $S \subseteq N$, $S \neq \emptyset$, if

(13) $$y \in X,$$

where X is the set of all imputations, and if

(14) $$y_i > x_i \quad \text{for all} \quad i \in S$$

and, if at the same time,

(15) $$\sum_{i \in S} y_i \leqq v(S).$$

(Condition (13) is usually not included in the definition of dominance, but for our purposes it is convenient to do so.) Thus, y must yield every player i in S a *higher* payoff $y_i > x_i$ than x would; and coalition S must also have the *power* to achieve these higher payoffs, which will be the case if their sum $\sum y_i (i \in S)$ does not exceed the quantity $v(S)$.

The statement "y dominates x" will mean that y dominates x through *some* coalition(s) $S \subseteq N$.

A given set V of imputations in game G is called a von Neumann-Morgenstern stable set (or solution) if it has the following two properties:

(16) If any imputation x *is* in V,
then x is *not* dominated by any imputation y in V; and

(17) If any imputation x is *not* in V,
then x *is* dominated by some imputation y in V.

Property (16) is called *internal stability* whereas property (17) is called *external stability*. It is easy to verify that if V is a stable set in this sense, then it will have external stability also in the following stronger form:

(18) If any semi-imputation x is not in V, whether it is an imputation or not, it will be dominated by some imputation y in V.

As we have seen, the definition of a game G given in characteristic-function form implies that the outcome of the game will be a semi-imputation (or that it will be an imputation if we assume efficient behavior on the part of the players). But this does not answer the question why the players should be interested in choosing an imputation from some particular stable set V of the game. This question can be answered only by making more specific assumptions about the bargaining process among the players.

3. The Bargaining Process Underlying the Definition of Stable Sets

In defining stable sets, von Neumann and Morgenstern have presumably envisaged a bargaining process of the following kind.

At first, one of the players proposes some payoff vector x. Then, some player coalition S may make a counter proposal, suggesting an alternative payoff vector y. But this counter proposal will not be taken seriously, unless y dominates x through this coalition S. That is, the s players i in S must be able to show that they have the *power* to achieve the payoffs y_i they are proposing for themselves, and that they also have an *interest* in demanding these payoffs y_i rather than accepting the payoffs x_i previously proposed for them. If y does dominate x through S, then y will now replace x as the payoff vector under consideration.

Then again, another coalition S' may propose still another payoff vector y', which will in turn replace y as the payoff vector under consideration if y' dominates y through coalition S', etc.

This bargaining process will come to an end if the last payoff vector z under consideration is no longer challenged by anybody (at least for some stipulated period of time), i.e., if no coalition T proposes any payoff vector z' dominating z through this coalition T.[1] If this situation arises, then this payoff vector z will become the final payoff vector of the game, and the players will receive the corresponding payoffs $z_1, \cdots, z_n$.

A very important additional assumption underlying the concept of a stable set is this. Suppose that in this bargaining process the players' expectations come to be focused on some particular stable set V, so that they expect all payoff vectors x in V to be *stable*, and all payoff vectors x' not in V to be *unstable*. Then, these very expectations will give them incentives to behave in such a way as to *justify* these expectations.[2] More particularly,

(A) All payoff vectors in V will be in fact *stable* in the sense that, once such a payoff vector x has come under consideration by the players, their discussion will *not* shift to any other payoff vector y thereafter, because no payoff vector y dominating x will be proposed by any coalition S. Thus, x will become the final payoff vector of the game.

(B) In contrast, all payoff vectors x' not in V will be in fact *unstable* in the sense that, even if such a payoff vector x' came to be proposed and to be considered by the players, their discussion *would* soon shift to an alternative payoff vector y proposed by some coalition S and dominating x through S. Thus, x' would not become the final payoff vector of the game.

How can we justify assumptions (A) and (B) in terms of the two defining properties (16) and (17) of stable sets? To justify (A), the following argument seems to suggest itself:

(A*) In view of (16), if x is in V, then any payoff vector y dominating x must lie outside V and therefore, according to assumption (B), any such y will be unstable. Thus, no coalition S will find it profitable to get x replaced by such a payoff vector y. Even though y would yield higher payoffs $y_i > x_i$ than x would yield to the s players i in S, they *could not retain* these higher payoffs y_i because y, being unstable, would soon be itself replaced by still another payoff vector y' proposed by some coalition S' and dominating y through S'.

To justify assumption (B), one may argue as follows:

(B*) In view of (17), if payoff vector x' is not in V, then it must be dominated by some payoff vector y which does lie in V and which therefore, according to assumption (A), will itself be stable. Hence, there will always be some coalition S that will find it profitable to get x' replaced by such a payoff vector y. This is so because, y being itself stable, the s players i in S will be able in this case to *retain* the higher payoffs $y_i > x_i$ corresponding to y.

By (A*), we need assumption (B) to justify assumption (A); and, conversely, by (B*), we need (A) to justify (B). But the question is, are (A*) and (B*) valid arguments? It seems to me that (B*) in itself is valid enough—provided that all payoff vectors y in V are in fact stable as required by assumption (A). But—at least in the absence of some alternative justification (see §10 below)—assumption (A) itself will be, in general, false because argument (A*) is open to the following objection.

[1] In many games (viz. in those with empty cores), *every* payoff vector z *could* be challenged by some players—but, it may not always be profitable for them to do so (see below).

[2] In contrast, let V^0 be some set of payoff vectors *not* possessing properties (16) and (17), and therefore *not* having the nature of a stable set. Nevertheless, suppose that the players would somehow come to expect all payoff vectors x in V^0 to be stable, and all payoff vectors x' not in V^0 to be unstable. Then, these very expectations would give them incentives to behave in such a way as to *falsify* these expectations.

Suppose that coalition S does get payoff vector x replaced by some unstable payoff vector y, and that the latter in turn is later replaced by a third payoff vector y'. Then, this payoff vector y' itself may very well be a stable imputation lying in set V, and may very possibly yield every player i in S a higher payoff $y'_i > x_i$ than the first payoff vector x would have yielded him. But, should this be the case, then every player i in S will have obtained a *lasting benefit* by getting x replaced with y (so as to permit another coalition S' in turn to get y replaced with the desirable payoff vector y'). Hence, a given payoff vector x may turn out to be *unstable*, even though it does belong to set V.

4. Indirect Dominance

More generally, a given payoff vector x in V may turn out to be unstable because some coalition $S = S^1$ may wish to take advantage of what we shall call a monotone chain of dominance relations. We shall speak of a *chain* of dominance relations in a given game G if there is a sequence of payoff vectors $y^1, \cdots, y^k, \cdots, y^K$ with $K \geqq 3$, and there is a corresponding sequence of coalitions $S^1, \cdots, S^k, \cdots, S^{K-1}$, with the property that

(19) Each payoff vector y^k $(k = 1, \cdots, K - 1)$ is dominated by the next payoff vector y^{k+1} through coalition S^k.

In view of (13), property (19) implies that all y^k with $k \geqq 2$ must be imputations. For convenience, we shall call any sequence of payoff vectors $y^1, \cdots, y^K$ itself a chain if it has property (19) with respect to some sequence of coalition $S^1, \cdots, S^{K-1}$.

We shall speak of a *monotone chain* of dominance relations, as well as of a monotone chain of payoff vectors $y^1, \cdots, y^K$, if these payoff vectors y^k also have the additional property that

$$(20) \qquad y_i^k < y_i^K \quad \text{for all} \quad i \in S^k, \quad \text{and for} \quad k = 1, \cdots, K - 1.$$

Here y_i^k and y_i^K denote the ith components of y^k and of y^K, respectively.

Finally, we shall say that a given payoff vector x is *indirectly dominated* by another payoff vector z if there exists a monotone chain $y^1, \cdots, y^K$ connecting $x = y^1$ with $z = y^K$. In contrast to the indirect dominance relation just defined, the relation defined by (13), (14), and (15) will sometimes be called *direct dominance*.

We have included condition (20) in our definition of indirect dominance, because the members i of any coalition S^k cannot be expected to cooperate in getting payoff vector x^k replaced by the next payoff vector x^{k+1}; unless the likely end result of this move, the last payoff vector x^K of the chain, will yield them higher payoffs $y_i^K > y_i^k$ than the payoff vector y^k to be replaced would yield them.

In this terminology, our objection can be restated as follows. Assumption (A) lacks general validity because a given imputation x in some stable set V may be *indirectly* dominated by another imputation z in the same stable set V. To be sure, as V will possess property (16), x cannot be *directly* dominated by another imputation z in V, but, in general, we cannot exclude the possibility of indirect dominance. Yet, if x is, in fact, indirectly dominated by some stable imputation z in V, this will tend to make x unstable, contrary to assumption (A).

This fact suggests the following definition. Let V be a stable set possessing internal stability in the following stronger form, going beyond property (16):

(21) If some imputation x is in V, then x is not dominated, either directly or indirectly, by any imputation z in V.

We shall call property (21) *indirect* internal stability. If any stable set V has this property, then it will be called *strictly stable.*[3]

We shall call a given game G itself *strictly stable* if *all* stable sets V in G are strictly stable.

Thus, we reach the following conclusion. Assumptions (A) and (B) are always justified in the case of strictly stable sets V. But they are, in general, unjustified in the case of stable sets lacking strict stability—or, more exactly, as we shall see in Section 10, they can be justified only by additional assumptions about the strategies the players will follow in the bargaining game $B(G)$.

5. Trivial indirect dominance relations and absolutely stable games

We shall now give some examples for strictly stable sets V and for strictly stable games G. But, first, we need a few more definitions.

Suppose that payoff vector x is indirectly dominated by payoff vector y. We shall call this indirect dominance relation *trivial* if x is at the same time also directly dominated by y. Otherwise, we shall call this indirect dominance relation *nontrivial.*

A given game G will be called *absolutely stable* if *every* indirect dominance relation of one imputation over another imputation in G is trivial. This terminology is motivated by the following lemma.

LEMMA 1. *Every absolutely stable game G is strictly stable.* (*That is, if all indirect dominance relations between imputations in G are trivial, then all stable sets V in G are strictly stable.*)

PROOF. Let V be a stable set in G, and let x and y be imputations in V. We have to show that x cannot be even indirectly dominated by y. Now, if x were in fact indirectly dominated by y, then it would also be directly dominated by y (because in G all indirect dominance relations between imputations are trivial). But this is impossible since V has property (16). Hence, x cannot be indirectly dominated by y.

THEOREM 1. *Games of the following two classes are always absolutely stable:*

(a) *Simple games, and*

(b) *Games where all effective coalitions are* 1 *or* $(n-1)$ *or n-person coalitions* (*which include all two- and three-person games*).

PROOF. In a simple game G, let $y^1, \cdots, y^K$ be a monotone chain of imputations with respect to the coalitions $S^1, \cdots, S^{K-1}$. We have to show that the indirect dominance of y^K over y^1 is trivial, i.e., that y^K also directly dominates y^1. By (20),

$$y_i^K > y_i^1 \quad \text{for all} \quad i \in S^1. \tag{22}$$

In a simple game, dominance can occur only through winning coalitions. Therefore, $v(S^1) = \cdots = v(S^{K-1}) = v(N)$. Since y^K is an imputation, we have

$$\textstyle\sum_{i \in S^1} y_i^K \leqq v(N) = v(S^1). \tag{23}$$

Consequently, by (22) and (23), y^K does directly dominate y^1 through S^1.

Next, suppose that $y^1, \cdots, y^K$ is a monotone chain in a game G' of class (b). In games of this class, dominance over an imputation can occur only through $(n-1)$-

[3] Our concept of *strictly stable* sets is somewhat similar to Vickrey's [1959] concept of *strong* solutions. Yet, the two concepts are not equivalent. For example, as we shall see (Theorem 1 in Section 5, below), all stable sets of any three-person game are always strictly stable. In contrast, as Vickrey has shown [*op. cit.*, pp. 216–19], some of these stable sets are *not* strong solutions.

person coalitions. Therefore, $|S^1| = \cdots |S^{K-1}| = n - 1$. Let $S^1 = N - \{j\}$ and $S^2 = N - \{m\}$. Without loss of generality, we can assume that $S^1 \neq S^2$, because if $S^1 = S^2$, then we can simply use the shorter sequence $y^1, y^3, \cdots, y^K$, which, in this case, will be itself a monotone chain with respect to $S^1, S^3, \cdots, S^{K-1}$. Therefore, $j \neq m$, and $j \in S^2$. Hence, by (20)

$$(24) \qquad y_j^K > y_j^2.$$

By (15)

$$(25) \qquad \sum_{i \in S^1} y_i^2 \leqq v(S^1).$$

Since y^K and y^2 are both imputations, in view of (24) and (25), we can write

$$(26) \qquad \sum_{i \in S^1} y_i^K = v(N) - y_j^K < v(N) - y_j^2 = \sum_{i \in S^1} y_i^2 \leqq v(S^1).$$

Consequently, by (22) and (26), y^K again directly dominates y^1 through coalition S^1. This completes the proof of Theorem 1.

Theorem 1 shows that, for games of classes (a) and (b), assumptions (A) and (B) of §3 are always valid. But, in general, for other games, this will not be the case.

II. FORMAL ANALYSIS OF THE BARGAINING GAME

6. Definition of the Bargaining Game

Let $0^* = (0, \cdots, 0)$ be the zero vector, i.e., an n-vector whose n components all are zero. Let X be the set of all imputations in game G. We define

$$(27) \qquad X^* = X \cup \{0^*\}.$$

Thus X^* is the set of imputations enlarged by including the zero vector.

We shall assume that the bargaining game $B(G)$ is divided into stages $1, \cdots, k, \cdots$. At each stage k, the state of the game can be characterized by some payoff vector $x^k \in X^*$ under active consideration by the players, to be called the *current payoff proposal* x^k for stage k.

We shall say that a given coalition $S \subseteq N$ makes an *effective* counter proposal at stage k if the s members of S unanimously propose some imputation y^k dominating the current payoff proposal x^k through this coalition S. If this happens, then this imputation y^k will become the payoff proposal $x^{k+1} = y^k$ for the next stage $(k + 1)$. In order to handle the case where two or more coalitions would like to make different effective counter proposals to x^k, we shall introduce an extra player to act as *chairman* during the negotiations, and to decide which particular coalition $S = S^k$ "should have the floor" at any stage k. This chairman will be called player C or player $(n + 1)$. The n players other than chairman C will be called *regular players*. The set of all $(n + 1)$ players will be called M. Thus

$$(28) \qquad M = N \cup \{n + 1\}.$$

Player C can be regarded as a personification of all factors that enable any particular coalition S at any stage k of the game to get organized, and to make an effective counter proposal y^k to the current payoff proposal x^k, *before* other coalitions $S' \neq S$ could do so.

Let $\mathcal{S}$ be the set of all $(2^n - 1)$ nonempty subsets of N. We shall assume that the bargaining game $B(G)$ is played as follows. At each stage k (except for the last stage $k = K$ if the game ends after a finite number of stages), player C will choose some coalition $S^k \in \mathcal{S}$ for making a counter proposal to x^k. Then, every player i in this coalition S^k will independently choose some imputation $y^{ki} \in X$. Finally, the payoff proposal x^{k+1} for the next stage $(k + 1)$ will be determined as follows:

(29) $x^{k+1} = y^k$ if all players i in S^k have chosen the *same* imputation $y^{ki} = y^k$, and if this imputation y^k dominates x^k through S^k; but

(30) $$x^{k+1} = x^k \quad \text{otherwise.}$$

The game will end at stage K if case (30) arises at stage $(K - 1)$, that is, if

(31) $$x^K = x^{K-1} = z.$$

If this happens, then the last payoff proposal $z = x^K$ will become the *final payoff vector* of the game, and the n regular players will receive the corresponding payoffs $z_1, \cdots, z_n$. Accordingly, at stage K none of the $(n + 1)$ players will have a move any more.

If case (30) never arises, then the game will go on indefinitely. In this event, we define the *final payoff vector* of the game as

(32) $$z = 0^* = (0, \cdots, 0),$$

which will yield every regular player $i \in N$ a payoff $z_i = 0$.

We want player C to have an incentive at every stage k to give the floor to one of those coalitions S he thinks are likely to make an effective counterproposal y^k to the current payoff proposal x^k. Therefore, we shall assume that at each stage k player C will receive a *premium* α^k to be defined as follows:

(33) $$\alpha^k = (\tfrac{1}{2})^k \quad \text{if} \quad x^{k+1} \neq x^k, \qquad \text{but}$$

(34) $$\alpha^k = 0 \quad \text{if} \quad x^{k+1} = x^k.$$

The *final payoff* of player C will be

(35) $$\alpha^* = \sum_{k=1}^{K} \alpha^k,$$

where K is the number of stages through which the game passes on that particular occasion, where K may be finite or infinite. (These definitions ensure that α^* will be finite even if $K = \infty$.)

The payoff proposal $x^1 = u$ for stage 1 will be called the *initial payoff vector*. We shall assume that in $B(G)$ the latter will always be the zero vector

(36) $$x^1 = u = 0^* = (0, \cdots, 0).$$

(That is, in the absence of any alternative agreement, all regular players i would receive zero payoffs $u_i = 0$.)

However, for mathematical convenience, we shall also consider bargaining games $B(G, u)$ with $u \in X^*$, where u need not be the zero vector but rather may be an imputation $u \in X$.

We shall also consider bargaining games $B^*(G, u, S)$ with $u \in X^*$ and with $S \in \mathcal{S}$, defined as bargaining games having a given vector u as initial payoff vector, and starting from a position where a given coalition $S^1 = S$ already has the floor (without any prior move by player C), so that in these games the first move belongs to the s players in this coalition S. Apart from these differences in the initial payoff vector u, and in the way the coalition S^1 having the floor at stage 1 is selected, we shall assume that all these games $B(G, u)$ and $B^*(G, u, S)$ will be played according to the *same* rules as $B(G)$ is. The games $B(G, u)$, including the game $B(G) = B(G, 0^*)$, will be called bargaining games of the *first kind*. The games $B^*(G, u, S)$ will be called bargaining games of the *second kind*.

Both kinds of bargaining games will be useful auxiliary concepts in our analysis. However, our main interest will be always in the bargaining game $B(G) = B(G, 0^*)$.

7. Strategies and Equilibrium Points

We shall define a strategy σ_i for any player $i \in M$ in such a way that σ_i will determine his behavior, not only in one specific bargaining game $B(G, u)$ or $B^*(G, u, S)$, but rather in *all* bargaining games B or B^* derived from a given cooperative game G.

Let A^k be the sequence $A^k = (S^1, \cdots, S^k)$, where $S^1, \cdots, S^k$ are the coalitions holding the floor at stages $1, \cdots, k$ (as a result of player C's choice—or, in the case of coalition S^1, possibly without any prior move by player C).

Let $b^0 = u = x^1$; and let b^k (for $k = 1, 2, \cdots$) be the ordered set $b^k = (y^{ki}, \cdots, y^{kj})$, where $S^k = \{i, \cdots, j\}$. Thus, for $k \geqq 1$, b^k consists of the imputations chosen by the s^k players in coalition S^k. Let B^k be the sequence $B^k = (b^0, b^1, \cdots, b^k)$.

Then, the pair $d^k = (A^{k-1}, B^{k-1})$ will provide full information about the initial situation, and about the $(n + 1)$ players' moves up to the point where player C has to choose some coalition S^k at stage k. Let D be the set of all possible pairs d^k. Accordingly, a strategy σ_{n+1} of player C (player $(n + 1)$) will be an arbitrary function from this set D to the set $\mathcal{S}$ of all nonempty coalitions. If player C uses a given strategy σ_{n+1}, then at any stage $k \neq K$ he will choose the coalition

$$S^k = \sigma_{n+1}(d^k). \tag{37}$$

On the other hand, the pair $e^k = (A^k, B^{k-1})$ will provide full information about the $(n + 1)$ players' moves up to the point where player C has already chosen some coalition S^k at stage k, and the next move is up to the s^k players in S^k. Let E be the set of all possible pairs e^k. Then, a strategy σ_i of any regular player $i \in N$ will be an arbitrary function from this set E to the set X of all imputations. If some regular player i uses a given strategy σ_i, then, at any stage k where $S^k \ni i$, he will choose the imputation

$$y^{ki} = \sigma_i(e^k). \tag{38}$$

(It is sufficient to assume that the function σ_i is defined for those pairs $e^k = (A^k, B^{k-1})$ only where the last element S^k of the sequence A^k satisfies $S^k \ni i$. But we can just as well assume that σ_i is defined for all possible pairs e^k.) The set of all strategies σ_i available to any player $i \in M$ will be called $\sum_i$.

Since $A^k = (S^1, \cdots, S^{k-1}, S^k) = (A^{k-1}, S^k)$, we shall sometimes make the dependence of y^{ik} on S^k more explicit by writing (38) in the form:

$$y^{ki} = \sigma_i(e^k) = \sigma_i(A^k, B^{k-1}) = \sigma_i(A^{k-1}, S, B^{k-1}), \tag{38a}$$

where $S = S^k$.

A strategy $(n + 1)$-tuple of the $(n + 1)$ players will be written as $\sigma = (\sigma_1, \cdots, \sigma_{n+1})$. The n-tuple which remains if the ith component σ_i is omitted from σ will be called σ^i. Thus $\sigma^i = (\sigma_1, \cdots, \sigma_{i-1}, \sigma_{i+1}, \cdots, \sigma_{n+1})$.

In any game $B(G, u)$ and $B^*(G, u, S)$, the payoff function of any player $i \in M$ will be written as[4]

[4] This will cause no ambiguity because it will be always clear from the context whether the payoff functions of game $B(G, u)$ or of game $B^*(G, u, S)$ [or of both games] are meant.

(39) $U_i = U_i(\sigma; u) = U_i(\sigma_1, \cdots, \sigma_{n+1}; u) = U_i(\sigma_i, \sigma^i; u) = U_i(\sigma_S, \sigma_{S^0}, \sigma_{n+1}; u)$.

Here σ_S is the s-tuple of strategies used by the s regular players in coalition S, while σ_{S^0} is the $(n - s)$-tuple of strategies used by the $(n - s)$ regular players in the complementary coalition $S^0 = N - S$.

We shall also use the vector notation

(40) $$U = U(\sigma; u),$$

where $U = (U_1, \cdots, U_n)$ is called the vector payoff function for the n regular players. Note that under our definition U does not include the payoff function U_{n+1} of player C.

In any game $B(G, u)$ or $B^*(G, u, S)$, a given strategy σ_i of any player $i \in M$ will be called a *best reply* to the strategy combination σ^i used by the other n players if

(41) $$U_i(\sigma_i, \sigma^i; u) = \max_{\tau_i \in \Sigma_i} U_i(\tau_i, \sigma^i; u).$$

If requirement (41) is satisfied for *all* $(n + 1)$ components of a given strategy $(n + 1)$-tuple σ, then σ is called an *equilibrium point* in game $B(G, u)$ or $B^*(G, u, S)$ (Nash, [5, 1951]).[5]

Let x^k be the current payoff proposal for stage k. Suppose the strategies σ_i of all players i in a given coalition S are such that, *were* coalition S be given the floor by player C at stage k, then S *would* make an effective counterproposal y^k to x^k. That is, suppose that for all players i in S we have $y^k = y^{ki} = \sigma_i(A^{k-1}, S, B^{k-1})$, where y^k is an imputation dominating x^k through this coalition S, whereas the two ordered sets A^{k-1} and B^{k-1} represent the past history of the game during stages 1 to $(k - 1)$. Then, we shall say that S is an *opposing coalition* against the current payoff proposal x^k at stage k (regardless of whether coalition S is or is not actually given the floor by player C).

LEMMA 2. *Suppose that the $(n + 1)$ players' strategies σ_i in a given game $B(G, u)$ form an equilibrium point, and that the game ends at stage K with $x^K = x^{K-1}$. This can happen only if, as a result of its members' strategies σ_i, $i \in S$, no coalition $S \subseteq N$ is an opposing coalition against the payoff proposal $x^K = x^{K-1}$.*

The lemma follows from the fact that, by (33) to (35), player C will always have an incentive to choose an opposing coalition S as coalition $S^{K-1} = S$ if such a coalition exists.

Suppose that the sequence of payoff vectors $y^1, \cdots, y^K$ and the sequence of coalitions $S^1, \cdots, S^{K-1}$ satisfy condition (19) as well as the condition

(42) $$y_i^K \geqq y_i^k \quad \text{for all} \quad i \in S^k, \quad \text{and for} \quad k = 1, \cdots, K - 1$$

(which is the same as condition (20), except that the strong inequality sign $>$ has been replaced by the weak inequality sign $\geqq$). Then we shall say that y^K *indirectly dominates* y^1 *in a weak sense.*

LEMMA 3. *Under the assumptions of Lemma 2, let $x^1, \cdots, x^K$ be the sequence of payoff proposals in a game $B(G, u)$ [or $B^*(G, u, S)$], with $x^1 = u$, and $x^K \neq u$.*

(a) *Then $x^K = x^{K-1}$ will directly dominate x^{K-2} and*

[5] In what follows, when we call a strategy n-tuple an "equilibrium point," we shall always mean that it is an equilibrium point with respect to the strategy spaces $\sum_i$ just defined; and we shall not mean merely that it is an equilibrium point with respect to some more restricted strategy spaces (such as the sets R_i of reactive strategies, or the sets R_i^* of quasi-reactive strategies, to be defined in Sections 8 and 10—see Lemma 7, below).

(b) $x^K = x^{K-1}$ *will indirectly dominate every payoff proposal* x^k *with* $1 \leqq k \leqq K - 3$, *at least in a weak sense.*

PROOF. Part (a) follows from (29). Part (b) follows from the fact that the sequence $x^1, \cdots, x^{K-1}$ will possess properties (19) and (42) with respect to the coalitions $S^1, \cdots, S^{K-2}$ chosen by player C at stages $1, \cdots, K - 2$. Now, property (19) is again implied by (29). On the other hand, property (42) follows from the assumption that σ is an equilibrium point. Any player i in a given coalition S^k can make the payoff vector x^k into the final payoff vector of the game, e.g., by choosing $y^{ki} = x^k$. Thus, if he nevertheless cooperates in making $x^K = x^{K-1} \neq x^k$ into the final payoff vector, this can only mean that $x_i^{K-1} \geqq x_i^k$, which establishes property (42).

Suppose again that the players' strategies σ_i in game $B(G, u)$ form an equilibrium point σ. Let $x^1, \cdots, x^k, \cdots$ be the sequence of payoff proposals. Let $S^1, \cdots, S^k, \cdots$ be the coalitions chosen by player C at stages $1, \cdots, k, \cdots$. Let z be the final payoff vector of the game. Suppose that *either*

$$z = x^1, \quad \text{or} \tag{43}$$

$$z_i > x_i^k \quad \text{for all } i \in S^k \text{, and for all } k \text{, except for } k = K \text{ and } k = K - 1 \text{, where } K \text{ is the last stage of the game} \tag{44}$$

(if the game ends after a finite number of stages). Then, we shall say that σ is a *conservative* equilibrium point. In other words, an equilibrium point σ is conservative if it makes each player $i \in S^k$ refuse to cooperate in making an effective counter proposal to the current payoff proposal x^k at any stage k, unless he will positively benefit from this move. (That is, in the absence of a positive incentive to the contrary he will always prefer the status quo.)

LEMMA 4. *Suppose the* $(n + 1)$ *players' strategies* σ_i *in a game* $B(G, u)$ *or* $B^*(G, u, S)$ *form a conservative equilibrium point. Then*

(a) *The phrase "at least in a weak sense" can be omitted from part* (b) *of Lemma* 3. *Moreover,*

(b) *The game will end after a finite number of stages.*

The lemma directly follows from the definition of a conservative equilibrium point.

Let σ be an equilibrium point in a game $B(G, u)$ or $B^*(G, u, S)$. Let R be a specific coalition $R \subseteq N$. Suppose we permit each player i in S to change his strategy σ_i in those situations e^k where he is acting as a member of coalition R, i.e., where $S^k = R$. But in all other situations e^k, we keep his strategy σ_i constant. That is, we permit him to shift only to strategies $\tau_i \in \sum_i$ satisfying

$$y^{ki} = \tau_i(e^k) = \sigma_i(e^k) \quad \text{for all } e^k \text{ such that } S^k \neq R. \tag{45}$$

At the same time, we keep constant the strategies σ_j of the $(n - r)$ regular players j not in R. Let σ_{R^0} be the $(n - r)$-tuple consisting of all these strategies σ_j. We also keep constant the strategy σ_{n+1} of player C.

Now suppose there does not exist any r-tuple τ_R of strategies τ_i satisfying (45) for all $i \in R$ and having the property that

$$U_i(\tau_R, \sigma_{R^0}, \sigma_{n+1}; u) > U_i(\sigma; u) \quad \text{for all } i \in R. \tag{46}$$

That is, there is no permissible strategy combination τ_R that would yield all players i in R higher payoffs than their payoffs at the equilibrium point σ (if the strategies σ_j of all other players j are kept constant). Under these conditions we shall say that this equilibrium point σ is *efficient* for coalition R in this game $B(G, u)$ or $B^*(G, u, S)$.

A given equilibrium point σ is called *coalitionally efficient* in a given game $B(G, u)$ or $B^*(G, u, S)$ if σ is efficient for *every* coalition $R \subseteq N$.[6]

At any given stage k, a coalition $S = S^k$ may act inefficiently because (1) different players i in S may vote for different imputations y^{ki}, with the result that none of these imputations y^{ki} will be accepted as the next payoff proposal x^{k+1}, or because (2) even though all players i in R do vote for the same imputation $y^{ki} = y^k$, voting for another imputation $y_0^{ki} = y_0^k$, may have better served their common interests. An equilibrium point σ may very well involve either kind of inefficient behavior, because any individual player i in R might be unable to obtain a better outcome for himself (and for coalition R as a whole) by changing his own strategy σ_i, so long as the other players $h \neq i$ in R persist in their inefficient strategies σ_h. Thus, a given strategy σ_i may very well be an equilibrium strategy and may still involve inefficient behavior. Therefore, we need the concept of a coalitionally efficient equilibrium point if we want to exclude these two types of inefficient behavior on the part of the members of the various coalitions R.

An equilibrium point σ in a given game $B(G, u)$ or $B^*(G, u, S)$ is called *admissible* if σ is both conservative and coalitionally efficient in $B(G, u)$ or in $B^*(G, u, S)$.

8. Reactive Strategies and Canonical Equilibrium Points

We shall assume that in every game $B(G, u)$ and $B^*(G, u, S)$ the players will use strategies of a particularly simple mathematical form, to be called reactive strategies, because they make the players react merely to the current situation in the game, regardless of what happened at earlier stages. More precisely, we define a *reactive strategy* ρ_i of any player $i \in M$ as a strategy that makes his behavior at any stage k depend only on the current payoff proposal x^k and on moves made at stage k itself, but makes it independent of moves made at earlier stages (except insofar as the payoff proposal x^k itself has been determined by such earlier moves).

Specifically, a reactive strategy ρ_{n+1} of player C is an arbitrary function from set X^* to the set $\mathcal{S}$ of all nonempty coalitions, making him at any stage k (except for the last stage $k = K$) choose some coalition S^k such that

$$S^k = \rho_{n+1}(x^k). \tag{47}$$

On the other hand, a reactive strategy ρ_i of any regular player $i \in N$ is an arbitrary function from the Cartesian product $X^* \times \mathcal{S}$ to the set X of all imputations, making him, at any stage k where $S^k \ni i$, choose some imputation y^{ki} such that

$$y^{ki} = \rho_i(x^k, S^k). \tag{48}$$

The set of all reactive strategies ρ_i available to any player $i \in M$ will be called R_i. Of course, $R_i \subset \sum_i$ for all $i \in M$.

The assumption that the players will use reactive strategies can be mathematically justified by Lemmas 6 and 7 below. But our more specific reason for making this assumption is this: following von Neumann and Morgenstern, we want the stability of any imputation x, to be decided solely by whether x does or does not belong to the stable set V chosen by the players—quite independently of the history of the negotiations among the players at earlier stages of the game. (In particular, we do not want the stability of x to be decided by some players' attempt to reward or to punish some other players for moves the latter made at earlier stages.[7])

[6] Our concept of a *coalitionally efficient* equilibrium point is similar to Aumann's [1959] concept of a *strong* equilibrium point. But coalitional efficiency is a less exacting requirement than strongness is.

[7] This assumption will be relaxed in Section 10, below.

Suppose the $(n + 1)$ players use some $(n + 1)$-tuple $\rho = (\rho_1, \cdots, \rho_{n+1})$ of reactive strategies in any game $B(G, u)$, i.e., in a bargaining game of the first kind. Then this $(n + 1)$-tuple ρ will uniquely determine a *transition function* ϕ, specifying the next payoff proposal x^{k+1} as a function of the current payoff proposal x^k, so that

$$x^{k+1} = \phi(x^k). \tag{49}$$

Thus, ϕ will be a function from the set X^* into itself. By the definition of reactive strategies, the transition function ϕ will remain the *same* at all stages k of the game $B(G, u)$, so long as the players use the same $(n + 1)$-tuple ρ of reactive strategies.

If any payoff vector x is a fixed point of ϕ, that is, if

$$\phi(x) = x, \tag{50}$$

then x is called a *stationary point* or a *pole* of the $(n + 1)$-tuple ρ of reactive strategies that gave rise to ϕ. Any payoff vector x not satisfying (50) will be called a *nonstationary point* of ρ. The set $V = V(\rho)$ of all stationary points x of ρ will be called the *stationary set* of ρ.[8]

Two $(n + 1)$-tuples $\rho = (\rho_1, \cdots, \rho_{n+1})$ and $\rho' = (\rho'_1, \cdots, \rho'_{n+1})$ of reactive strategies will be called *co-polar* if they have exactly the same stationary sets $V(\rho) = V(\rho')$.

Two $(n + 1)$-tuples ρ and ρ' of reactive strategies are called *equivalent* if they yield the very same final payoff vector $U(\rho; 0^*) = U(\rho'; 0^*)$ in game $B(G) = B(G, 0^*)$.

LEMMA 5. *Suppose the $(n + 1)$ players use some $(n + 1)$-tuple ρ of reactive strategies in game $B(G, u)$. Then*

(a) *The game will end at stage K if and only if the payoff proposal x^{K-1} for stage $(K - 1)$ is a stationary point of ρ.*

(b) *If the final payoff vector z is not a stationary point of ρ, then z must be the zero vector $z = 0^*$.*

PROOF. Part (a) follows from (31), while part (b) follows from (32).

LEMMA 6. *Suppose the n players other than player i use the reactive strategies $\rho_1, \cdots, \rho_{i-1}, \rho_{i+1}, \cdots, \rho_{n+1}$ in a game $B(G, u)$ or $B^*(G, u, S)$. Then player i will have a best reply ρ_i that is itself also a reactive strategy. This is true regardless of whether i is a regular player or is player C.*

PROOF. We have to show that player i, by using a reactive strategy ρ_i can always achieve a final payoff at least as high as he could achieve by using any nonreactive strategy σ_i. First, let i be a regular player. We can distinguish two cases:

CASE I. If i used σ_i, then the game would go on indefinitely. Then, by (32), he would obtain the final payoff $z_i = 0$. Hence, he could only gain by switching to an alternative strategy such as ρ_i.

CASE II. If i used σ_i, then the game would end at stage K. His switching to ρ_i can make a difference only if the same combination (x^k, S^k) arises at different stages k,

[8] However, our statements about transition functions, and in particular equations (49) and (50), do not in general extend to games of the form $B^*(G, u, S)$, i.e., to bargaining games of the second kind. For example, suppose that a given imputation x is a nonstationary point of ρ, with $\phi(x) = y \neq x$. This means that if the players use the strategy $(n + 1)$-tuple ρ in game $B(G, x)$, then we shall have $x^1 = x$ but $x^2 = y \neq x$. This in turn implies that there exists an opposing coalition T against x, which will make the effective counter proposal y to x at stage 1.

In contrast, in game $B^*(G, x, S)$, if S itself is not an opposing coalition to x, then we shall have $x^1 = x^2 = x$. This is so because at stage 1 coalition S will have the floor, and this will prevent T (or any other opposing coalition) from making an effective counterproposal to x. This example also shows that Lemma 2 in general does not apply to bargaining games of the second kind.

say, at stages $k = k_1, \cdots, k_m$. Suppose that strategy σ_i would yield him the final payoff z_i, and would make him choose imputations $y^{k_i} = y(k_1), \cdots, y(k_m)$ at stages $k_1, \cdots, k_m$, respectively. We claim that i can achieve the same payoff z_i also by using a reactive strategy ρ_i which will make him choose the *last* of these imputations, $y^{k_i} = y(k_m)$, at *every* stage k characterized by the combination (x^k, S^k). Since all other players are assumed to use reactive strategies, if i switches over to this strategy ρ_i, the result will be simply that the old stages $k_1, \cdots, (k_m - 1)$ will be omitted, and the old stages $k_m, \cdots, K$ will now become stages $k_1, \cdots, (k_1 + K - k_m)$. The final payoff vector of the game will remain z, except that z now will not be achieved at stage K, but rather already at stage $(k_1 + K - k_m)$.

Next, let i be player C. We shall again distinguish two cases.

CASE I′. If C used σ_i, then at every stage k a *different* payoff proposal x^k would be considered by the regular players. In this case, it cannot make any difference if C switches to a reactive strategy ρ_i which prescribes the same moves for C at every stage k as σ_i does.

CASE II′. If C used σ_i, then the same payoff proposal $x^k = x^{k'}$ would emerge at least at two different stages k and k' with $k' > k$. Suppose that σ_i would make player C choose coalition S^k at stage k. Then, since the regular players are assumed to use reactive strategies, player C can ensure that the game will go on indefinitely, yielding him the highest possible final payoff $\alpha^* = 1$, if he switches over to a reactive strategy ρ_i which will make him choose the *same* coalition S^k at *every* stage k characterized by this payoff proposal x^k. This is so because, as a result of strategy ρ_i, this same payoff proposal x^k will now re-emerge at stages $k, k', (2k' - k), (3k' - 2k), \cdots$, indefinitely.

LEMMA 7. *Suppose that the $(n + 1)$-tuple $\rho = (\rho_1, \cdots, \rho_{n+1})$ would be an equilibrium point in a given game $B(G, u)$ or $B^*(G, u, S)$, if the $(n + 1)$ players were restricted to using reactive strategies. Then ρ will be an equilibrium point even if every player $i \in M$ is free to use any strategy $\sigma_i \in \sum_i$ without restriction.*

Lemma 7 follows directly from Lemma 6.

An equilibrium point ρ where all $(n + 1)$ players use reactive strategies ρ_i is called a reactive equilibrium point.[9]

We feel that both the use of strategies representing an admissible equilibrium point, and the use of reactive strategies, are important rationality requirements. But, then, full rationality requires that the players should follow these two rationality requirements, not only in game $B(G)$ as a whole, but also in all of its subgames.[10] Even if, given the players' actual strategies $\rho_1, \cdots, \rho_{n+1}$, some particular subgame will never be reached by the players, they must always act on the assumption that, *were* this subgame ever to be reached, all players *would* act in it in accordance with these two rationality requirements. These considerations suggest the following definition.

An equilibrium point ρ is called *canonical* if it is an admissible reactive equilibrium point in game $B(G)$ itself and in all of its subgames (both in those subgames the players will actually reach, and in those they will not reach, with their equilibrium strategies ρ_i).[11]

[9] See footnote 4.

[10] For definition of a subgame, see Kuhn [1953, p. 204].

[11] Selten [1965] has argued that full rationality requires that the players should use strategies representing a *perfect* equilibrium point, defined as a strategy n-tuple that is an equilibrium point both in the main game as a whole and in all of its subgames. Our concept of a canonical equilibrium point is based on a similar philosophy.

It is easy to verify that this definition is equivalent to the following one. An equilibrium point ρ is called *canonical* if it is an admissible reactive equilibrium point in *all* games $B(G, u)$ with $u \in X^*$, and in *all* games $B^*(G, u, S)$ with $u \in X^*$ and with $S \in \mathcal{S}$.[12]

LEMMA 8. *Every canonical equilibrium point ρ must have at least one stationary point z which is an imputation.*

PROOF. Suppose that all players use the strategies ρ_i in game $B(G, u)$ with $u \neq 0^*$. By part (b) of Lemma 4, the game will end after a finite number of stages. Hence, since $u \neq 0^*$, in view of (29) to (31), the final payoff vector $z = U(\rho; u)$ can only be an imputation. Moreover, by Lemma 5, z will be a stationary point of ρ.

LEMMA. 9. *Every stationary point x of a canonical equilibrium point ρ must be an imputation.*

PROOF. We have to show that the zero vector 0^* cannot be a stationary point of ρ. We shall consider two cases.

CASE I. All imputations are stationary points of ρ. Let $z^* = (z_1^*, \cdots, z_n^*)$ be the imputation defined by $z_1^* = \cdots = z_n^* = v(N)/n$. This stationary imputation z^* obviously dominates 0^* through the all-player coalition N.

CASE II. At least one imputation y is a nonstationary point of ρ. Suppose the players use the strategies ρ_i in game $B(G, y)$. Then, by Lemmas 4(b) and 5, the game will eventually move from $x^1 = y$ to some stationary imputation $z = x^K = x^{K-1}$, after a finite number of stages. By Lemma 4(a), z will directly or indirectly dominate y. Therefore, by (14), (15), and (19), there must exist some coalition $S \subseteq N$ such that $z_i > 0$ for all players i in S, and such that $z(S) \leqq v(S)$, where $z(S) = \sum z_i \ (i \in S)$.

Thus, in either case, 0^* is dominated by a stationary imputation z (with $z = z^*$ or $z \neq z^*$) through some coalition S (with $S = N$ or $S \neq N$). Therefore, if $x^1 = 0^*$ is the current payoff proposal for stage 1, then, by coalitional efficiency, at least this coalition S will be an opposing coalition against 0^*. Consequently, by Lemma 2, 0^* will be nonstationary, as desired.

LEMMA 10. *Suppose the payoff vector x is dominated by some stationary point y of a canonical equilibrium point ρ. Then x itself must be a nonstationary point of ρ.*

PROOF. If $x \notin X$, then x must be nonstationary by Lemma 9. Thus we only have to consider the case where $x \in X$. Let S be the coalition through which y dominates x. Suppose the players use the strategies ρ_i in game $B(G, x)$. Clearly, $x^1 = x$. We have to show that $x^2 \neq x$. By Lemma 2, this will be the case if there is any opposing coalition against $x^1 = x$. Yet, by coalitional efficiency, S must be an opposing coalition, since if it proposes $y^1 = y$, then y will become the final payoff vector (as y is stationary) and, by (14), y will yield higher payoffs than x would to all players in S.

LEMMA 11. *Let G be an absolutely stable game, and let ρ be a canonical equilibrium point in $B(G)$. Suppose the imputation x is a nonstationary point of ρ. Then x is dominated by some stationary point y of ρ.*

[12] By requiring admissibility, not only in the games $B(G, u)$, but also in the games $B^*(G, u, S)$, we ensure that the concept of an "opposing coalition" will have its intended meaning. More particularly, we ensure that at each stage k any coalition S that would find it advantageous to be an opposing coalition against the current payoff proposal x^k *if it expected to be given the floor by player C*, would remain an opposing coalition even if it knew that player C's strategy ρ_{n+1} was such as to give the floor to another coalition $T \neq S$ at this stage k.

PROOF. Suppose the players use the strategies ρ_i in game $B(G, x)$, and obtain the final payoff vector $U(\rho; x) = z = y$. By Lemmas 4(b) and 5, y will be a stationary point of ρ. As x itself is nonstationary, $x \neq y$. To prove the lemma, all we have to show is that y dominates x. But this follows from Lemmas 3 and 4(b), and from the fact that in G all indirect dominance relations between imputations are trivial.

LEMMA 12. *Let G again be absolutely stable, and let ρ be a canonical equilibrium point in $B(G)$. Then the stationary set V of ρ will be a von Neumann-Morgenstern stable set.*

PROOF. We have to show that V has properties (16) and (17). But the former follows from Lemma 10, while the latter follows from Lemma 11.

LEMMA 13. *Let V be a von L Neumann-Morgenstern stable set in an absolutely stable game G. Let x be any interior imputation in V. Then there exists a canonical equilibrium point ρ in $B(G)$ such that*

(a) *ρ yields this imputation $x = U(\rho; 0^*)$ in game $B(G)$, and*
(b) *ρ has this strictly stable set $V = V(\rho)$ as its stationary set.*

PROOF. We shall prove the lemma by constructing a canonical equilibrium point ρ with properties (a) and (b). But before doing so, we shall construct the transition function ϕ associated with ρ.

Let $V^0 = X^* - V$. Let $V^{00} = V^0 - \{0^*\} = X - V$. For each payoff vector $y \in V^{00}$, we shall choose some imputation $z = z(y)$ such that $z \in V$ and z dominates y through some coalition $S = S(y)$. As V has property (18), this can always be done. Now we set

(51) $$\phi(y) = y \quad \text{for all} \quad y \in V;$$

(52) $$\phi(y) = x \quad \text{for} \quad y = 0^*;$$

(53) $$\phi(y) = z(y) \quad \text{for all} \quad y \in V^{00}.$$

Next, we construct the equilibrium strategy ρ_{n+1} of player C as follows.

(54) $$\rho_{n+1}(y) = N \quad \text{if} \quad y = 0^* \quad \text{or if} \quad y \in V, \qquad \text{but}$$

(55) $$\rho_{n+1}(y) = S(y) \quad \text{if} \quad y \in V^{00}.$$

Finally, we construct the equilibrium strategy ρ_i of each regular player $i \in N$ as follows.

(56) $$\rho_i(S, y) = z(y) \quad \text{if}$$

(56a) $z(y)$ dominates y through this coalition S.

(57) $$\rho_i(S, y) = w \quad \text{if (56a) fails, but if}$$

(57a) there exists at least one stationary imputation $w = \phi(w) \neq z(y)$ dominating y through S.

(If there exist two or more such imputations w, then we can choose any one of them for use in (57).)

(58) $$\rho_i(S, y) = y \quad \text{if both (56a) and (57a) fail.}$$

It is easy to check that the strategy n-tuple $\rho = (\rho_1, \cdots, \rho_{n+1})$ just constructed is a canonical equilibrium point and possesses properties (a) and (b), as desired.

A given set V of payoff vectors will be called a *canonical set* if $V = V(\rho)$ is the stationary set of some canonical equilibrium point ρ. By Lemmas 8 and 9, any canonical set must always be a nonempty set of imputations.

Let x be a point in a canonical set V. We shall say that x is a *reachable* point of V if there exists a canonical equilibrium point ρ having $V = V(\rho)$ as its stationary set, and yielding $x = U(\rho; 0^*)$ as final payoff in game $B(G) = B(G, 0^*)$. By Lemma 13, if V is a stable set in an absolutely stable game, and if x is an *interior* imputation, then x is always a reachable point. However, in general this need not be the case if x is an *exterior* imputation.

For example, let G be a three-person game, with $v(N) = v((123)) = 1$ and $v(S) = 0$ for all $S \neq N$. G has only one stable set, viz. the set $V = X$ of all imputations (which is also the core of the game). By Theorem 1, G is absolutely stable. Let $x = (1, 0, 0)$. This imputation x is obviously an *unreachable* point of V (since it does not dominate the zero vector 0^* through any coalition whatever).

Intuitively this means that, *if* this stationary imputation x ever became the payoff proposal $x^k = x$ considered by the players at any stage k, then all further bargaining would stop, and x would immediately become the final payoff vector of the game (assuming that the players' strategies would form a canonical equilibrium point ρ with $V = V(\rho)$ as its stationary set). But, in actual fact, if the game starts with the zero vector 0^* as initial payoff vector, then x can *never* become a payoff proposal $x^k = x$ considered by the players at any stage k.

We now state the following theorem.

THEOREM 2. *Let G be an absolutely stable game. Then*

(a) *Every canonical set V in G will be a von Neumann-Morgenstern stable set and conversely.*

(b) *The set W^* of all canonical equilibrium points ρ in game $B(G)$ can be partitioned into families $V^* = \psi(V)$ of co-polar equilibrium points having the same stable set V as stationary set.*

(c) *Any such family $V^* = \psi(V)$ can be further partitioned into subfamilies $T^* = \psi(x)$ of equivalent equilibrium points ρ yielding the same imputation $x = U(\rho; 0^*,) \in V$ as final payoff vector. There will be a nonempty subfamily $T^* = \psi(x)$ for every interior imputation x in the stable set V associated with this family $V^* = \psi(V)$. (But in general this will not be true for exterior imputations x' in V, because the latter will often be unreachable points of V.)*

Theorem 2 follows from Lemmas 12 and 13 (see also the discussion on unreachable points, following Lemma 13).

9. Canonical Sets in Arbitrary Games

In any game G lacking absolute stability, the canonical sets of the game in general will *not* be von Neumann-Morgenstern stable sets, and conversely. Any given imputation x will be stable (in the sense of assumptions (A) and (B) of Section 3) if and only if x belongs to the stationary set $V = V(\rho)$ of the canonical equilibrium point ρ chosen by the players. In contrast, if x belongs to any particular von Neumann-Morgenstern stable set V', this, in general, will not assure its stability, owing to the destabilizing effect of indirect dominance relations.[13]

[13] At the present time, it is not known whether Lemma 13 can be extended to *strictly* stable sets V belonging to games G that lack absolute stability. Therefore, it is not known whether every *strictly* stable set is always a *canonical* set or not. In the terminology of Theorem 3 (see below), of course every strictly stable set V will possess *internal* effective stability. But we do not know whether we can always find a canonical equilibrium point ρ such that V will have *external* effective stability in terms of the transition function ϕ generated by ρ.

Accordingly, *we propose to re-define a stable set as a canonical set*, i.e., as a stationary set $V = V(\rho)$ of some canonical equilibrium point ρ.

By Theorem 2, in an absolutely stable game, every canonical set V has both internal and external stability, as defined by (17) and (18). In order to obtain a general theorem about the stability properties of canonical sets for *all* games, we shall now introduce the concept of effective dominance.

Suppose that in game $B(G)$ the $(n + 1)$ players' stategies $\rho_1, \cdots, \rho_{n+1}$ form a canonical equilibrium point ρ. Let ϕ be the transition function generated by ρ. Then we shall say that a given payoff vector $z = (z_1, \cdots, z_n)$ *effectively dominates* some payoff vector $x = (x_1, \cdots, x_n)$ in terms of this transition function ϕ if either of the following two conditions is satisfied:

(59) *Either z directly* dominates x,

(60) *Or z indirectly* dominates x through some monotone chain

$y^1, \cdots, y^K$ with $y^1 = x$ and $y^K = z$, such that

(60a) $$y^{k+1} = \phi(y^k) \quad \text{for} \quad k = 2, \cdots, K - 1.$$

If z indirectly dominates x, but if the monotone chain(s) $y^1, \cdots, y^K$ connecting $y^1 = x$ with $y^K = z$ does not (or do not) have property (60a), then we shall say that *z ineffectively dominates x*. The importance of property (60a) will become apparent from the following lemma.

Lemma 14. *Suppose that a given payoff vector x is effectively dominated by a stationary point z of some canonical equilibrium point ρ, in terms of the transition function ϕ generated by ρ. Then x itself will be a nonstationary point of ρ.*

Proof. Lemma 10 takes care of case (59). Therefore we shall assume that x and z come under case (60). Let S^1 be the coalition through which imputation y^2 dominates $y^1 = x$ in the monotone chain $y^1, \cdots, y^K$ mentioned under (60). Now suppose that in game $B(G, x)$ the players use the strategies ρ_i corresponding to ρ. Clearly, we shall have $x^1 = x$. By Lemma 2, $x^1 = x$ will not be stationary if there is any opposing coalition against it. Yet, by coalitional efficiency, S^1 must be such an opposing coalition since if it proposes y^2, then, by (60a) and by the stationariness of z, this move will make z the final payoff vector of the game. Moreover, by (20), z will yield every player i in S^1 a higher payoff $z_i > x_i$ than x would. Consequently, x will be nonstationary.

Obviously, if x is indirectly dominated by z, but if this dominance is *ineffective*, i.e., if (60a) fails, then the proof of Lemma 14 does not go through.

Lemma 15. *Suppose that x is a nonstationary point of some canonical equilibrium point of ρ. Then x is effectively dominated by some stationary point z of ρ, in terms of the transition function ϕ generated by ρ.*

Proof. Suppose that in game $B(G, x)$ the players use the strategies ρ_i corresponding to ρ. Let $z = U(\rho; x)$ be the resulting final payoff vector. By Lemmas 4(b) and 5, z will be a stationary point of ρ. As x itself is nonstationary, $z \neq x$. Let $x^1, \cdots, x^K$ be the sequence of payoff proposals in the game. Clearly, $x^1 = x$ and $x^K = x^{K-1} = z$. By Lemmas 3 and 4(a), x will be either directly or indirectly dominated by z. Moreover, as the sequence $x^1, \cdots, x^K$ has been generated by the strategy $(n + 1)$-tuple ρ, we shall have $x^{k+1} = \phi(x^k)$ for $k = 1, \cdots, K - 1$. Therefore condition (60a) will be satisfied. Consequently, the direct or indirect dominance of x by z will be effective dominance.

THEOREM 3. (a). *Suppose that* $V = V(\rho)$ *is the stationary set of some canonical equilibrium point* ρ. *Then* V *will have the following two properties:*

(61) *If some payoff vector* x *is in* V, *then* x *is not dominated effectively by any payoff vector* z *in* V, *in terms of the transition function* ϕ *generated by* ρ.

(62) *If some payoff vector* x *is not in* V, *then* x *is effectively dominated by some payoff vector* z *in* V, *in terms of the transition function* ϕ.

Property (61) *will be called internal effective stability whereas property* (62) *will be called external effective stability.*

(b). *Conversely, suppose that some set* V *of payoff vectors has both internal and external effective stability in terms of a transition function* ϕ *generated by some canonical equilibrium point* ρ. *Then* V *will be the stationary set of* ρ.

PROOF. In part (a), property (61) follows from Lemma 14, while property (62) follows from Lemma 15. As to part (b), let V^* be the stationary set of ρ. Suppose that V has properties (61) and (62). We have to show that $V = V^*$. We shall first show that $V^* \subseteq V$, i.e., that any payoff vector x not in V is nonstationary. Since V has property (62), any x not in V is effectively dominated by some stationary point z in V. Yet, by Lemma 14, this means that x itself will be nonstationary.

Next, we shall show that $V \subseteq V^*$, i.e., that no nonstationary payoff vector y can belong to V. Let y be such a nonstationary point. Then, by Lemma 15, y is effectively dominated by some stationary point z. As z is stationary, we have $z \in V^* \subseteq V$. Consequently, y cannot belong to V since, by assumption, V has property (61).

Hence, $V^* \subseteq V$ and $V \subseteq V^*$, so that $V = V^*$, as desired.

Thus, according to Theorem 3, the basic difference between canonical sets and von Neumann-Morgenstern stable sets is this. The former have both internal and external stability in terms of *effective* dominance relations, which include both direct dominance relations and *effective* indirect dominance relations. In contrast, the latter have both external and internal stability merely in terms of *direct* dominance relations alone, and their definition disregards indirect dominance relations altogether.

THEOREM 4. *Let* G *be any game given in characteristic function form, such that the set of canonical equilibrium points in the corresponding bargaining game is nonempty.* (*G need not be absolutely stable or strictly stable.*)

(a) *Then the set* W^* *of all canonical equilibrium points in game* $B(G)$ *can be partitioned into families* $V^* = \psi(V)$ *of copolar equilibrium points, each family having the same canonical set* V *as stationary set.*

(b) *Any such family* $V^* = \psi(V)$ *can be further partitioned into subfamilies* $T^* = \psi(x)$ *of equivalent equilibrium points* ρ, *each subfamily yielding the same imputation* $x = U(\rho; 0^*) \in V$ *as final payoff vector. There will be a nonempty subfamily* $T^* = \psi(x)$ *for every interior imputation* x *in the canonical set* V *associated with this family* $V^* = \psi(V)$. (*But in general this will not be true for exterior imputations* x' *in* V, *because the latter will often be unreachable points of* V.)

PROOF. Part (a) and the first sentence of part (b) follow from the definitions of canonical sets, of co-polarity, and of equivalence, as well as from Lemma 9. The second sentence of part (b) can be verified by a construction procedure similar to the one used in the proof of Lemma 13. As to the last sentence (in parentheses), see the discussion of unreachable points, following Lemma 13.

10. A Model Yielding von Neumann-Morgenstern Stable Sets

As we have seen, in games lacking absolute stability the stable sets (canonical sets) generated by our model, in general, will not be von Neumann-Morgenstern stable sets. We now propose to show that the stable sets our model yields are von Neumann-Morgenstern stable sets in every game if we assume that the players' bargaining strategies will incorporate a deliberate policy of "neutralizing" indirect dominance relations.

Suppose that at stage k of some bargaining game $B(G, u)$, coalition $S = S^k$ unanimously proposes some imputation y^k dominating the current payoff proposal $x = x^k$, so that this imputation y^k replaces x^k as the payoff proposal $x^{k+1} = y^k$ for stage $(k + 1)$. Then we shall say that coalition S *vetoed* the payoff vector x at stage k.

Such a vetoing move can serve either of two purposes. Coalition S may wish to make y itself into the final payoff vector of the game. In this case, we shall say that S tries to take advantage of the *direct* dominance of y over x. Alternatively, coalition S may wish to make some other payoff vector $x^K = z \neq y$ into the final payoff vector at some later stage K of the game. In this latter case, we shall say that S tries to take advantage of the *indirect* dominance of z over x. In the former case, in view of (15), coalition S cannot obtain a joint payoff exceeding $v(S)$ by this vetoing move. But in the latter case, there is no such automatic restriction on the joint payoff that coalition S can eventually achieve by this move.

However, it is possible that the players will adopt a deliberate policy restricting the joint payoff that any coalition S can achieve by a vetoing move, even with the help of indirect dominance relations. More specifically, each regular player $i \in N$ may adopt the following restrictive policy. If a given coalition $S = S^h$ vetoed the payoff proposal $x = x^h$ at some stage h, then at all stages $k > h$ player i will refrain from proposing any imputation $y = y^{ki}$ such that

$$\sum_{i \in S} y_i > v(S). \tag{63}$$

To make this policy consistent with player i's own self-interest, we may assume that he will follow this rule only if the whole shift from payoff vector x to payoff vector y would *not* benefit him personally, i.e., if

$$y_i \leqq x_i. \tag{64}$$

Of course, a bargaining strategy σ_i incorporating this restrictive policy cannot be a reactive strategy because it will make player i's behavior dependent on the vetoing moves of various coalitions S at earlier stages of the game. But we may assume that player i will use a "quasi-reactive strategy," under which he basically follows some reactive strategy ρ_i and departs from the latter only to the extent necessary for implementing this restrictive policy.

Suppose that a given coalition $S = S^h$ vetoed the payoff proposal $x = x^h$ at stage h. Let β_i^h be the set of all imputations y satisfying (63) with respect to $S = S^h$, and also satisfying (64) with respect to $x = x^h$ for a given regular palyer i. Finally, let

$$\gamma_i^k = \cup_{h<k} \beta_i^h. \tag{65}$$

The restrictive policy we have described amounts to requiring that any regular player i should refrain from proposing any imputation $y = y^{ki}$ in this set γ_i^k.

Thus, we can define a *quasi-reactive strategy* ρ_i^* for any regular player i as follows. Let $\rho_i \in R_i$ be a reactive strategy of player i. Then, in view of (38) and (65), we define

$$y^{ik} = \rho_i^*(x^k, S^k, \gamma_i^k) = \rho_i(x^k, S^k) \quad \text{if} \quad \rho_i(x^k, S^k) \notin \gamma_i^k, \tag{66}$$

but

$$y^{ik} = \rho_i^*(x^k, S^k, \gamma_i^k) = x^k \quad \text{if} \quad \rho_i(x^k, S^k) \in \gamma_i^k. \tag{67}$$

In case of player C (player $(n+1)$), since he has no role in enforcing the restrictive policy, we define his *quasi-reactive strategies* ρ_{n+1}^* to be identical with his reactive strategies ρ_{n+1}.

The set of all quasi-reactive strategies ρ_i^* available to any player $i \in M$ will be called R_i^*. Of course, $R_i^* \subset \sum_i$ for all $i \in M$.

Let $\rho^* = (\rho_1^*, \cdots, \rho_{n+1}^*)$ be an $(n+1)$-tuple of quasi-reactive strategies. We define the stationary set $V = V(\rho^*)$ of ρ^* as the set of all payoff vectors x satisfying

$$U(\rho^*; x) = x \quad \text{in game } B(G, x). \tag{68}$$

We define co-polarity and equivalence between different $(n+1)$-tuples of quasi-reactive strategies in the same way as we did in the case of $(n+1)$-tuples of reactive strategies.

If an $(n+1)$-tuple $\rho^* = (\rho_1^*, \cdots, \rho_{n+1}^*)$ of quasi-reactive strategies is an equilibrium point in a given game $B(G, u)$ or $B^*(G, u, S)$, then ρ^* is called a *quasi-reactive equilibrium point* in this game $B(G, u)$ or $B^*(G, u, S)$.[14]

Lemma 16. *Suppose the $(n+1)$ players' strategies $\rho_1^*, \cdots, \rho_{n+1}^*$ in some game $B(G, u)$ or $B^*(G, u, S)$ form a conservative quasi-reactive equilibrium point. Let $x^1, \cdots, x^K$ be the sequence of payoff proposals in the game, with $x^1 = u$ and with $x^K = x^{K-1} = z$. Suppose that $x^K \neq x^1$. Then z will directly dominate all payoff vectors x^k with $1 \leqq k \leqq K-2$.*

Proof. For x^k with $k = K-2$, the lemma follows from (29). So we have to consider only payoff vectors x^k with $1 \leqq k \leqq K-3$. Let $S = S^k$ and $T = S^{K-2}$ be the coalitions selected by player C at stages k and $(K-2)$, respectively. In view of the restrictive policy followed by the players, only two cases are possible: (a) *Either* z will satisfy the condition

$$\textstyle\sum_{i \in S} z_i \leqq v(S), \tag{69}$$

(b) *Or* z will satisfy the alternative condition

$$z_i > x_i \quad \text{for all} \quad i \in T, \tag{70}$$

with respect to $x = x^k$.

As ρ^* is a conservative equilibrium point, in view of (45), we can write

$$z_i > x_i \quad \text{for all} \quad i \in S. \tag{71}$$

By (29), $z = x^{K-1}$ must dominate x^{K-2}. Hence, by (13) and (15), we have

$$z \in X \quad \text{and} \tag{72}$$

$$\textstyle\sum_{i \in T} z_i \leqq v(T). \tag{73}$$

Therefore, in case (a), by (69), (71), and (72), z will dominate x through coalition S; whereas, in case (b), by (70), (72), and (73), z will dominate x through coalition T. Thus, in either case, z will directly dominate x, as desired.

Let ρ^* be an admissible quasi-reactive equilibrium point in all games $B(G, u)$ with $u \in X^*$, and in all games $B^*(G, u, S)$ with $u \in X^*$ and with $S \in \mathcal{S}$. Then ρ^* will be called a *quasi-canonical equilibrium point*. The stationary set $V = V(\rho^*)$ of any quasi-canonical equilibrium point will be called a *quasi-canonical set*.

[14] See footnote 4.

Lemma 16 directly implies:

THEOREM 5. *Let G be any game given in characteristic-function form. (Thus, G need not be absolutely stable or strictly stable.) Then, every quasi-canonical set V in G will be a von Neumann-Morgenstern stable set.*

It is not known at present whether the converse is true, i.e., whether every von Neumann-Morgenstern stable set V of any game G can always be obtained as the stationary set of a quasi-canonical equilibrium point in the corresponding bargaining game $B(G)$.

To conclude, we have argued that, in view of the destabilizing effect of indirect dominance relations, in defining stable sets for a given game G, the von Neumann-Morgenstern definition should be replaced by an alternative definition, based on the canonical equilibrium points of the bargaining game $B(G)$. Yet, if we assume that the players will deliberately adopt a "restrictive policy" discouraging the use of indirect dominance relations, then the stable sets this bargaining game $B(G)$ yields will always be von Neumann-Morgenstern stable sets.

11. Postscript

If we are willing to go one step further, and drop the requirement that stable sets must be defined in terms of direct dominance relations—or at least in terms of chains of such relations—then the following alternative definition suggests itself for stable sets.

Let us say that y *semi-dominates* x through coalition S if x and y satisfy (13) and (15), regardless of whether they also satisfy (14). Then, in (19), we may replace the word "dominated" by the word "semi-dominated," and likewise in (29), we may replace the word "dominates" by the word "semi-dominates." Let us call the resulting modified bargaining game $B^0(G)$. We may now define a stable set as the stationary set $V = V(\rho^0)$ of a canonical equilibrium point ρ^0 in this *modified* bargaining game $B^0(G)$.

Intuitively, these changes in our definitions would amount to permitting any coalition S^k (selected by player C at any given stage k) to propose an imputation $y = y^k$ possibly *less favorable* to its own members than the current payoff proposal $x = x^k$ is—provided that coalition S^k has the power to enforce y (in accordance with (15)). Of course, if the members of S^k are rational individuals, then they will not initiate a reduction of their own *provisional* payoffs from x_i to $y_i < x_i$, unless they expect that this move will increase their *final* payoffs z_i at the end of the game. But it seems reasonable that the rules of the bargaining game should permit any player to make such temporary sacrifices for the sake of an ultimate gain.[15]

[15] As Robert Weber has pointed out to me, another change in the rules of the bargaining game might also be desirable. One might assume that at each stage k, before player C gave the floor to any coalition, all coalitions would have to *declare* how they would act if they were given the floor; and player C would choose one particular coalition in the light of this information. Under these rules, a given player i in a certain coalition S might find it advantageous to cooperate in making an effective counterproposal to the current payoff proposal $x^k = x$, even though he would think that the final payoff vector z resulting from this move would be *less favorable* to him than x itself was, with $z_i < x_i$. He might find this more advantageous because he might come to the conclusion that otherwise player C would give the floor to another coalition $T \neq S$ and $T \ni i$, and this would result in a final payoff vector z' even less favorable to him, with $z'_i < z_i$. In real-life bargaining situations as described in Section 3, it is often rational to accept cuts in one's payoff on the basis of such considerations. Therefore, one might argue that our formal bargaining model should also give proper scope for this kind of behavior. (However, such a change in the rules of our bargaining game would represent a major change in our model, and would no longer leave our main results unaffected.)

It is easy to verify that these changes in our definitions would leave our main conclusions unaffected. In particular, all of our theorems would retain their validity.[16]

References

1. AUMANN, ROBERT J., "Acceptable Points in General Cooperative n-Person Games," in A. W. Tucker and R. D. Luce (eds.), *Contributions to the Theory of Games*, Vol. IV, Princeton, N. J., 1959, pp. 287-324.
2. —— AND PELEG, BEZALEL, "von Neumann-Morgenstern Solutions to Cooperative Games Without Side Payments," *Bulletin of the American Mathematical Society*, Vol. 66 (1960), pp. 173–179.
3. KUHN, H. W., "Extensive Games and the Problem of Information," in H. W. Kuhn and A. W. Tucker (eds.), *Contributions to the Theory of Games*, Vol. II, Princeton, N. J., 1953, pp. 193–216.
4. MASCHLER, MICHAEL, "n-Person Games with Only 1, $n - 1$, and n-Person Permissible Coalitions," *Journal of Mathematical Analysis and Applications*, Vol. 6 (1963), pp. 230–256.
5. NASH, JOHN F., "Noncooperative Games," *Annals of Mathematics*, Vol. 54 (1951), pp. 286–295.
6. SELTEN, REINHARD, "Spieltheoretische Behandlung eines Oligopolmodels mit Nachfrageträgheit" (Game-Theoretical Treatment of an Oligopoly Model with Demand Inertia), Parts I–II, *Zeitschrift für die Gesamte Staatswissenschaft*, Vol. 121 (1965), pp. 301–324 and 667–689.
7. VICKREY, WILLIAM, "Self-policing Properties of Certain Imputation Sets," in A. W. Tucker and R. D. Luce (eds.), *Contributions to the Theory of Games*, Vol. IV, Princeton, N. J., 1959, pp. 213–246.
8. VON NEUMANN, JOHN, AND MORGENSTERN, OSKAR, *Theory of Games and Economic Behavior*, Princeton, N. J., 1947.

---

[16] In this paper, we have discussed stable sets only for games given in characteristic-function form. In a forthcoming paper, we shall extend our definition for stable sets—more particularly, the definition suggested in this Postscript—to games given in normal form.

CHAPTER 11

A NEW GENERAL SOLUTION CONCEPT FOR BOTH COOPERATIVE AND NONCOOPERATIVE GAMES

I. Some limitations of classical game theory

1. Cooperative games vs. noncooperative games

A basic distinction in classical game theory is that between cooperative and noncooperative games.[1] This distinction was introduced by NASH [1951], who defined *cooperative* games as games where the players can communicate and can make enforceable agreements, in contrast to *noncooperative* games, where neither communication nor enforceable agreements are allowed. Of course, a binary distinction based on *two* different criteria is logically unsatisfactory. We cannot define one category as the class of objects possessing *both* properties A and B while defining the other category as the class of objects possessing *neither* property. For the question then immediately arises, What about objects having the first property but lacking the second, and what about objects having the second property but lacking the first?

Accordingly, the modern approach is to use a one-criterion distinction; *cooperative* games are defined as those permitting enforceable agreements while *noncooperative* games are defined as those not permitting them. How much communication is permitted among the players is, of course, also important; but it is a less fundamental issue. For example, consider a Prisoners' Dilemma game:

	X	Y
A	10, 10	−10, 11
B	11, −10	1, 1

If enforceable agreements are possible then two rational players will no doubt immediately agree to use the strategy pair (A, X) since this outcome is much better for both of them than the other symmetric outcome (B, Y) would be. But if agreements are unenforceable then the players cannot do any better

[1] I want to thank the National Science Foundation for supporting this research through Grant SOC77–06394 to the Center for Research in Management Science, University of California, Berkeley. I would like to thank also Professor Wilhelm Krelle and the University of Bonn for their kind hospitality.

than use the strategy pair (B, Y). This is so because if they agreed to use (A, X) instead, then each player would know that his opponent would be most likely to *break* this agreement (and he himself, also, would be most likely to do the same). The reason is that, even if player 1 expected player 2 to keep to the agreement and use strategy X, his own (player 1's) best reply to this would be to use strategy B, i.e., to violate the agreement. Of course, player 2 would be in a similar position. This inherent instability of strategy pair (A, X) can be overcome only if the two players can irrevocably *commit* themselves to using strategies A and X by means of an enforceable agreement.

On the other hand, it makes little difference whether the players can *talk* to each other or not. Even if they can talk and can negotiate an agreement, this will be of little help if such an agreement is unenforceable and, therefore, has little chance of being kept. An ability to negotiate agreements is useful only if the rules of the game make such agreements fully binding and enforceable. (In real life, this may be the case because an outside government agency, or public opinion, etc., will enforce such agreements; or because the players are, and *know* each other to be, spontaneously unwilling of their own accord, to break agreements, say, as a matter of moral principle.)

As Nash has already pointed out, since in a noncooperative game agreements are unenforceable, rational players will always use a strategy combination that is *self-enforcing*, i.e., a strategy combination with the property that each player's strategy is a *best reply* to the other players' strategies. Such a strategy combination is called an *equilibrium point*.

For instance, in our example, the strategy pair (B, Y) is an equilibrium point; if player 1 uses strategy B, then player 2's best reply is to use strategy Y; and, conversely, if player 2 uses strategy Y, then player 1's best reply is to use strategy B. Thus, B and Y are *mutually* best replies to each other. Such a strategy pair is *self-enforcing*, in the sense that, if either player expects the other player to abide by it, then he will have an incentive to do the same.

In contrast, all other strategy pairs in this game are *self-destabilizing:* even if the players expected each other to use such a strategy pair, *this very expectation* would give at least one player an incentive to act *contrary* to this expectation. (For instance, the strategy pair (A, X) is self-destabilizing because, as we have seen, if the two players expected each other to use this strategy pair then this expectation would immediately give them an incentive not to use strategies A and X but rather to use strategies B and Y instead.)

Besides inventing the concept of an equilibrium point, Nash [1951] has also shown that every finite game (i.e., every game with a finite number of players, and with a finite number of pure strategies for each player) has at least one equilibrium point.

2. Limitations of the classical theory of cooperative games

After NASH had introduced the concept of equilibrium points, the theory of noncooperative games has made little progress for a number of years, and the main interest of game theorists shifted to the theory of cooperative games. A number of solution concepts for such games were proposed: the VON NEUMANN – MORGENSTERN [1944] stable sets, NASH's [1950, 1953] solution for two-person bargaining games, the SHAPLEY [1953] value, the core [GILLIES, 1959], the AUMANN – MASCHLER [1964] bargaining sets, and several others.

Individually, each of these solution concepts is of great theoretical interest. But, taken as a group, they fail to provide an analytically clear and coherent theory of cooperative games. Indeed, the different solution concepts have very little theoretical connection with one another and even less can they be interpreted as special cases of one general theory.

These shortcomings of the classical theory of cooperative games are, of course, basically *conceptual* limitations, important primarily for the logician, the methodologist, and the philosopher, and are perhaps of lesser importance to the social scientist, whose main interest is in possible *applications* of game-theoretical ideas to economics, political science, or sociology. But these conceptual limitations do give rise to major difficulties also in empirical applications.

First of all, classical game theory offers quite a number of very different cooperative solution concepts, but gives little guidance to the social scientist as to *which* particular solution concept to use in analyzing any specific real-life social situation. Nor does it tell him *why* so many different solution concepts are needed in the first place.

Another difficulty lies in the fact that classical game theory is restricted to *fully cooperative* games, where all agreements are always fully enforceable, and to *fully noncooperative* games, where no agreements are ever enforceable at all. It does not cover *partially cooperative* games, where some types of agreements may be fully enforceable, while other types may be either unenforceable or enforceable only with specified probabilities; where agreements by some groups of players may be enforceable but agreements by other groups may not be; where agreeements made at some stages of the game may be enforceable while agreements made at other stages may lack this property, etc. Yet, this intermediate category of partially cooperative games has great importance in many empirical applications.

Within the class of partially cooperative games, special importance attaches to cooperative games with a *sequential structure*. These are games involving two or more successive stages, in which agreements among the players may be built up gradually in several consecutive steps. Unlike strictly cooperative games, in which any agreement is always final, such sequential games will often permit renegotiation and modification of earlier agreements under certain conditions in later stages of the game. In real life, games of this type are extremely important, yet they are altogether beyond the reach of classical game theory.

All these limitations of classical cooperative game theory have their roots in one basic shortcoming of this theory. In almost every cooperative game, bargaining negotiations among the players play an all-important role. Yet, the classical theory of cooperative games completely excludes the players' bargaining moves and countermoves from its formal analysis by postulating that the latter (called "pre-play negotiations") always occur *before* the game is actually played and, therefore, are not a part of the game itself. This amounts to voluntarily relinquishing any serious attempt to understand how the outcome of the game in fact depends on the nature of the bargaining process among the players.

3. Games with incomplete information

One of the most serious limitations of classical game theory is its inability to handle games with incomplete information. We say that a given game involves *complete information* if all players know the nature of the game—in the sense of knowing the extensive form of the game (i.e., the game tree), or at least knowing the normal form of the game (i.e., the payoff matrix).

A game with complete information can be either a game with *perfect* information or a game with *imperfect* information: it will be the former if the players not only know the nature of the game as such but also know all *previous moves* (made by other players or by chance) at any stage of the game; and it will be the latter if the players know the nature of the game but have less than full information about some or all previous moves during the game.

On the other hand, a game involves *incomplete information* if the players have less than full information about each other's strategy possibilities and/or each other's payoff functions. (The latter problem may arise because the players do not have full knowledge about each other's strategic objectives in the game, or because they do not know each other's attitudes to risk taking as expressed by their cardinal utility functions.) At the same time, the players may also fail to know how much information the other players have about any given player's strategy possibilities and about his payoff function, etc.

Classical game theory cannot handle games with incomplete information at all (but it does cover both games with *perfect* and with *imperfect* information, as long as these have the nature of games with *complete* information). This is obviously a very serious limitation because virtually all real-life game situations involve incomplete information on the part of the players. (In particular, it very rarely happens that the participants of any real-life social situation have reasonably full information about each other's utility functions. Uncertainty about the strategies actually available to the other players is also very common.)

It turns out, however, that we can bring a game with incomplete information within the scope of game-theoretical analysis by using a *probabilistic model* to represent the incomplete information that the players have about various parameters of the game [HARSANYI, 1967–68]. More specifically, let G be a game with incomplete information. Then, analysis of this game G can be reduced to analysis of a new game G* involving random moves, called the probabilistic model for G. In this new game G*, the fact that (some or all of) the players have limited information about certain basic parameters of the game is mathematically represented by the assumption that these players have limited information about the outcomes of these *random moves*.

From a technical point of view, this probabilistic-model game G* will be a game with *complete information*. (But it will be a game with *imperfect information* because of the players' having less than full information about the random moves occurring in the game.) Thus, our approach essentially amounts to reducing the analysis of a game with *incomplete* information, G, to the analysis of a game with *complete* (yet imperfect) information, G*, which, being a game with complete information, is of course fully accessible to the usual analytical tools of game theory.

By constructing suitable probabilistic models of various types, we can produce games with any desired distribution of knowledge and of ignorance among the players, and can study how alternative informational assumptions will change the nature of the game. We can also study how any given player can *infer* some pieces of information originally denied to him, by observing the moves of those players who already posses this information. We can also investigate how each player can optimally *convey* information to some other players, or can optimally *withhold* this information from them, in accordance with his own strategic interests within the game [see HARSANYI, 1977b].

I must add, however, that use of such probabilistic models in general provides only a *partial* solution for the problem of how to analyze games with incomplete information. This is so because, as soon as a probabilistic-model game G* has been constructed as a more convenient mathematical representation for the originally given game with incomplete information, G, the problem immediately arises what *solution concept* to use for this newly constructed game G* itself.

In actual fact, if the game G we start with is a *noncooperative* game with incomplete information, then this question has an easy and natural answer. In this case, the probabilistic-model game G* derived from G will be itself also a *noncooperative* game (though of course one with complete information), and can be analyzed in terms of its *equilibrium points:* thus, the concept of equilibrium points can be extended to games with incomplete information without any difficulty [HARSANYI, 1967–68, pp. 320–329].

The situation, however, is very different if the game G we are trying to analyze is a *cooperative* game with incomplete information. In this case, we will find that typically the probabilistic-model game G* derived from G will not admit of analysis in terms of *any* of the cooperative solution concepts of conventional game theory.

For example, it turns out that the NASH solution for two-person bargaining games, which is such an attractive solution concept for such games in the case of *complete* information, cannot be used at all to define solutions for two-person bargaining games with *incomplete* information or for the probabilistic-model games derived from them: if we try to use the NASH solution for this purpose then we obtain completely nonsensical results [HARSANYI, 1967–68, pp. 329–334]. Other classical cooperative solution concepts give equally unsatisfactory results when applied to incomplete-information games. This lack of solution concepts applicable to games with incomplete information is another serious weakness of the classical theory of cooperative games.

4. *Difficulties with the concept of equilibrium points*

Compared with the classical theory of cooperative games, the classical theory of noncooperative games presents a much more satisfactory picture. First of all, it has a much higher degree of *theoretical unity* because it is wholly based on one specific solution concept, that of equilibrium points. It is also a *more complete* theory than the theory of cooperative games is because it tries to cover *all* aspects of any given game, and does not exclude the players' bargaining moves from the scope of its analysis, in the way the theory of cooperative games does.[2] Furthermore, as I have already mentioned, the concept of equilibrium points — and, therefore, the classical theory

[2] I am assuming that our definition of noncooperative games includes games where inter-player communication is permitted (see Section 1 above).

of noncooperative games—can be easily extended to games with *incomplete information.*

Finally, equilibrium points are one of the very few game-theoretical solution concepts that have direct application both to games in extensive form and to games in normal form, which enables the theory of noncooperative games to deal with both game forms in terms of a uniform theoretical framework. (This has many desirable consequences. One of them is the fact that the classical theory of noncooperative games, unlike the theory of cooperative games, can handle games possessing a sequential structure, without any difficulty.)

Yet, even though the concept of equilibrium points has many strong points, it also has at least three important weaknesses:

1. Almost every nontrivial game has *many* (sometimes even infinitely many) *essentially different* equilibrium points. Hence, a theory which could only predict that the outcome of a noncooperative game should be an equilibrium point, without specifying *which* equilibrium point this actually were to be, would be an extremely weak and uninformative theory. This difficulty I will call the *multiplicity problem.*

2. Secondly, any mixed-strategy equilibrium point is, or at least appears to be, fundamentally *unstable* (see Section 6 below) and, therefore, to be unsuited to be the solution of a game. This gives rise to what I will call the *instability problem:* how are we to define a solution for a noncooperative game that has only mixed-strategy equilibrium points?

3. A third difficulty connected with equilibrium points was discovered by REINHARD SELTEN [1965, 1975]. He pointed out that many equilibrium points require some or all of the players to use highly *irrational* strategies (see Section 7). He proposed to call such equilibrium points *imperfect* equilibrium points, as distinguished from *perfect* equilibrium points, which involve no irrational strategies. The problem posed by the fact that many games contain imperfect equilibrium points I will call the *imperfectness problem.*

This completes my survey of the main problems posed by classical game theory, both by its cooperative and by its noncooperative branches. I now propose to show how the new solution theory developed by REINHARD SELTEN and by me tries to resolve these problems.

II. The new solution theory

5. *Analysis of cooperative games by means of noncooperative bargainig models*

The new solution concept developed by REINHARD SELTEN and by me is primarily a solution concept for noncooperative games. But it actually grew out of our research concerning cooperative games. When it became clear to us that the NASH solution in its original form could not be used as a solution concept for two-person bargaining games with incomplete information [HARSANYI, 1967–68, pp. 329–334], REINHARD SELTEN and I first tried to overcome this difficulty by proposing an *ad hoc* modification of the NASH solution, specifically designed to resolve this particular problem [HARSANYI and SELTEN, 1972]. But soon after this we have come to the conclusion that thinking up new *ad hoc* cooperative solution concepts whenever the need arises is not really a satisfactory approach, and that a radically new theoretical departure is required.

We have decided to follow NASH's [1951, p. 285] suggestion that analysis of any *cooperative game* G should be based on a formal *bargaining model* B (G), involving bargaining moves and countermoves by the various players, and resulting in an agreement about the outcome of the game. Formally, this bargaining model would have the nature of a *noncooperative game* in extensive form (or possibly in normal form). The solution of the cooperative game G would then be defined in terms of the equilibrium points of this noncooperative bargaining game B (G).

At the same time, we have been aware of the fact that NASH's approach cannot possibly work unless we can overcome the three—already mentioned—weaknesses in the concept of equilibrium points itself, viz. the multiplicity problem, the instability problem, and the imperfectness problem. Overcoming the multiplicity problem has been particularly crucial: experience shows that bargaining games (bargaining models) are especially prone to having an overabundance of essentially different equilibrium points.

This fact can be illustrated by considering a very simple two-person bargaining game, where two players have to agree on how to divide $ 100, the money being lost to them if they cannot agree. (I will assume that both players have linear utility functions for money.) This game can be represented by the following bargaining model. Each player has to name a real number, representing his payoff demand. The numbers named by players 1 and 2 will be called x_1 and x_2, respectively. If $x_1 + x_2 \leqq 100$, i.e., if the two players' payoff demands are mutually compatible, then both will obtain their payoff demands, i.e., they will obtain the payoffs $u_1 = x_1$ and $u_2 = x_2$. In contrast, if $x_1 + x_2 > 100$, i.e., if their payoff demands are incompatible, then they will receive zero payoffs $u_1 = u_2 = 0$ (since this will be taken to mean that they could not reach an agreement).

If the players are freee to divide the $ 100 in all mathematically possible ways, then this game will have infinitely many essentially different equilibrium points in pure strategies since all possible pairs (x_1, x_2) satisfying $x_1 + x_2 = 100$ and $x_1, x_2 \geqq 0$ will be equilibrium points. But even if we restrict the players to payoff demands representing integer numbers of dollars, the game will still have 101 essentially different equilibrium points, from (0,100), (1,99), . . ., to (100,0). Clearly, a theory telling us no more than that the outcome can be *any one* of these equilibrium points will not give us much useful information. We need a theory selecting *one specific* equilibrium point as the solution of the game. In fact, the main purpose of REINHARD SELTEN's and my solution concept is to provide a mathematical criterion that always selects one specific equilibrium point as the solution. In other words, its main purpose is to overcome the multiplicity problem. (But, as I will try to show, our theory also overcomes the two other problems posed by the concept of equilibrium points, viz. the instability problem and the imperfectness problem.)

Used in conjunction with our solution theory, analysis of cooperative games by means of noncooperative bargaining models, as suggested by NASH, does provide a full remedy for the various problems posed by the classical theory of cooperative games. First of all, it yields a uniform approach to analyzing all classes of cooperative games. Even though different cooperative games may have to be analyzed in terms of very different bargaining models, the solution of each bargaining model, and therefore the solution of each cooperative game, will be defined in terms of the very same basic mathematical criteria, as specified by our solution theory.

Indeed, the suggested approach will permit a unification, not only of cooperative game theory, but rather of game theory as a whole, including *both* the theories of cooperative and of noncooperative games. This is so because the problem of defining a solution for a *cooperative* game G will always be reduced to the problem of defining a solution for the *noncooperativ* game B (G).

A further advantage of this approach is that it shows exactly how the outcome of any given cooperative game will depend on the actual nature of the bargaining process among the players, including such factors as who can talk to whom, and who can talk to whom *first*, before anybody else can; what the rules are for concluding agreements, or for withdrawing from agreements already concluded, or again for making one's tentative agreements final and irrevokable; how easily coalitions can be formed, dissolved, or recombined; what threats can be made by whom against whom, and to what extent such threats are irrevokable, etc.

In constructing bargaining models we can take advantage of the very great flexibility that bargaining games in extensive form provide for us – a flexibility not available to the classical theory of cooperative games because of its insistence on using the much more restrictive normal form (or the even more rigid characteristic-function form). For example, we can easily construct bargaining models that represent cooperative games wholly inaccessible to classical theory, such as partially cooperative games, or cooperative games possessing a sequential structure, or cooperative games with incomplete information, etc.

In what follows, I will first describe how our theory resolves the instability and the imperfectness problems. Then, I will discuss our solution concept itself, which we use primarily to overcome the multiplicity problem.

6. A new justification for use of mixed-strategy equilibrium points

To illustrate the problem posed by games that have only mixed-strategy equilibrium points, consider the following game:

	B_1	B_2
A_1	45, 30	0, 90
A_2	30, 75	60, 45

The only equilibrium point of this game is the mixed-strategy equilibrium point (A^*, B^*), where $A^* = (1/3, 2/3)$ and $B^* = (4/5, 1/5)$. In other words, players 1's equilibrium strategy A^* requires him to use his two pure strategies A_1 and A_2 with the probabilities $1/3$ and $2/3$, respectively. In contrast, player 2's equilibrium strategy B^* requires him to use his two pure strategies B_1 and B_2 with the probabilities $4/5$ and $1/5$.

We can better understand the problem if we add a new row and a new column to the payoff matrix, corresponding to the equilibrium strategies A^* and B^*:

	B_1	B_2	B^*
A_1	45 / 30	0 / 90	36 / 42
A_2	30 / 75	60 / 45	36 / 69
A^*	35 / 60	40 / 60	36 / 60

As can be seen from this enlarged payoff matrix, if player 1 expects player 2 to use his equilibrium strategy B^*, then player 1 himself will have no real incentive to use his own equilibrium strategy A^*. This is so because he will obtain the same payoff $u_1 = 36$, regardless of whether he does use his mixed equilibrium strategy A^*, or uses either of his two pure strategies A_1 and A_2, or uses any other mixed strategy whatever. Likewise, player 2 will have no real incentive to use his equilibrium strategy B^* even if he does expect player 1 to use his own equilibrium strategy A^*. This is so because player 2 will obtain the same payoff $u_2 = 60$, regardless of whether he uses his equilibrium strategy B^*, or uses either of his two pure strategies B_1 and B_2, or uses any mixed strategy whatever.

This is what we mean by saying that the equilibrium point (A^*, B^*) is—seemingly—*unstable:* even if it does not provide any incentive for either player *not* to use his equilibrium strategy, it does not provide any incentive, either, that would make it positively attractive for him to *use* his equilibrium strategy.

I now propose to argue that this instability of such mixed-strategy equilibrium points is only *apparent*. First of all, even if the players have as complete information about the payoff matrix of the game as they can possibly have, each player will always have some irreducible minimum of *uncertainty* about the other player's payoffs. For example, even though the payoff matrix shows player 2's payoff associated with the strategy pair (A_1, B_1) to be $U_2(A_1, B_1) = 30$, player 1 will never be able to exclude the possibility that, at this very moment, this payoff may be in fact $30 - \varepsilon$ or $30 + \varepsilon$, where ε is a small positive number. This is so because every person's utility function is subject at least to small unpredictable random fluctuation as a result of changes in his mood, or a possible sudden urge to use one of his pure strategies in preference to his other pure strategy, etc.

This means that a realistic model of any given game will not be one with *fixed* payoffs but rather one with *randomly fluctuating* payoffs (though these fluctuations may be very small). Mathematical analysis shows that such a game will have no mixed-strategy equilibrium points at all. Rather, all its equilibrium points will be in pure strategies, in the sense that neither player will ever intentionally randomize between his two pure strategies: instead, he will always find that one of his two pure strategies will yield him a higher payoff, and this is the pure strategy that he will actually use.

At the same time, the random fluctuations in the two players' payoffs will interact in such a way that player 1 will find strategy A_1 to be more profitable than strategy A_2 approximately $^1/_3$ of the time, and will find A_2 more profitable than A_1 approximately $^2/_3$ of the time. As a result, even though he will make no attempt to randomize, he will in fact use his two pure strategies approximately with the probabilities actually prescribed by his equilibrium strategy $A^* = (^1/_3, ^2/_3)$. By the same token, even though player 2 will make no attempt to randomize, he will in fact use his two pure strategies B_1 and B_2 approximately with the probabilities prescribed by his equilibrium strategy $B^* = (^4/_5, ^1/_5)$. [For detailed discussion and for mathematical proofs, see HARSANYI, 1973.]

To conclude, when a given game is interpreted as a game with *fixed* payoffs, then it will not provide any incentives for the players at a mixed-strategy equilibrium point to use their pure strategies with the probabilities prescribed by their equilibrium strategies. But if the game is, more realistically, reinterpreted as a game with *randomly fluctuating* payoffs, then it will provide the required incentives, so that the instability problem associated with mixed-strategy equilibrium points will disappear.

7. Elimination of imperfect equilibrium points from the game

To illustrate the imperfectness problem, we have to consider a game in extensive form because, in the normal form, the distinction between perfect and imperfect equilibrium points becomes unclear. I will use the following example:

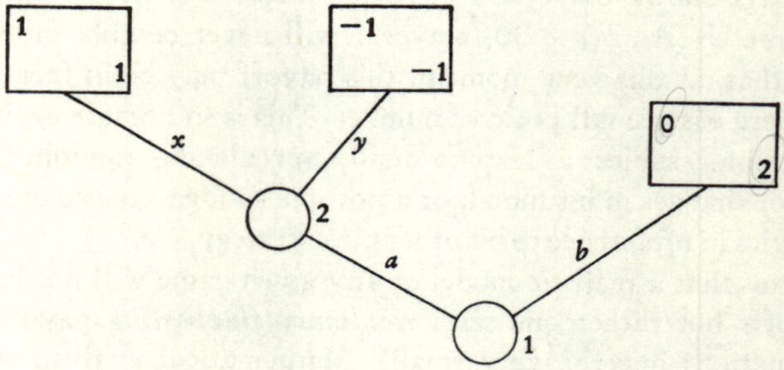

In this game, player 1 has the first move: he can choose between move *a* and move *b*. If he chooses *b* then the game will immediately end with the payoffs $u_1 = 0$ and $u_2 = 2$. But if he chooses *a*, then player 2 will also have a move: he will be able to choose between moves *x* and *y*. If he chooses *x* then the payoffs will be $u_1 = u_2 = 1$ while of he chooses *y* then the payoffs will be $u_1 = u_2 = -1$.

For people more accustomed to considering games in normal form, the normal form of this game is:

	X	Y
A	1 / 1	−1 / −1
B	0 / 2	0 / 2

Thus, player 1 has two pure strategies, viz. strategy A ("Choose move *a*") and strategy B ("Choose move *b*"). Player 2 also has two pure strategies, viz. strategy X ("Choose move *x* if player 1 has chosen *a*"), and strategy Y ("Choose move *y* if player 1 has chosen *a*").

Obviously, the game has two pure-strategy equilibrium points, viz. the strategy pairs $E = (A, X)$ and $E^* = (B, Y)$. E is a perfect equilibrium point. But I will now show that E* is an imperfect equilibrium point, involving irrational strategies.

In fact, strategy Y is surely irrational because it requires player 2 to choose move *y*, which will yield him as well as player 1 the payoff $u_1 = u_2 = -1$, even though by choosing move *x* both he and player 1 could obtain the payoff $u_1 = u_2 = 1$. Strategy B is equally irrational: player 1 should know that if he chose move *a* then player 2 would surely choose move *x*, which would yield him the payoff $u_1 = 1$; therefore, player 1 should not choose move *b*, which would yield him only $u_1 = 0$.[3]

How is it possible that an equilibrium point should involve such irrational strategies? In particular, how can an equilibrium point require a player to choose a move like *y* when choosing this move is inconsistent with maximizing his own payoff?

[3] The reader may object that we do not need the concept of imperfect equilibrium points to show that $E^* = (B, Y)$ is an undesirable equilibrium point, because it is sufficient to point out that E* uses a (weakly) dominated strategy, viz. strategy Y. This is, of course, a correct observation. In fact, in two-person games *all* imperfect equilibrium points always involve dominated (or equivalent) strategies. But this is *not* true for three-person or even larger games. (The reason why I did not use a three-person example was that its analysis would take up much more space.)

The answer is that a move like *y* will in fact reduce a given player's expected payoff only if this move occurs with a *positive probability.* Yet, if the two players act in accordance with equilibrium point E* = (B, Y), then player 2 *will never come into a position of having to implement* this irrational move *y*, i.e., move *y* will occur with *zero probability,* and, therefore, will not actually reduce player 2's expected payoff.

More generally, an equilibrium strategy by definition must maximize the relevant player's expected payoff if the other players' strategies are kept constant; and this means that no equilibrium strategy can possibly prescribe an irrational (i.e., a non-payoff-maximizing) move at any point of the game tree that will actually be *reached* with a positive probability if all players use their equilibrium strategies. But an equilibrium strategy *may* prescribe an irrational move for a given player at some point of the game tree that will be reached with *zero probability.* Imperfect equilibrium points are precisely those equilibrium points that prescribe a move contrary to payoff maximization at some point of the game tree that will be reached with zero probability.

This analysis, however, suggests a method of eliminating imperfect equilibrium points from any given game. We first replace every player by as many *agents* as the number of his information sets in the extensive form of the game. Each agent's pure strategies will be simply the moves that the main player had at the relevant information set. His payoff function will be the same as the main player's. Then, we construct a normal-form game whose players are these agents: this game will be called the *agent normal form.*[4]

Given the agent normal form, we can now eliminate all imperfect equilibrium points as follows. Suppose that agent i ($i = 1, \ldots, n$) has K_i pure strategies, denoted as $a_i^1, \ldots, a_i^k, \ldots, a_i^{K_i}$. I will make the following assumption. Whenever agent i tries to use any given pure strategy a_i^k, he will succeed in doing so only with probability $(1 - \varepsilon)$, where ε is a small positive number; whereas with probability ε he will make a *mistake,* using each of his $(K_i - 1)$ unintended pure strategies with probability $\varepsilon/(K_i - 1)$. The resulting game, involving these imposed mistake probabilities, is called the *perturbed agent normal form* (PANF).

As a result of these positive mistake probabilities, in this PANF game *all* information sets of the game tree will always be *reached* with positive probabilities. Consequently, this PANF game will no longer contain any

[4] If we want to eliminate all imperfect equilibrium points from the game then we have to work with the *agent normal form* rather than with the ordinary *normal form* because in the latter it is in general impossible to distinguish perfect and imperfect equilibrium points from each other.

imperfect equilibrium points. Moreover, if we let the mistake-probability parameter ε go to zero, the equilibrium points of the PANF game will converge to perfect equilibrium points of the original game.

To prevent our solution concept from ever selecting an imperfect equilibrium point as the solution of any given game, we define our solution as follows. We first apply our solution concept to the PANF games G_ε characterized by given ε values. Then, we compute the solutions s_ε for these games G_ε. Finally, we define the solution of the original game G as that particular perfect equilibrium point $s = (s_1, \ldots, s_n)$ which is the *limit* of these solutions s_ε when the parameter ε goes to zero.

8. *The tracing procedure*

Our solution theory consists of two parts. One is a mathematical procedure called the *tracing procedure*. The other is a theory of the *prior probability distributions* used in this tracing procedure. This latter part of our theory also deals with the *hierarchical structure* of the game, based on the various *component games* into which the game can be divided, since the prior probability distributions used in the tracing procedure are themselves defined in terms of this hierarchical structure of the game. For lack of space, I can describe our solution theory only in very rough outlines. A more detailed discussion of the tracing procedure can be found in HARSANYI [1975]. The prior probability distributions used by our theory, and the hierarchical structure of the game, are described in HARSANYI [1977a]. Some applications of our theory are discussed in HARSANYI [1977c and 1979]. Other aspects of our solution theory will be discussed in forthcoming publications by REINHARD SELTEN and by me.

I will use the following notations. The pure strategies of player i ($i = 1, \ldots, n$) in game G will be written as $a_i^1, \ldots, a_i^k, \ldots, a_i^{K_i}$. Any mixed strategy s_i of player i will have the form of a probability vector $s_i = (s_i^1, \ldots, s_i^k, \ldots, s_i^{K_i})$, where s_i^k ($k = 1, \ldots, K_i$) is the probability that s_i assigns to the pure strategy a_i^k. A strategy combination s of the n players will have the form $s = (s_1, \ldots, s_n)$. A strategy combination of the $(n - 1)$ players other than player i will be written as $\bar{s}_i = (s_1, \ldots, s_{i-1}, s_{i+1}, \ldots, s_n)$. I will also write $s = (s_i, \bar{s}_i)$. Player i's payoff function will be denoted as $U_i(s) = U_i(s_i, \bar{s}_i) = U_i(s_1, \ldots, s_n)$.

The *tracing procedure* is an attempt to extend Bayesian decision theory from *one-person* decision problems, where it has been so eminently successful, to the *n-person* decision problems arising in n-person noncooperative games. We use the following model. Each player j expresses his expectations about the strategies that any other player *i* is likely to use in the game, in the form of a subjective probability distribution $p_i = (p_i^1, \ldots, p_i^k, \ldots, p_i^{K_i})$ over player i's pure strategies, where p_i^k $(k = 1, \ldots, K_i)$ denotes the subjective probability that player *j* assigns to the hypothesis that player *i* will use the pure strategy a_i^k. Since all players $j \neq i$ will base their subjective probability distributions p_i over the pure strategies a_i^k of any given player *i* on the *same* information concerning the nature of the game, all of them will use the *same* subjective probability distribution p_i. This subjective probability distribution p_i is called the *prior probability distribution* over player *i*'s pure strategies.

Since any prior distribution p_i is a probability distribution over player *i*'s pure strategies, mathematically it will have the nature of a *mixed strategy* by player *i*. But it will have a very different intuitive interpretation from an ordinary mixed strategy. While an ordinary mixed strategy s_i involves the assumption that player *i* himself will *randomize* among his different pure strategies, a prior probability distribution p_i involves no such assumption. Rather, it merely expresses *uncertainty* in the other players' minds about the actual strategy that player *i* will use in the game.

I will write $p = (p_1, \ldots, p_n)$, $\bar{p}_i = (p_1, \ldots, p_{i-1}, p_{i+1}, \ldots, p_n)$, and $p = (p_i, \bar{p}_i)$. Thus, $\bar{p}_i$ is a probability vector summarizing player *i*'s expectations about the behavior of all the other $(n - 1)$ players in the game whereas p summarizes the expectations of all *n* players about one another's behavior.

Suppose we know p. How can we predict the outcome of the game? The simplest answer would be this. Since the expectations of each player *i* about his fellow players' behavior are expressed by the relevant probability vector $\bar{p}_i$, he will use that particular strategy s_i^+ that is his *best reply* to $\bar{p}_i$, i.e., he will use a strategy $s_i = s_i^+$ that maximizes his expected payoff

$$(8.1) \qquad U_i(s_i, \bar{p}_i).$$

Accordingly, one may try to argue that the solution of game G should be defined as the strategy combination $s^+ = (s_1^+, \ldots, s_n^+)$, consisting of these best-reply strategies $s_1^+, \ldots, s_i^+, \ldots, s_n^+$. I will describe the solution theory based on this assumption as the *naive Bayesian approach*.

Regrettably, this simple-minded naive Bayesian approach must be rejected because the strategy combination s^+ in general *will not be an equilibrium point*.

Yet, even though this naive Bayesian approach is an unsatisfactory solution theory, it can be used as a *starting point* of a more satisfactory theory. In fact, the tracing procedure always starts with this *best-reply* strategy combination s^+, but then continuously and systematically transforms s^+

until it *converges* to an equilibrium point $s^* = (s_1^*, \ldots, s_n^*)$ of the game, which will then be chosen as the solution.

The tracing procedure is based on a one-parameter family of auxiliary games G^t, with $0 \leqq t \leqq 1$. In any given game G^t, each player i will have the payoff function

$$(8.2) \qquad U_i^t (s_i, \bar{s}_i) = (1 - t)\, U_i (s_i, \bar{p}_i) + t\, U_i (s_i, \bar{s}_i).$$

Clearly, for $t = 1$, game $G^t = G^1$ will have the same payoff functions

$$(8.3) \qquad U_i^1 (s_i, \bar{s}_i) = U_i (s_i, \bar{s}_i)$$

as the original game G has. Therefore, we can write $G^1 = G$. On the other hand, for $t = 0$, game $G^t = G^0$ will have payoff functions of form (8.1) since

$$(8.4) \qquad U_i^0 (s_i, \bar{s}_i) = U_i (s_i, \bar{p}_i).$$

Thus, this auxiliary game G^0 will be a game of a very special form since it will make the payoff U_i^0 of each player i depend only on his *own* strategy s_i and will make it independent of the other players' strategies (since $\bar{p}_i$ does not depend on $\bar{s}_i$). Consequently, G^0 can be factored into n mathematically independent maximization problems, one for each player i.

More generally, for $0 \leqq t \leqq 1$, the payoff function U_i^t of each player i will be a convex combination of U_i^0 and of $U_i^1 = U_i$.

For each game G^t, let E^t be the set of all equilibrium points of G^t. By NASH's [1951] theorem (see Section 1), each of these sets E^t will be nonempty. Let P be the *graph* of the correspondence $t \rightarrow E^t$, $0 \leqq t \leqq 1$. For any fixed t, this graph P will have as many points as the number of equilibrium points in E^t. Accordingly, in nondegenerate cases, in region $t = 0$, graph P will have only *one* point x^0, because each player i will have a *unique* (pure) strategy $s_i = s_i^+$ maximizing the payoff function $U_i^0 = U_i (s_i, \bar{p}_i)$, and as a result the equilibrium-point set E^0 will contain only the *one* point $s^+ = (s_1^+, \ldots, s_n^+)$. Moreover, one can show that, in nondegenerate cases, graph P will contain exactly *one* continuous path L, leading from this one point x^0 in region $t = 0$ to some point x^1 in region $t = 1$. This point x^1 will always correspond to a specific equilibrium point $s^* = (s_1^*, \ldots, s_n^*)$ of game $G^1 = G$, and this equilibrium point s^* can be chosen as the *solution* of game G. The mathematical procedure selecting s^* as solution is called the *tracing procedure* because it is based on tracing (i.e., following) path L from its starting point x^0 (corresponding to the strategy combination s^+) to its end point x^1 (corresponding to the solution s^*).

This version of the tracing procedure, called the *linear* tracing procedure, in all nondegenerate cases always yields a unique solution s^*. In degenerate cases, however, it will not yield a unique result because graph P may contain more than one point x^0 in region $t = 0$ (in which case it will always contain infinitely many points x^0) and, even if x^0 is unique, there may be several (and even infinitely many) continuous paths leading from x^0 to the region $t = 1$. But one can show that a modified version of the tracing procedure, called the *logarithmic* tracing procedure, will *always* yield a unique solution, even in highly degenerate cases.

The solution defined by the tracing procedure always has the form $s^* = T(G, p)$. That is to say, the solution s^* obtained will depend not only on the game G we are analyzing but also on the vector $p = (p_1, \ldots, p_n)$ of priors used as a starting point. Indeed, in most cases, s^* will very *crucially* depend on p. Therefore, the theory of how to select the vector p of priors is a crucially important part of our solution theory.

Intuitively, the tracing procedure has the following interpretation. Its purpose is to model the *solution process*, i.e., the reasoning process by which rational players choose a specific equilibrium point s^* as the solution of any given noncooperative game G.

At the beginning of this solution process, the players' *expectations* about each other's strategy choices will correspond to the prior probability distributions $p_1, \ldots, p_n$; whereas their tentative strategy choices themselves will correspond to the strategies $s_1^+, \ldots, s_n^+$, which represent their best-reply strategies to these expectations. Then, the solution process will consist in gradually and systematically *modifying* both these expectations and these tentative strategy choices until both of them *converge* to the solution strategies $s_1^*, \ldots, s_n^*$, corresponding to a unique equilibrium point $s^* = (s_1^*, \ldots, s_n^*)$ of the game.

9. *The hierarchical structure of the game*

For any given game G, our theory makes use of two kinds of component games, viz. formations and cells. A *formation* F is a smaller game derived from game G in the following way. F is played be the same n players as G is, but each player i is now restricted to some nonempty *subset* A_i^F of the pure-strategy set A_i he would have in G. He can also choose any mixed strategy s_i which uses only pure strategies a_i^k included in A_i^F. These pure and mixed strategies will be called his *F-permissible* strategies. We also require that the new game F should satisfy the following *closure* requirement: If any given player i expects all the other players to use F-permissible strategies, then player *i* should have a positive incentive likewise to use an F-permissible strategy.

On the other hand, a *cell* C is a game played by some *subset* N^C of the player set N of game G. The players included in N^C are called the *C-players*. A cell C must satisfy the following closure requirement: If any given C-player *i* wanted to decide which strategy s_i of his was a best reply to the strategies he expects the other players would use, then it should be sufficient for him to consider the strategies he expected his fellow C-players would use, regardless of the strategies he expected the non-C-players would use.

Thus, a formation is a self-sufficient strategy-restricted component game whereas a cell is a self-sufficient player-restricted component game.

Our theorey defines the solution $s^*(G)$ of the entire game in terms of the solutions $s^*(F)$ and $s^*(C)$ of the formations F and the cells C of G. On the other hand, the solution $s^*(F)$ of any formation F, and the solution $s^*(C)$ of any cell C, is itself defined in terms of the solutions of the subformations and the subcells of F or of C, etc., until we reach the smallest formations and the smallest cells which cannot be any further decomposed into subformations and subcells.

This inductive definition procedure will always involve only a finite number of steps because any finite game G can contain only finite numbers of formations and of cells (even if we include the subformations and subcells of all subformations and subcells of all orders).

10. *The prior probability distributions*

Our theory uses two types of prior probability distributions, called *bicentric* priors and *uniform* priors. Under our theory, the players will use *bicentric* priors $\check{p}_i$ whenever they want to decide which of two equilibrium points $s = (s_1, \ldots, s_n)$ and $t = (t_1, \ldots, t_n)$ should be the solution of the game, in case *their choice could be restricted* to these two equilibrium points. For lack of space, I cannot state the formal definition of the bicentric probability vectors $\check{p}_i = (\check{p}_i^1, \ldots, \check{p}_i^{K_i})$. Let me merely state the following fact. Consider the quantities

$$(10.1) \qquad v_i = U_i(s_i, \overline{s_i}) - U_i(t_i, \overline{s_i})$$

and

$$(10.2) \qquad w_i = U_i(t_i, \overline{t_i}) - U_i(s_i, \overline{t_i}) \qquad \text{for } i = 1, \ldots, n.$$

We may interpret the quantity v_i as the *net payoff incentive* that player *i* will have to stick to his s-strategy s_i rather than shift to his t-strategy t_i in case all the other $(n-1)$ players j likewise stick to their s-strategies s_j. On the other hand, the quantity w_i is the *net payoff incentive* that player i will have to stick to his t-strategy t_i rather than shift to his s-strategy s_i in case all the other $(n-1)$ player j likewise stick to their t-strategies t_j.

Now, in a fairly large and very important class of games, the bicentric prior $\check{p}_i$ induced by the two equilibrium points s and t will be equivalent to an expectation that each player i will use his s-strategy s_i and his t-strategy t_i with probabilities *proportional* to these net payoff incentives v_i and w_i, which amounts to assigning the prior probability

$$(10.3) \qquad \text{Prob}(s_i) = \frac{v_i}{v_i + w_i}$$

to the hypothesis that player i will use strategy s_i while assigning the prior probability

$$(10.4) \qquad \text{Prob}(t_i) = \frac{w_i}{v_i + w_i}$$

to the alternative hypothesis that he will use strategy t_i.

Our theory uses the vector $\check{p} = (\check{p}_1, \ldots, \check{p}_n)$ of bicentric prior probability distributions for defining *risk-dominance* relations. Let F° be the smallest formation of the game which contains both equilibrium points s and t. We say that equilibrium point s *risk-dominates* equilibrium point t if the tracing procedure, when applied to $\check{p}$, yields s as solution, in preference over t, i.e., if we have $T(F^\circ, \check{p}) = s$. Conversely, we say that t risk-dominates s if we have $T(F^\circ, \check{p}) = t$. We call this relation "risk-dominance" because, if (say) s risk-dominates t, this will have the intuitive interpretation that, if the players have to choose between the two equilibrium points s and t, then they will choose s on the ground that they will consider t to be a *more risky*, and therefore a less attractive, choice than s is.

As I have already mentioned, under our theory the players will use *bicentric* priors $\check{p}_i$ only when they want to choose between *two* alternative equilibrium points as possible solutions for the game. In all other cases, they will use *uniform* priors $\hat{p}_i$. While a bicentric prior $\check{p}_i$ typically assigns *different* prior probabilities to different strategies of any given player i in accordance with the payoff incentives this player has to use each particular strategy, a uniform prior $\hat{p}_i$ always assigns the *same* prior probability to each strategy. (This is so because uniform priors are used in situations where the other players have no reason to regard player i as being more likely to use one strategy than to use another.)

Both uniform and bicentric priors play essential roles in our theory. But, in an important sense, bicentric priors, and the risk-dominance relations based on them, play a more fundamental role. Indeed, one may even say that our theory defines the solution of a game basically in terms of the *risk-dominance relations* that exist among the various equilibrium points.

11. Conclusion

To sum up, REINHARD SELTEN and I have developed our solution theory basically to overcome certain deficiencies of classical game theory. The primary application of our theory is to noncooperative games. But it does yield solutions also for all classes of cooperative games because any cooperative game can always be analyzed in terms of a suitable noncooperative bargaining model.

Lack of space does not permit me to describe some of the actual solutions our theory defines for various specific classes of games. [But see HARSANYI, 1977c, and 1979. This latter paper also discusses the actual *computation procedure* for a one-parameter family of two-person bargaining games with incomplete information as an illustrative example.] In all cases investigated so far, the solutions we obtain are in very good agreement with our intuitive expectations.

References

AUMANN, ROBERT J. and MASCHLER, MICHAEL, The Bargaining Set fort Cooperative Games, in *Annals of Mathematics Studies No. 52*, Princeton University Press: Princeton, 1964; pp. 443–76.

GILLIES, DONALD B., Solutions to General Non-Zero-Sum Games, in *Annals of Mathematics Studies No. 40*, Princeton University Press: Princeton, 1959; pp. 47–85.

HARSANYI, JOHN C., Games with Incomplete Information Played by 'Bayesian' Players, Parts I–III, *Management Science,* 14 (1967–68), pp. 159–82, 320–34, and 486–502.

HARSANYI, JOHN C., Games with Randomly Disturbed Payoffs: A New Rationale for Mixed-Strategy Equilibrium Points, *International Journal of Game Theory,* 2 (1973), pp. 1–23.

HARSANYI, JOHN C., The Tracing Procedure: A Bayesian Approach to Defining a Solution for n-Person Noncooperative Games, *International Journal of Game Theory,* 4 (1975), pp. 61–94.

HARSANYI, JOHN C., A Solution Concept for Noncooperative Games, *International Journal of Game Theory,* 5 (1977a), pp. 211–25.

HARSANYI, JOHN C., Time and the Flow of Information in Noncooperative Games, in ALBACH, HORST, *et al.* (editors), Quantitative Wirtschaftsforschung: Festschrift für W. Krelle, J. C. B. Mohr (Paul Siebeck): Tübingen, 1977b; pp. 255–67.

HARSANYI, JOHN C., A Solution Theory for Noncooperative Games and Its Implications for Cooperative Games, *Working Paper* CP-401 (1977c), Center for Research in Management Science, University of California, Berkeley.

HARSANYI, JOHN C., Analysis of a Family of Two-Person Bargaining Games with Incomplete Information, *Working Paper* CP-406 (1979), Center for Research in Management Science, University of California, Berkeley.

HARSANYI, JOHN C. and SELTEN, REINHARD, A Generalized Nash Solution for Two-Person Bargaining Games with Incomplete Information, *Management Science,* 18 (January 1972, Part II), pp. 80–106.

NASH, JOHN F., The Bargaining Problem, *Econometrica,* 18 (1950), pp. 155–62.

NASH, JOHN F., Noncooperative Games, *Annals of Mathematics,* 54 (1951), pp. 286–95.

NASH, JOHN F., Two-Person Cooperative Games, *Econometrica,* 21 (1953), pp. 128–140.

SELTEN, REINHARD, Spieltheoretische Behandlung eines Oligopolmodells mit Nachfrageträgheit, *Zeitschrift für die Gesamte Staatswissenschaft,* 121 (1965), pp. 301–24 and 667–89.

SELTEN, REINHARD, Reexamination of the Perfectness Concept for Equilibrium Points in Extensive Games, *International Journal of Game Theory,* 4 (1975), pp. 25–55.

SHAPLEY, LLOYD S., A Value for n-Person Games, in *Annals of Mathematics Studies No. 28,* Princeton University Press: Princeton, 1953; pp. 307–17.

VON NEUMANN, JOHN and MORGENSTERN, OSKAR, *Theory of Games and Economic Behavior,* Princeton University Press: Princeton, 1944.

SECTION E

GAME-THEORETICAL ANALYSIS OF MORAL DECISIONS

CHAPTER 12

RULE UTILITARIANISM, RIGHTS, OBLIGATIONS AND THE THEORY OF RATIONAL BEHAVIOR

Abstract. The paper first summarizes the author's decision-theoretical model of moral behavior, in order to compare the moral implications of the act-utilitarian and of the rule-utilitarian versions of utilitarian theory. This model is then applied to three voting examples. It is argued that the moral behavior of *act-utilitarian* individuals will have the nature of a *noncooperative* game, played in the *extensive mode*, and involving action-by-action maximization of social utility by each player. In contrast, the moral behavior of *rule-utilitarian* individuals will have the nature of a *cooperative* game, played in the *normal mode*, and involving a firm *commitment* by each player to a specific moral strategy (viz. to the strategy selected by the rule-utilitarian choice criterion) – even if some individual actions prescribed by this strategy, when considered in isolation, should fail to maximize social utility.

The most important advantage that rule utilitarianism as an ethical theory has over act utilitarianism lies in its ability to give full recognition to the moral and social importance of individual *rights* and personal *obligations.* It is easy to verify that action-by-action maximization of social utility, as required by act utilitarianism, would *destroy* these rights and obligations. In contrast, rule utilitarianism can fully recognize the moral validity of these rights and obligations precisely because of its *commitment* to an overall moral strategy, independent of action-by-action social-utility maximization.

The paper ends with a discussion of the *voter's paradox* problem. The conventional theory of rational behavior cannot avoid the paradoxical conclusion that, in any large electorate, voting is always an irrational activity because one's own individual vote is extremely unlikely to make any difference to the outcome of any election. But it can be shown that, by using the principles of rule-utilitarian theory, this paradox can easily be resolved and that, in actual fact, voting, even in large electorates, may be perfectly rational action. More generally, the example of rule utilitarianism shows what an important role the concept of a *rational commitment* can play in the analysis of rational behavior.

1. INTRODUCTION

In earlier publications (Harsanyi, 1953, 1955, 1977a, b, 1978a, b, c), I have argued that the modern theory of *rational behavior* – as developed by economists (as well as by decision theorists and game theorists) – can make important contributions to our understanding of the nature of *morality.*[1] Now I would like to carry this argument a little further and discuss the important role that the concept of *rational commitment* plays in rule-utilitarian theory. I also propose to show how rule-utilitarian theory, in turn, can help us to resolve certain problems in the theory of rational behavior itself and, in particular, how it can help us to resolve the *voter's paradox* problem.

When economists take an active interest in the foundations of moral philosophy, they are following a long tradition, going back to Adam Smith himself. In fact, an important branch of economic theory, namely welfare economics, is primarily a study of how to apply certain *moral* principles (Pareto-optimality, distributive justice, individual freedom, etc.) to economic activities. What is more, a major school of thought in moral philosophy itself, that of *utilitarianism*, was developed by a number of distinguished scholars (James Mill, John Stuart Mill, Sidgwick, Edgeworth, and others) who combined active work in moral philosophy with active work in economics. Indeed, utilitarianism can be interpreted as an attempt to base the whole of moral philosophy on the economic principle of utility maximization – or, more exactly, on the requirement of maximizing *social utility*, the latter being defined as the sum, or as the arithmetic mean, of all individual utilities.

Again, when utilitarianism came under heavy criticism because of its apparent insensitivity to individual *rights* and personal *obligations*, it was an economist, Roy Harrod (1936), who pointed out that these difficulties can be easily overcome if the *act-utilitarian* theory of the 19th-century utilitarians is replaced by a *rule-utilitarian* approach.[2] Act utilitarianism is the theory that the utilitarian criterion of social-utility maximization should be directly applied to each individual *action* (or act): In any given situation, a morally right action is that particular one that maximizes social utility. In contrast, rule utilitarianism is the theory that the utilitarian choice criterion should be applied, in the first instance, to alternative possible *moral rules*, rather than directly to alternative possible actions: In any given situation, a morally right action is one that conforms to the "correct moral rule" applicable to this kind of situation; whereas the correct moral rule is defined as that particular behavioral rule which will maximize social utility if it is followed by everybody in this kind of situation.

As I have argued elsewhere (Harsanyi, 1977a), the actual moral implications of either version of utilitarian theory can be ascertained only by a study of rigorous decision-theoretical models of moral behavior. Such a study will show that the rule-utilitarian approach does overcome the notorious difficulties of the act-utilitarian approach, and does yield a very attractive, truly rational and deeply humane, standard of morality.

In this paper I propose to show that the rule-utilitarian conceptual framework not only helps us to clarify the nature of morality, but also can help us to resolve important problems in the theory of *rational behavior*: in particular, it can help us to resolve the *voter's paradox* problem and, more generally, it can help us to appreciate the important role that *rational commitments* play in rational behavior.

2. ACT UTILITARIANISM VS. RULE UTILITARIANISM

To clarify the moral implications of the two utilitarian theories, I have proposed the following decision-theoretical model (Harsanyi, 1977a). Society consists of $(n + m)$ moral agents (individuals). Out of these individuals, agents $1, \ldots, n$ are *utilitarians*, while agents $(n + 1), \ldots, (n + m)$ are *nonutilitarians*, whose behavior is governed by self-interest or by some nonutilitarian morality (say, by traditional morality), or by some mixture of the two. Each agent has a finite set, S, of pure strategies, which is the same for all agents. By 'strategy' I mean the usual decision-theoretical and game-theoretical strategy concept: thus, a strategy is a mathematical function from the set of all possible 'situations' to the set of all possible 'actions', specifying, for each situation, the specific action that the agent will take in that situation. (But it is required that any strategy should always specify the *same* action for two different situations if the agent has insufficient information to distinguish between the two.) The strategy of a given agent k will be called s_k.

The strategies s_k of the m nonutilitarian agents k $(k = n + 1, \ldots, n + m)$ are regarded as *given*, i.e., as being determined outside the model. In contrast, the strategies s_i of the n utilitarian agents i $(i = 1, \ldots, n)$ are determined within the model, by the requirement that they should maximize the social-utility (or social-welfare) function W. But the *constraints* of this maximization problem will differ according as the n utilitarian agents are act utilitarians or are rule utilitarians.

First take the *act-utilitarian* case. In this case, each utilitarian agent i will regard, not only the nonutilitarian agents' strategies, but also the other $(n - 1)$ utilitarian agents' strategies, as *given*. Therefore, he will try to solve the following maximization problem.

Maximize $W = W(s_1, \ldots, s_i, \ldots, s_n; s_{n+1}, \ldots, s_{n+m})$, subject to:

$$(2.1) \qquad s_j = s_j^0 = \text{const.} \quad \text{for } j = 1, \ldots, i-1, i+1, \ldots, n;$$

and

$$(2.2) \qquad s_k = s_k^* = \text{const.} \quad \text{for } k = n+1, \ldots, n+m.$$

In contrast, in the *rule-utilitarian* case, each utilitarian agent will act on the assumption that all the other $(n-1)$ utilitarian agents will follow the *same* moral rule and, therefore, will use the *same* strategy as he does himself. Thus, he will try to solve the following maximization problem.

Maximize $W = W(s_1, \ldots, s_i, \ldots, s_n; s_{n+1}, \ldots, s_{n+m})$, subject to:

$$(2.3) \qquad s_1 = \ldots = s_i = \ldots = s_n;$$

and

$$(2.4) \qquad s_k = s_k^* = \text{const.} \quad \text{for } k = n+1, \ldots, n+m.$$

Even though the maximand is the same function W in both cases, the two maximization problems are quite *different* maximization problems because Equations (2.1) and (2.3) are very different constraints. Accordingly, in many cases the two maximization problems will have very different solutions, as I will illustrate by three voting examples (see below).

Mathematically, the behavior of the n utilitarian agents in both models can be regarded as an n-person *game*. (It is an n-person game, rather than an $(n+m)$-person game because only the n utilitarian agents are active players.) But it is a very special game because all n players are trying to maximize the *same* social-utility function W, which means that we are dealing with a game where all n players have the *same* payoff function W.

More specifically, if the n players follow the *act-utilitarian* approach then the resulting game will be formally a *noncooperative* game since each player will choose his strategy independently of all other agents in trying to maximize the function W. In contrast, if the players follow the *rule-utilitarian* approach then the resulting game will be a *cooperative* game since, according to requirement (2.3), each player will choose his strategy on the assumption that all other players will use the same strategy, so that the different players' strategies will *not* be independent. But their strategies *will* be independent in another sense: they will be chosen in general without any prior consultation between the players.[3]

We can express this fact by saying that the rule-utilitarian game is a cooperative game of a somewhat unusual type. It is *not* a cooperative game with *negotiated* strategy coordination as most cooperative games discussed in the game-theoretical literature are. Rather, it is a cooperative game with *spontaneous* strategy coordination, in which this coordination is typically achieved without actual communication between the players, merely by the fact that all players will follow the *same* choice criterion in choosing their strategies.

To illustrate the differences between the act-utilitarian and the rule-utilitarian approaches, I will briefly discuss three voting examples. (For fuller mathematical analysis, see Harsanyi, 1977a, Mathematical Note II.)

EXAMPLE 1. 1000 voters have to decide the fate of a socially very desirable policy measure *M*. All of them favor the measure. Yet it will pass only if all 1000 voters actually come to the polls and vote for it. But voting entails some minor costs in terms of time and inconvenience. The voters cannot communicate and cannot find out how many other voters actually voted or will vote.

Under these assumptions, if the voters are *act utilitarians* then each voter will vote only if he is reasonably sure that *all* other 999 voters will vote. Therefore, if even one voter doubts that all other voters will vote then he will stay home and the measure will fail. Thus, defeat of the measure will be a fairly likely outcome.

This voting process by act-utilitarian voters, as a noncooperative game, will have two game-theoretical equilibrium points: one will occur if everybody votes while the other will occur if nobody votes. This second, socially undesirable, equilibrium point is much more stable than the first, socially desirable, equilibrium point because it will be destabilized as soon as even one player becomes doubtful about whether everybody will vote.

In contrast, if the voters are *rule utilitarians* then all of them will be certain to vote, ensuring passage of the measure. This is so because, for rule-utilitarian voters, their choice reduces to a choice between everybody voting and nobody voting; and since the former alternative will yield a much higher social utility they will always choose that alternative

EXAMPLE 2. The situation is as in Example 1, except that now only 800 affirmative votes are needed to pass the measure. The voters can use only *pure* strategies.

In this case, an *act-utilitarian* voter will vote only if he is reasonably sure that *exactly* 799 other voters will vote. (If fewer of them vote then his vote will not be sufficient to pass the measure; whereas, if more of them vote then his vote will not be needed.) But this means that he will be most unlikely to vote.

Now, the noncooperative game representing voting by act-utilitarian voters will have an astronomical number of equilibrium points, which fall into two classes. Class 1 consists of one unique equilibrium point where nobody votes. On the other hand, Class 2 consists of all strategy combinations where exactly 800 voters do vote and exactly 200 voters do not. Thus, it contains as many equilibrium points as the number of possible ways of selecting 800 elements out of a set of 1000 elements, which is, of course, a very large number. Since it is very hard to ensure that exactly 800 voters should vote if no communication is permitted, equilibrium points of Class 2 are very unlikely to occur. A much more likely outcome is the one equilibrium point of Class 1, which makes all voters refrain from voting, with socially very undesirable results.

In contrast, if the voters are *rule utilitarians* then again everybody will vote.

Thus, neither rule utilitarianism nor act utilitarianism is able in this case to ensure the socially efficient outcome where exactly 800 voters would vote. This is, of course, not at all surprising: in the absence of communication between the voters this outcome can arise only as a result of a most unlikely accidental coincidence. But, given this constraint, rule utilitarianism has a clear advantage over act utilitarianism. It will ensure passage of the socially desirable measure, even though it will do so somewhat inefficiently, with more votes cast than legally needed. Yet, this outcome will be much preferable to a defeat of the measure, which would be the likely outcome if the voters followed the act-utilitarian approach.

EXAMPLE 3. The situation is as in Example 2, but now the voters can also use *mixed* strategies.

In this case, once more, an act-utilitarian voter will vote only if he is reasonably sure that exactly 799 other voters will vote, which will make him very unlikely to vote.

Now, the noncooperative game representing voting by act-utilitarian voters will have an even larger number of equilibrium points than it had in Example 2. The two classes of pure-strategy equilibrium points described for Example 2 will remain equilibrium points. But now they will be joined by a very large number of mixed-strategy equilibrium points as well. Again, the equilibrium point most likely to arise as an actual outcome will be the pure-strategy equilibrium point at which all voters choose nonvoting with probability one.

Among the very large number of mixed-strategy equilibrium points that the game now has, two are of special theoretical interest because they are *symmetric* equilibrium points where all players use the *same* mixed strategies. I will call them equilibrium points $\bar{\sigma}^*$ and $\bar{\sigma}^{**}$.

At $\bar{\sigma}^*$, every player will use a mixed strategy σ^* which assigns a probability p^*, *slightly smaller* than $800/1000 = 4/5$, to voting and assigns the remaining probability $(1 - p^*)$ to nonvoting. On the other hand, at $\bar{\sigma}^{**}$, every player will use a mixed strategy σ^{**} which assigns a probability p^{**}, *slightly greater* than $4/5$, to voting and assigns the remaining probability $(1 - p^{**})$ to nonvoting. (For proof, see Harsanyi, 1977a, Mathematical Note II.)

Finally, if the voters are *rule utilitarians,* then, in this case the common strategy they use will be a mixed strategy. More particularly, it can be shown (*loc. cit.*) that it will be the same mixed strategy σ^{**} that would be used at equilibrium point $\bar{\sigma}^{**}$ of the act-utilitarian game. It is easy to verify that if the players use this mixed strategy σ^{**} then measure M will pass with a probability very close to (but slightly less than) *one*. Moreover, the outcome obtained in this way will represent a true social *optimum*: it will achieve the highest social utility that can be achieved at all without actual communication among the players.

These three voting examples show that the act-utilitarian and rule-utilitarian choice criteria will often lead to different outcomes. Moreover, in the absence of communication, by using the rule-utilitarian approach, people can achieve a much higher degree of spontaneous strategy coordination than they could by using the act-utilitarian approach. I will call this fact the *coordination effect*.

Yet, in spite of the theoretical importance of this coordination effect, in my opinion, the superiority of rule utilitarianism over act utilitarianism rests primarily on other considerations. It rests primarily on the fact that rule utilitarianism can provide a much more satisfactory theory of individual *rights* and personal *obligations* than act utilitarianism can provide (see below).

3. RULE UTILITARIANISM AND MORAL COMMITMENT

In a given game, any player may follow two basically different approaches in playing the game. One is to make a *separate* decision on each particular move he undertakes: this may be called the *extensive mode* of playing. The other is to choose a comprehensive *strategy* at the beginning of the game, with a firm *commitment* to follow this strategy all the time even if at some later stages of the game he were to feel tempted to deviate from this strategy: this may be called the *normal mode* of playing the game.

A game where all players act in the extensive mode is called a game in *extensive form*, while a game where all players act in the normal mode is called a game in *normal form*. But it is often analytically convenient to analyze a game where the players in fact act in the extensive mode, by studying the normal form of the game. This is, of course, a perfectly legitimate procedure as long as we clearly remember that we are actually dealing with a game played in the extensive mode, in which the players are *not* committed to specific strategies chosen in advance.[4] Accordingly, when a game theorist studies a game in normal form, he must always clearly state whether this game modelled in normal form is meant to represent a real-life game situation where the players in fact will act in the *normal mode*, or is meant to represent a game situation where in fact they will act in the *extensive mode*, so that the normal-form representation is used merely for analytical convenience.

Using this terminology, it is clear that the moral behavior of *act-utilitarian* moral agents will represent a *noncooperative* game played in the *extensive mode* since they will choose each individual action one by one so as to maximize social utility by this particular action, assuming that all other moral agents' behavior is given. In contrast, the moral behavior of *rule-utilitarian* moral agents will represent a *cooperative* game played in the *normal mode* since their behavior will be governed by firm *commitment* to some moral rules – and, therefore, to some moral *strategy* – chosen in advance in accordance with the rule-utilitarian choice criterion (*viz*., with the criterion that this moral strategy should maximize social utility if all other rule-utilitarian moral agents follow the same strategy).[5]

The cooperative nature and the normal-mode decision structure of the rule-utilitarian game can be illustrated by means of Example 2 of Section 2. As we have seen, in this example, if all voters follow the rule-utilitarian approach then this will ensure passage of the proposed measure – in fact, this will ensure that all 1000 voters will cast affirmative votes, even though 800 votes would suffice for passage. The point is that, even though there will be too many votes cast, rule-utilitarian theory cannot excuse *any one* voter from voting because if it did then it would have to do the same for *every* voter, so that the measure would most certainly fail.

Yet, voting by every voter is *not* a game-theoretical equilibrium point and, therefore, it could not be the planned outcome of a noncooperative game played by rational voters. (No equilibrium point can involve more than 800 voters voting in this case.) It is only because the game played by rule-utilitarian voters is a *cooperative* game that it can have such a non-equilibrium-point outcome as voting by all the voters.

Moreover, this outcome can occur only because the game is played in the *normal mode*, involving a firm *commitment* by each voter to the rule-utilitarian voting strategy. In effect, if a given voter were not under such a commitment, then he would obviously choose *not* to vote since nonvoting would be the social-utility maximizing action for him. This is so because he would know that 999 other voters would vote for the measure anyhow; therefore, if he himself did not vote, then the measure would still pass, yet he would have saved the *costs* (in time and in inconvenience) of himself going to the polls. To be sure, as we have seen, as a *joint strategy*, voting by all voters is the social-utility maximizing strategy, given that, in the absence of communication, only joint strategies under which all players do the same thing are practicable. Yet, even if voting by all voters is the optimal joint strategy, it will be implemented only if each voter feels *committed* to this joint strategy, and does not feel free to depart from it on the ground that by not voting he could actually increase social utility, as long as the other voters did vote.

As we will see, it is generally true that the social benefits of the rule-utilitarian approach crucially depend on each individual feeling *committed* to the joint strategy chosen on the basis of the rule-utilitarian choice criterion, and on his firm adherence to this joint strategy even if some action deviating from this strategy, when considered in isolation, would yield a *higher* social utility. To put it differently, the rule-utilitarian approach requires each individual to *renounce* the right to choose his individual actions by the act-utilitarian criterion of *action-by-action maximization* of social utility.

Under the modern theory of rational behavior, rationality always involves some form of utility maximization. For a utilitarian decision maker, devoted to the common interests of society as a whole, rational behavior will always involve maximization of social utility. Therefore, it may appear that, of the two versions of utilitarian theories, act utilitarianism is the more rational approach because it requires every individual to choose *each* individual *action* of his by the criterion of social-utility maximization, whereas, rule utilitarianism restricts social-utility maximization essentially to *one* decision problem, that of choosing a comprehensive *joint strategy*. But such a view would amount to misapplication of the concept of rationality. Whether one approach is more rational than another does not depend on the *number* of different occasions where it makes use of utility maximization. Rather, it depends on its *effectiveness* in maximizing utility, i.e., on its ability to achieve a *higher* level of utility.

Yet, in this respect, the rule-utilitarian approach is incomparably superior to the act-utilitarian approach. Our three voting examples have shown that in many cases society will be much better off, i.e., will reach a much higher level of social utility, if its members follow the rule-utilitarian approach than if they follow the act-utilitarian approach; and, as we will see, it is quite generally true that the rule-utilitarian approach will yield socially much more beneficial results than the act-utilitarian approach would – primarily because the rule-utilitarian approach has a much better appreciation for the social importance of individual *rights* and personal *obligations.*

Thus, the rule-utilitarian approach does restrict the freedom of action of each individual by asking him to make a firm *commitment* to a moral strategy chosen in accordance with the rule-utilitarian choice criterion – even if this strategy is to be often inconsistent with action-by-action maximization of social utility. Nevertheless, when an individual commits himself to a rule-utilitarian moral strategy he is making a *rational commitment* because by doing so he will contribute to achieving a higher level of social utility.

4. INDIVIDUAL RIGHTS AND OBLIGATIONS

It is this commitment to a comprehensive moral strategy, independent of action-by-action maximization of social utility, that enables rule utilitarianism to recognize the existence of morally protected individual rights and obligations.

According to common-sense morality, people have many individual rights, and have many specific obligations to certain other people, which cannot be overridden merely on the basis of social-expediency considerations, except perhaps under very exceptional conditions. On the other hand, act utilitarianism is inherently unable to provide a proper recognition for these rights and obligations because it must take the position that any action violating such rights and obligations is morally permissible – and is, indeed, morally obligatory – if it here and now maximizes social utility. Yet, such a view deprives these rights and obligations of all effective moral force.

For example, act utilitarianism must take the view that the legal owner will lose all his moral claim to his private property if this property is desired by a needy person who presumably would have a *higher* utility for it. Likewise, it must take the view that a person who has made a promise will be under no moral obligation to keep it, and the recipient of his promise will have no moral right to demand its fulfilment, should the utility to the former of breaking his promise be somewhat *greater* than would be the utility to the latter of actually obtaining the promised benefits. Again, act utilitarianism must take the position that the parents of any given child will have neither the moral right, nor the moral obligation, of spending their time and their money on the support and the education of their own child, if there are any other children (or any adults) in a somewhat *greater* need for support or for educational help, etc.

Of course, by being logically compelled to take such positions, act utilitarianism essentially *destroys* all individual rights and all personal obligations. The right to private property, the right to demand fulfilment of promises, the right to parental support, as well as the obligation to respect other people's private property, the obligation to keep one's promises, and the obligation to support one's children, all become empty words if they can and, indeed, have to be violated on every occasion where this might produce slightly more total utility than disutility.

In contrast, rule utilitarianism has no logical difficulty in taking the position that individual rights and personal obligations have full moral validity – even if they do not pass the act-utilitarian test of action-by-action maximization of social utility – as long as they pass the rule-utilitarian test, i.e., as long as a general moral strategy committed to full *observance* of these rights and obligations by all honest people will in the long run yield a higher social utility than a moral strategy *disregarding* some or all of these rights and obligations would yield.

Thus, to the extent that such rights and obligations actually pass this test of social utility, rule utilitarianism – in sharp contrast to act utilitarianism – will fully recognize the moral validity of these rights and obligations, much in the same way as common-sense morality does. At the same time, rule utilitarianism goes beyond common-sense morality in two philosophically important respects. It not only recognizes the *existence* of such morally binding rights and obligations, but also provides a *rational justification* for them in terms of their social utility. Moreover, it also provides a rational criterion for discovering the legitimate *exceptions* to actual observance of these rights and obligations, which common-sense morality cannot do. More specifically, suppose we want to find out whether it is morally permissible to disregard a given individual right, or a given personal obligation, under such and such conditions. Then, rule-utilitarian theory tells us that, in order to answer this question, we have to ask whether society will be better off in the long run by adopting a moral strategy *permitting* its members to disregard this right or this obligation under the specified conditions, or whether society will be better off by adopting a moral strategy *not* permitting any such exception from observance of this right or this obligation.

For example, any reasonable moral code must permit *some* exceptions to the general obligation of keeping our promises, at least in cases of special hardship. On the other hand, it must not permit too many easy exceptions because this would greatly reduce the social value of promises by greatly diminishing the probability that these promises would actually be kept. Even though in practice it may be hard to judge at what particular point the socially optimal balance will be reached between these two considerations pulling in opposite directions, the social-utility maximization principle, at least in theory, does provide a conceptually clear criterion for finding this social optimum.

By recognizing the moral importance of individual rights and personal obligations, but by making this recognition strictly dependent on the social utility of these rights and obligations, rule utilitarianism takes up a middle ground between act utilitarianism (which essentially fails to recognize them) and many other ethical theories (which do recognize them but which fail to provide any rational justification for them), such as *deontological*, *contractarian*, and *entitlement*-based ethical theories (cf. Prichard, 1965; Ross, 1963; Rawls, 1971; Nozick, 1974). Since these theories reject the utilitarian test of social utility, they have to leave the network of moral rights and moral obligations without any rational foundation.

For instance, deontologists such as Prichard and Ross could never explain *where* our alleged moral duties actually come from, and *why* a rational individual should feel bound to observe them. Contractarians like Rawls are in an even weaker position: even if the postulated 'social contract' were a historical fact, it would be hard to see *why* any rational individual should feel bound by a contract that his very remote ancestors had concluded presumably many thousands of years ago. But, of course, we know that this postulated 'social contract' is sheer fiction, which makes it even harder to see why it should have any moral force. Indeed, more fundamentally, even in the case of real contracts we have ourselves concluded, the principle that contracts are to be kept needs some *rational justification*, and such rational justification cannot be provided in terms of any contractarian theory. Rather, such a justification can be provided only in utilitarian terms, i.e., in terms of the *social utility* of a moral rule requiring the observance of contracts, except in some very special cases. Again, entitlement theorists like Nozick cannot explain where 'entitlements' actually come from. Did some mysterious godhead invest us with all the unlikely-sounding 'entitlements' postulated by these entitlement theorists? If so, how can we ever find out that we have been so invested? And, in any case, why should any rational person pay any attention to these 'entitlements'?

Given that the justification of individual rights and obligations lies in their social utility, I now propose to consider what advantages, if any, the existence of such rights and obligations has for society.

One advantage lies in what may be called their *expectation effect* and their *incentive effect*. The former consists in the fact that the existence of such rights and obligations makes it easier for us to form reasonably specific *expectations* about other people's future behavior, and this fact gives us a stronger feeling of security, and makes it easier for us to plan our future activities. On the other hand, the incentive effect consists in the fact that the existence of such rights and obligations greatly increases our *incentives* to engage in socially desirable activities. For example, the existence of private property rights increases people's incentives to hard work, to saving, and to investment. The existence of an obligation to keep one's promises increases people's incentives to agree on certain joint activities, and to perform services for other people in exchange for promised future rewards, etc.

Another advantage lies in the fact that this network of rights and obligations gives rise to a certain *division of labor* among the members of society as a whole, and among the members of each particular social institution, by creating a large number of mutually complementary *social roles*, and by assigning specific rights and specific obligations to the occupants of these roles. This fact may be called the *division-of-labor effect*. The resulting social division of labor has extremely great social utility.

For example, we could imagine a society in which *all* adults would have common responsibility for the support and the education of *all* the children. But, in actual fact, in all known societies the main responsibility for the well-being of any particular child is always vested in one particular adult or in a small group of such adults, whether these are the child's natural parents, or some other relatives, or some specially appointed guardians, etc. This division of labor among adult members of the society in looking after the children has both *informational* and *motivational* advantages. If each adult had to take an equal interest in all the children, then he or she could not keep informed about the specific needs of any child, and could not develop close emotional ties with any of them. In contrast, if each adult is responsible only for a few children, then he or she can pay very close attention to them, and can give each of them very close emotional support.

To conclude, the existence of a network of individual rights and obligations has great social advantages because of its *expectation effect*, its *incentive effect*, and its *division-of-labor effect*. The superiority of the rule-utilitarian approach over the act-utilitarian approach lies primarily in the fact that it is able to recognize the moral significance of these rights and obligations, which the act-utilitarian approach is logically unable to do.[6]

5. RULE UTILITARIANISM AND THE VOTER'S PARADOX

Economic theory has a long tradition of making use of the concept of rational behavior. Political science has been using this concept for a much shorter time, but has already achieved important insights by doing so. Yet, application of the concept of rational behavior to political activities has also yielded a curious paradox: It seems that voting cannot be analysed at all in terms of the concept of rational behavior because, in any large electorate, voting seems to be a highly *irrational* activity. This is so because the probability that one's own vote should actually decide the election is virtually nil. Thus, even if the election is concerned with extremely important issues, and even if the costs of voting, in terms of time and inconvenience, are actually quite small, any cost-and-benefit calculus will come out very clearly *against* voting. Yet, many people do vote, and do not seem to feel at all that they engage in an irrational activity. This fact is often described as the *voter's paradox*.

In actual fact, many of the more articulate voters will even offer a specific argument to defend the rationality of their voting. They will argue: "Admittedly, my own individual vote cannot make any difference. But imagine that *all* people (or most people) like me would not vote. That surely would make a significant difference in many cases!"

Under the customary model of rational behavior, this must be considered as being an invalid – and, in fact, a badly confused – argument. One would have to say to the voter who uses this argument: "Your argument would be valid if you had the power to decide whether all people (or most people) like you should vote or not. But, in actual fact, you can only decide whether you yourself will vote or not, and your decision will have no effect whatever on most other people you describe as people like you. Therefore, you cannot rationally make your decision on whether to vote or not to vote depend on what would happen if all people like you did not vote. All you can rationally ask is what difference it would make if you personally did not vote."

Yet, in my opinion, this voter's argument is a perfectly valid argument. The fault does not lie with this argument, but rather lies with the unduly narrow model of rational behavior which would suggest rejection of this argument. The fault lies in the fact that the customary model of rational behavior does not take account of the possibility of a *rational commitment* to a comprehensive joint strategy, such as the rule-utilitarian joint strategy.

Once the possibility of such a rational commitment is recognized, the argument in question makes perfectly good sense.

In the three voting examples discussed in Section 2, for simplicity I have disregarded all possible morally relevant *differences* that may exist among the voters. Accordingly, I have concluded that, in Example 2, the moral strategy suggested by the rule-utilitarian approach would require *every* voter to vote. But, more realistically, if there *are* relevant differences among the voters in age, in health, in how busy they are, etc., then the correct rule-utilitarian strategy may actually take the form, "People above a certain age, people who are sick, or people who are very busy, etc., need not vote, because we can win the election even without their votes. But all other people should definitely vote." (Of course, in the real world – unlike what has been assumed to happen in our voting examples – there will be considerable disagreements about the merits of any given policy measure. Therefore, a separate moral strategy of this kind must be formulated *both* for the supporters *and* for the opponents of any given measure.)

Accordingly, it makes very good sense to ask the question, "What would happen if people like me did not vote?" For it is precisely by asking such questions that (say) 70-year old voters can rationally decide whether the election will be so close that they will also have to vote, or whether people of that (or even higher) age can safely stay home in this election, etc.

Thus, we can conclude that voting in a large electorate need not represent irrational behavior at all; and that its rational justification actually lies in the often-rejected common-sense argument based on considering what would happen if *all* persons of a given category (as well as people who had even *better* excuses not to vote) refrained from voting.

More generally, our discussion of rule utilitarianism has shown that behavior based on a *rational commitment* must be classified as truly rational behavior, and that rational commitment by various individuals to specific strategies can be legitimately used as an explanatory variable in analyzing rational behavior.[7]

6. CONCLUSION

In this paper I have argued that moral behavior by *act-utilitarian* moral agents has the nature of a *noncooperative* game played in the *extensive mode*, and based on action-by-action maximization of social utility. In contrast, moral behavior by *rule-utilitarian* moral agents has the nature of a *cooperative* game played in the *normal mode*, and involving a firm commitment by each agent to an overall joint strategy chosen on the basis of the rule-utilitarian choice criterion.

We have seen that this cooperative nature, and this normal-mode decision structure, of the rule-utilitarian game enables the rule-utilitarian approach to achieve a more effective strategy coordination among different individuals than the act-utilitarian approach would be able to achieve. This fact I have called the *coordination* effect.

Yet, I have argued that the superiority of the rule-utilitarian approach over the act-utilitarian approach rests primarily on the ability of the former, and on the inability of the latter, to give proper recognition to the social importance of individual *rights* and *obligations*, and on the resulting *expectation* effect, *incentive* effect, and *division-of-labor* effect. As we have seen, the act-utilitarian approach would *destroy* all individual rights and obligations by its insistence on action-by-action maximization of social utility. In contrast, the rule-utilitarian approach can fully recognize the moral validity of such rights and obligations because it requires a firm *commitment* by each individual to a suitably chosen overall strategy, which will make his behavior independent of action-by-action maximization of social utility.

Finally, I have argued that the rule-utilitarian conceptual framework enables us to resolve the voter's paradox. It enables us to show that voting, even in large electorates, may be a perfectly rational activity. The example of rule-utilitarian theory also indicates the important role that the concept of a *rational commitment* can play in the analysis of rational behavior.

University of California, Berkeley, and *University of Bonn*

NOTES

[1] I want to thank Professor Wilhelm Krelle and the University of Bonn for their kind hospitality while I was writing this paper. I am also grateful to the National Science Foundation for supporting earlier parts of this research through Grant SOC 77-06394 to the Center for Research in Management Science, University of California, Berkeley. Finally I want to thank Professor Robert J. Aumann, of the Hebrew University, Jerusalem, for interesting discussions which I have found very helpful in clarifying my ideas, in spite of our obvious disagreements on the foundations of moral philosophy.

[2] Harrod did not actually use the terms 'act utilitarianism' and 'rule utilitarianism'. These terms were introduced only later by Brandt (1959, pp. 380 and 396). As the 19th-century utilitarians did not distinguish between the two versions of utilitarian theory, it is a rather controversial issue whether their views were closer to act utilitarianism or to rule utilitarianism.

[3] We have seen that each rule-utilitarian player will choose his strategy on the assumption that the other players, also, will choose the same strategy. But this assumption will not be based on any direct information about the strategy choices of the other players. Rather, it will be based merely on the fact that the other players will choose their strategies by using the same rule-utilitarian decision rule, i.e., by maximizing the same function W, subject to the same constraints (2.3) and (2.4).

[4] For instance, in analyzing noncooperative games, if we are dealing with a game played in the *extensive mode* then, as Reinhard Selten (1965, 1975) has pointed out, not all equilibrium points of the normal form of the game will qualify as rational outcomes, but only the *perfect* equilibrium points of the game will do so. In contrast, if the game is actually played in the *normal mode* then all equilibrium points will potentially represent rational outcomes.

[5] In fact, *any* cooperative game will always involve at least *some* behavior in the normal mode if it ends with an agreement among the players. For an agreement will always lead to a *commitment* by the players to some agreed strategies.

[6] As we have seen, the act-utilitarian approach would destroy the moral force of these individual rights and obligations. In doing so, it would destroy the *expectation* and the *incentive* effects, depending on these rights and obligations. For instance, if there are many readily available exceptions to the obligation of keeping one's promises then people cannot formulate reasonably firm expectations on the assumption that promises will be kept; and they will not have the incentive to engage in activities whose profitability depends on such expectations. Act utilitarianism would also destroy the *division-of-labor* effect by invalidating the special obligations one person is supposed to have to some specific other persons. For instance, as we have seen, it would prevent any parent from giving priority to looking after his own children's needs over looking after the needs of other children or the needs of any other individuals.

[7] Of course, within a model of rational behavior, a person's commitment to a specific strategy can explain his behavior only if this commitment is in fact a *rational commitment*. In game theory, in analyzing games in normal form, it is often too readily assumed that the players will feel committed to specific strategies – even though the conditions actually do not permit them to make such commitments, or even though it would be against their strategic interests to make, and to abide by, any such commitments.

REFERENCES

Brandt, R. B., *Ethical Theory*, Prentice-Hall, Englewood Cliffs, N.J., 1959.

Harrod, R. F., 'Utilitarianism Revised', *Mind* **45** (1936), 137–156.

Harsanyi, J. C., 'Cardinal Utility in Welfare Economics and in the Theory of Risk-Taking', *Journal of Political Economy* **61** (1953), 434–435. Reprinted in Harsanyi (1976).

Harsanyi, J. C., 'Cardinal Welfare, Individualistic Ethics, and Interpersonal Comparisons of Utility' *Journal of Political Economy* **63** (1955), 309–321, reprinted in Harsanyi (1976).

Harsanyi, J. C., *Essays on Ethics, Social Behavior, and Scientific Explanation*, Reidel, Dordrecht, Holland, 1976.

Harsanyi, J. C., 'Rule Utilitarianism and Decision Theory', *Erkenntnis* **11** (1977a), 25–53.

Harsanyi, J. C., 'Morality and the Theory of Rational Behavior', *Social Research* **44** (1977b), 623–656.

Harsanyi, J. C., 'Bayesian Decision Theory and Utilitarian Ethics', *American Economic Review, Papers and Proceedings* **68** (1978a), 223–228.

Harsanyi, J. C., 'Bayesian Decision Theory, Rule Utilitarianism, and Arrow's Impossibility Theorem', *Working Paper* CP-405, Center for Research in Management Science, University of California, Berkeley, 1978b.

Harsanyi, J. C., 'Rationality, Reasons, Hypothetical Imperatives, and Morality', *Working Paper* CP-407, Center for Research in Management Science, University of California, Berkeley, 1978c.

Nozick, R., *Anarchy, State, and Utopia*, Oxford, Blackwell, 1974.

Prichard, H. A., *Moral Obligation*, Oxford, Clarendon Press, 1965.

Rawls, J., *A Theory of Justice*, Cambridge, Mass., Harvard University Press, 1971.

Ross, W. D., *Foundations of Ethics*, Oxford, Clarendon Press, 1963.

Selten, R., 'Spieltheoretische Behandlung eines Oligopolmodells mit Nachfrageträgheit', *Zeitschrift für die gesamte Staatswissenschaft* **121** (1965), 301–324, 667–689.

Selten, R., 'Reexamination of the Perfectness Concept for Equilibrium Points in Extensive Games', *International Journal of Game Theory* **4** (1975), 25–55.

INDEX

THEORY AND DECISION LIBRARY

An International Series in the Philosophy and Methodology
of the Social and Behavioral Sciences

1. Günther Menges (ed.), *Information, Inference, and Decision.* 1974, viii + 195 pp.
2. Anatol Rapoport (ed.), *Game Theory as a Theory of Conflict Resolution.* 1974, v + 283 pp.
3. Mario Bunge (ed.), *The Methodological Unity of Science.* 1973, viii + 264 pp.
4. Colin Cherry (ed.), *Pragmatic Aspects of Human Communication.* 1974, ix + 178 pp.
5. Friedrich Rapp (ed.), *Contributions to a Philosophy of Technology. Studies in the Structure of Thinking in the Technological Sciences.* 1974, xv + 228 pp.
6. Werner Leinfellner and Eckehart Köhler (eds.), *Developments in the Methodology of Social Science.* 1974, x + 430 pp.
7. Jacob Marschak, *Economic Information, Decision and Prediction. Selected Essays.* 1974, three volumes, xviii + 389 pp.; xii + 362 pp.; x + 399 pp.
8. Carl-Axel S. Staël von Holstein (ed.), *The Concept of Probability in Psychological Experiments.* 1974, xi + 153 pp.
9. Heinz J. Skala, *Non-Archimedean Utility Theory.* 1975, xii + 138 pp.
10. Karin D. Knorr, Hermann Strasser, and Hans Georg Zilian (eds.), *Determinants and Controls of Scientific Developments.* 1975, ix + 460 pp.
11. Dirk Wendt, and Charles Vlek (eds.), *Utility, Probability, and Human Decision Making. Selected Proceedings of an Interdisciplinary Research Conference, Rome, 3–6 September, 1973.* 1975, viii + 418 pp.
12. John C. Harsanyi, *Essays on Ethics, Social Behavior, and Scientific Explanation.* 1976, xvi + 262 pp.
13. Gerhard Schwödiauer (ed.), *Equilibrium and Disequilibrium in Economic Theory. Proceedings of a Conference Organized by the Institute for Advanced Studies, Vienna, Austria, July 3–5, 1974.* 1978, 1 + 736 pp.
14. V. V. Kolbin, *Stochastic Programming.* 1977, xii + 195 pp.
15. R. Mattessich, *Instrumental Reasoning and Systems Methodology.* 1978, xxii + 396 pp.
16. H. Jungermann and G. de Zeeuw (eds.), *Decision Making and Change in Human Affairs.* 1977, xv + 526 pp.
18. A. Rapoport, W. E. Stein, and G. J. Burkheimer, *Response Models for Detection of Change.* 1978 (forthcoming)
19. H. J. Johnson, J. J. Leach, and R. G. Mühlmann (eds.), *Revolutions, Systems, and Theories; Essays in Political Philosophy.* 1978.
20. Stephen Gale and Gunnar Olsson (eds.), *Philosophy in Geography.* 1979, xxii + 470 pp.

21. Maurice Allais and Ole Hagen (eds.), *Expected Utility Hypotheses and the Allais Paradox: Contemporary Discussions of Decisions Under Uncertainty With Allais' Rejoinder.* 1979, vii + 681 pp. + indexes.
22. Teddy Seidenfeld, *Philosophical Problems of Statistical Inference: Learning from R. A. Fisher.* 1979, xiv + 246 pp.
23. L. Lewin and E. Vedung (eds.), *Politics as Rational Action.* 1980, xii + 274 pp.
24. J. Kozielecki, *Psychological Decision Theory.* 1982, xvi + 395 pp. + index (forthcoming).
25. I. I. Mitroff and R. O. Mason, *Creating a Dialectical Social Science, Concepts, Methods, and Models.* 1981, ix + 189 pp.
26. V. A. Lefebvre, *Algebra of Conscience, A Comparative Analysis of Western and Soviet Ethical Systems.* 1982, xxvii + 194 pp.
27. L. Nowak, *Property and Power, Towards a Non-Marxian Historical Materialism.* 1982, (forthcoming).
28. J. C. Harsanyi, *Papers in Game Theory, Papers 1956–1980.* 1982, x + 253 pp. + index.
29. B. Walentynowicz (ed.), *Polish Contributions to the Science of Science.* xii + 279 pp. + index.